7.11

VIIIB	IB	IIB	IIIA	IVA	VA	VIA	VIIA	Noble gases
								2 He 4.003
			5 B 10.811	6 C 12.011	7 N 14.007	8 O 15.999	9 F 18.998	10 Ne 20.183
			13 Al 26.982	14 Si 28.086	15 P 30.974	16 S 32.064	17 Cl 35.453	18 Ar 39.948
28 Ni 58.71	29 Cu 63.54	30 Zn 65.37	31 Ga 69.72	32 Ge 72.59	33 As 74.922	34 Se 78.96	35 Br 79.909	36 Kr 83.80
46 Pd 106.4	47 Ag 107.87	48 Cd 112.40	49 In 114.82	50 Sn 118.69	51 Sb 121.75	52 Te 127.60	53 I 126.90	54 Xe 131.30
78 Pt 195.09	79 Au 196.97	80 Hg 200.59	81 Tl 204.37	82 Pb 207.19	83 Bi 208.98	84 Po (210)	85 At (210)	86 Rn (222)

67 Ho 164.93	68 Er 167.26	69 Tm 168.93	70 Yb 173.04	71 Lu 174.97
99 Es (254)	100 Fm (253)	101 Md (256)	102 No (253)	103 Lr (257)

MODELS IN CHEMICAL SCIENCE

MODELS IN
CHEMICAL SCIENCE

an introduction to
general
chemistry

GEORGE S. HAMMOND
California Institute of Technology

JANET OSTERYOUNG
Colorado State University

THOMAS H. CRAWFORD
University of Louisville

HARRY B. GRAY
California Institute of Technology

W. A. Benjamin, Inc.   New York   1971

MODELS IN CHEMICAL SCIENCE

Copyright © 1971
by W. A. Benjamin, Inc.
All rights reserved
Standard Book Number 8053–3670–2
(Clothbound Edition)
Library of Congress Catalog
Card Number 75–145670
Manufactured in the
United States of America

12345K4321

The authors and publisher are
pleased to acknowledge
the assistance of
Wladislaw Finne, who
designed the text and
the cover.
W. A. Benjamin, Inc.
New York, New York 10016

PREFACE

We have tried in this book to provide an introduction to chemical science that stresses the role of models in stimulating interplay between theory and experiment. Everyone in science uses models in thinking about experimental data. We scientists tend to focus not on the real thing, but on the behavior of our own mental constructs. By highlighting the model approach, along with a general irreverence for the permanency of theoretical dogma, we hope that students will realize that science is a remarkable human endeavor, not the product of some mechanical super-intelligence.

The most valuable model of modern chemistry is the picture of molecular structure based upon the interaction of nuclei and electrons. The ideas contributing to the model range from simple notions concerning electrostatic attraction and repulsion, to sophisticated formalisms based on particle quantum mechanics. We try to show ways in which useful generalizations can be derived from both the simple and the complex. Thus, in Chapter 6, we show the way in which the pattern of known chemical composition of compounds of the light elements can be built up by a simple counting process. Modern teachers of chemistry have tended to soft-pedal this primitive, but powerful, method because the same conclusions can be derived from the much more complex reasoning that we introduce in Chapter 7 to deal with problems such as molecular geometry, which are not well understood in terms of simple theory. Models based upon quantum mechanics and those based upon the idea that the noble gases have especially stable electronic configurations are complementary, and presentation of both types should encourage students to seek models that work, without expecting them to be certified as having some ultimate reality.

The first seven chapters are directed almost exclusively toward the development of concepts of molecular structure. In Chapter 8 (Solids and Liquids) and Chapter 9 (Solutions), the orientation is still largely structural, although concepts of reactivity are introduced in connection with the processes of vaporization, melting, and dissolution. In Chapter 9, we preview the concept of dynamic equilibrium, which is the main theme of Chapter 10. In the latter chapter, we attempt to present ideas concerning chemical equilibrium, beginning with a primitive model of competing dynamic processes. By including the thermodynamic formulation of the equilibrium process, we have made a straightforward compromise that we believe to be necessary in an introductory book. We discuss the reasons for needing some function, such as the Gibbs free energy, to use

67023

as a criterion for the establishment of equilibrium, and then state that the function can be defined in terms of heat content and entropy.

Chapters 11–14 are devoted to an introduction to chemical reactions. The treatment is not enough really to satisfy the authors' desire to expose beginning students to the marvelous phenomena of chemical reactivity, but the coverage is compatible with criteria of brevity for the book.

Throughout the book, we have used both organic and inorganic compounds to illustrate topics such as molecular structure, bonding, equilibrium, and rates and mechanisms of chemical reactions. This is an example of the fact that we have written a *general chemistry* text.

We believe that a goal in teaching any subject should be to place the material in context. Students should learn something of the origins of the subject, because history is a great help in understanding the dynamic character of science. Perhaps even more important is some appreciation of the way in which chemical science, or any other discipline, fits into the context of current human understanding and activity. We have made no attempt to treat either kind of relationship extensively. Historical development is illustrated mainly by brief accounts of the development of molecular theory and the particulate structure of matter. Neither treatment is exceptional, because we feel that these stories have been told frequently in a meaningful way. We only hope that students will realize that the present state of theory of molecular structure is not the end of the evolutionary chain.

Relevance to the present is treated mainly in the postscripts to the fourteen chapters. We try to show ways in which chemistry is related to other sciences, to some of the complex problems of our society, and to some of our own quasiphilosophical views of life. We make no claim that the relationships are established rigorously. However, we are fascinated by the tremendous outreach of chemical science and have tried to share this feeling with our student audience.

George S. Hammond
Janet Osteryoung
Thomas H. Crawford
Harry B. Gray

Aspen, Colorado
September 1970

CONTENTS

14 Structures and Reactions of Compounds of Carbon and Silicon   346

MODELS IN CHEMICAL SCIENCE

INTRODUCTION
WHAT IS CHEMISTRY?

Many questions occur to students as they begin any new course of study. What is it? What can I get from the course besides fulfillment of some curriculum requirement? Is the course hard, and how can I study it most efficiently? Is the teacher real, or just a fellow pushing a party line? In the case of chemistry, as with most other subjects, no final and definitive answers can be given. The value of the course, and even what it really is, will be different for each person. A textbook author or a teacher can give answers only as he sees them. In this spirit, we will write briefly about chemistry.

The Random House Dictionary[1] defines chemistry as "The science that deals with or investigates the changes and properties of substances and various elementary forms of matter." The definition is acceptable and probably above par for dictionary definitions. However, if we really stop to think about the words, the meaning is unpleasantly vague. Some of us are not entirely sure what is "science" and what is not; we can ask "What changes?" "What properties?" and so on. As a matter of fact, the definition begins to make real sense only after we have learned enough about chemistry to feed some prejudices into interpretation of the words. For example, if we look up the word "substance" in the same dictionary, we find no less than thirteen definitions. We learn that substance can mean all kinds of things, from "reality" to "wealth." The only definition that obviously fits is "A species of matter of definite chemical composition." This is not much help if we want to know what chemistry is, because the relationship between the definitions is circular. Some of the other definitions make "substance" seem about the same as "matter." However, we know that a horse is matter, or at least material, and every child knows that the study of horses is biology or animal husbandry; the study of horses is not chemistry. Likewise, an electron is matter, but everyone knows that electrons belong to the physicist, unless the electron happens to be a part of some complicated form of matter, then the electron may become chemical.

All of this reflection and playing with words leads to the uncomfortable conclusion that chemistry is something done or discussed by people who call themselves chemists. Furthermore, history shows that by this criterion, chemistry changes with time. We can find many examples of research that only a few years ago would have been called physics or biology, but now are called chemistry because now the work is done by chemists. This state of affairs may seem disturbing to students, but it is a reflection of the fact that science, as with all areas of human understanding, is dynamic. Chemistry is not a cleanly defined, finished package; it is a diffuse field of study and performance that is moving. Part of our task as teachers and writers is to give

---

[1] *Random House Dictionary of the English Language* (Random House, New York, 1966).

a picture of where the subject came from and where it is today. We should also attempt the harder job of guessing where it may go in the future.

Chemistry Is A Part of Science

Chemistry is a part of the great field of learning called science, and even science is not well defined. However, most people will agree that science involves the *systematic* study of what things are and how they behave. Arguments now rage as to whether sociology and economics are sciences, largely because a method that seems systematic to one person appears to another to be entirely nonsystematic. In fact, some rigid purists attempt to dispense with a discussion by maintaining that nothing can be considered systematic unless the results can be expressed in the language of mathematics. This strikes us as an intolerably narrow and deceptive point of view. Mathematics does provide a beautifully effective way to express well-defined relationships and ideas. However, valuable systematic thought can exist and be expressed in other ways. In this book, we will present some topics in mathematical form because it is convenient and useful to do so. But we wish to emphasize that there are no mathematical statements in the book that cannot be translated into English. In fact, the mathematics in virtually any textbook of science can, in principle, be translated into English, although translation of the mathematics in advanced texts would often become ridiculously awkward. We will also present many ideas in nonmathematical form, again because that is the most convenient way to express what we want to say. Some of the ideas presented verbally can be turned into mathematical statements, but this cannot be done in every case, at least not by the authors of this book. Furthermore, we will issue a warning that a common pitfall in science is the desire to shove ideas into mathematical language prematurely. The urge to add "class" by producing mathematical statements can lead to distortion of basic concepts so that perfectly good theory is turned into bad theory.

Scientific Thinking Involves Models

Scientific thinking almost always involves models or pictures that exist in the human mind to help us think about the way nature behaves. The theoretical psychologist develops models for human behavior that he uses in thinking about the way in which a real human individual behaves. A physicist working with high-energy particles creates his own mental models for the ways in which fundamental particles behave, with a part of the model being the idea that fundamental particles exist.

Understanding the models of science is crucial to understanding science. It is also important to remember the fine distinction between the models and physical reality. A good model can be so successful in organizing our thoughts and predicting the behavior of real systems that we come to regard the model as real. Usually this confusion does not cause any problems. However, trouble may arise when two different scientists discussing the same phenomena, using slightly different models, engage in a bitter conflict in which they try to prove, or disprove, the reality of their models. Sometimes such strife has beneficial results because it stimulates people to do experiments that can provide valuable information as to the utility of the different models in predicting observable happenings. But the controversy can degenerate into an endless search for the truth in conflicting fairy tales. We must remember that scientific thinking is really a branch of symbolic logic.

Differentiation among the various branches of science can be made on the basis of the models commonly used by people in the different scientific disciplines. In these terms, chemists are likely to be people who think in terms of molecular models. We have a concept of the structure of matter that depends upon our picture of tiny bits of matter in which atoms are bound together to form molecules. We think of every molecule of a pure substance as being exactly the same as the billions of billions of other molecules of that substance in a visible sample of the material. When we think about chemical reactions, we think about the ways in which the molecules can stretch and bend to form new molecules.

We also think about molecules in various ways. Some chemists may visualize a molecule by thinking about a physical model made from balls and sticks. Another chemist may think of a molecule as an aggregate of heavy atomic nuclei embedded in a cloud of rapidly moving electrons. Others may think of molecules in terms of mathematical functions devised to model the behavior of electrons and nuclei. The common unit in thinking is really the molecule, so we can usually recognize a chemist by his reaction to the word "molecule." Some chemists spend their lives working with solid materials in which the pattern of atomic structure is endless; thus, there are really no discrete molecules. However, a chemist will think of the complex structure as a near relative of a molecule.

Chapter 1 in this book will give some history of the way in which the molecular model of matter developed. Every piece of physical evidence that we know of seems to be consistent with the idea that discrete molecules exist. This, then, is an example of a model that works so well that there doesn't seem to be much point in worrying about the question of whether molecules are real. However, we cannot make such claims for many of the things said about molecules. We will discuss ideas about the shapes of molecules, the ways in which they are bound together, and what happens to molecules in various observable changes in matter. In short, we will present ideas and facts in terms of the chemist's intellectual unit, the molecule. These ideas are not of equal stature; some are well tested and others more speculative. We hope that students will bear with us in the excursion, with the realization that we are dealing with ideas, not statements of ultimate reality.

Is the Scientific Method A Myth?

Science represents man's attempt to understand himself and his universe. In addition to thinking in terms of models, the scientist observes what happens. If the observation is specially directed, it may be called an experiment. The classical form of an experiment is to devise a system in which we control factors fairly carefully and watch for certain kinds of results. However, when an astronomer does an experiment, he has no control over the thing he is observing; he simply makes the observations in a carefully planned manner.

Science, according to our definition, is at least as old as the recorded history of man. Whoever invented the wheel must have made numerous observations of the ways in which objects move when pushed or pulled. There is a myth that is rather popular with both scientists and nonscientists. This myth is known as "the scientific method." The mythology maintains that scientists work by a scheme usually attributed to Sir Francis Bacon, who wrote some interesting essays about science in the seventeenth century. Briefly, the method is described as consisting of three key steps:

1   Observe the behavior of nature carefully.
2   Seek for a theory (or model) that seems to correlate the observations.
3   Devise experiments to test the validity, or generality, of the theory.

Although the Baconian method has the sound of system and elegance, it does not do justice to the ingenuity, or recognize the humanness, of scientists. A skillful scientist does anything that works, including the use of "the scientific method."

Most research in chemistry is not aimed directly at the creation of a new theory. Research chemists have two common motives: (1) to study known phenomena in detail so that they can be fitted into the body of existing theory; (2) to study well-chosen new examples of a general phenomenon so that existing theory can be developed in greater detail. The dominant urge is, therefore, to make existing theory more detailed, and, consequently, more powerful in its capacity for predictive use. Occasionally, the results of an investigation seem completely contradictory to existing theory. When this happens, a new theory, or at least new twists to an old theory, may be born. The birth is likely to be slow and painful. Chemists cling to established theory as tenaciously as an infant clings to its security blanket. This trait may seem less than noble, but conservatism probably serves the long-range benefit of science. Most of the research findings that initially disagree with existing theory usually do fit, or at least require only a minor extension of, the old theory when the experimental data are reexamined carefully. Consequently, reckless abandonment of an old theory at the first provocation could lead to utter chaos in science.

A casual observer of the scientific scene may be shocked by the behavior of scientists. They quarrel and compete in a highly personal way in attempts to prove that "I am right and you are wrong." Furthermore, they are inclined to forget that the arguments are always phrased in terms of models that must necessarily differ in some ways from the physically real things that are discussed and studied. However, the irrational behavior is just the behavior of people. Science is a very human activity carried on by individual humans. When we realize this rather obvious fact, we must be impressed by the stature to which scientific understanding and information have grown. The structure is a magnificent accomplishment when viewed as a construct of human minds, although science may look like a poor job when viewed as the product of an omniscient intelligence.

In this book, we have presented a discussion of chemistry by reporting our thoughts about chemical subjects. Much of the reporting is done in the language of chemistry, although an attempt has been made to translate some of the ideas into English. If every concept were presented in English, the book would become at least ten times its present length, and students would be ill-equipped to understand anything expressed in chemical language. Since most students will not become professional chemists, one of our objectives in writing the book is to present enough ideas and language to provide students with the tools needed to retain contact with the growing, changing field of chemistry.

Chemical Technology Affects All Our Lives

Definition of the term "technology" is at least as difficult as the definition of science. We usually say that technology is "applied science" and that it involves the direction of ideas, methods, and facts of science toward some specific goal. The goal must be tangible, because application of science to the goal of satisfying man's curiosity is not considered technology. The manufacture of steel using knowledge of the chemistry of iron ores, the design of supersonic aircraft using knowledge of aerodynamics, and the creation of space vehicles using knowledge of chemistry, physics, and several engineering sciences, are all commonly called technology. However, the practice of modern medicine, which depends heavily on the findings of medical science, is not usually called technology.

We live in a technological age, and ours is a technological society. No matter how we may define terms, we all recognize that applied science has had a powerful effect on the character of our lives. A few years ago, technological advance was commonly accepted as almost an index to human progress. Skeptics were dismissed as odd people, suffering from pangs of jealousy. The situation has now changed considerably, and technology is no longer regarded as an obvious blessing to man. Critics are now enjoying a heyday as they point out problems, such as the threat of nuclear warfare and pollution of the environment, that are related directly to technological development. They also call attention to many human problems that are not being solved, despite the efforts of the technological society. We see the excesses in antitechnological thinking as no more realistic than the blind adulation of technology a few years ago. Few people really want to return to the conditions of cavemen. We now know that technology, which has removed us from the conditions of primitive life, cannot alone guarantee that men will not return to these conditions. This conclusion disappoints and frustrates people who had begun to regard science and technology as a kind of benign force moving us toward Utopia. Scientists themselves have contributed to the utopian dream by their enthusiasm, and sometimes arrogance, about the value of their work.

We see modern science as a dynamic and valuable *human* enterprise. The technological outgrowth of science is so great that it is somewhat frightening. But it is the concern of all men that we use technology to provide a life of high quality for the people of the earth without destroying ourselves in the attempt. Some scientists would like to be removed from the technology that may grow from their work. Their fantasyland is no more strange than others to which various people retire when overcome by the total immensity of human problems. Most scientists, like most people, feel concerned with the problems of our time, and they will be involved in working for solutions whenever they can see a way of doing so effectively. However, we have no magic wand, and we suffer with our fellowmen in the realization that problems will never disappear from the lives of people.

Applied chemistry is found throughout technology. The chemical industry is directed toward the manufacture of goods by performing chemical reactions with available raw materials. In the United States, the principal sources of raw materials are petroleum products, coal, nitrogen from the atmosphere, and mineral deposits. From these starting materials come synthetic fibers, transistors, paints, plastic goods of all kinds, medicines, mind-altering drugs, surf boards, ink, paper, fertilizers, and myriad other products. If chemical manufacturing were to cease, people would die in countless numbers before we could readjust to the changes required.

Applied chemistry also appears in many other branches of technology. Environmental engineering is based upon knowledge of the chemical reactions that occur in the atmosphere and the bodies of water on the earth. Electrical engineering, which has brought the new era of global communication, is critically dependent on chemistry for the production of materials having the properties needed in all kinds of electronic devices. Food supplies can be augmented by agricultural chemistry, and solution of the world population problem, if it is found without resort to a devastating war, will surely be partly a chemical solution.

In a medium-sized textbook we cannot possibly deal adequately with all of the philosophical background and the outreach of chemistry. But this is nothing new! Every teacher who really wants to teach has an instinctive desire to tell his students nearly everything he knows. We know that we ourselves barely know enough to survive and fear that our students cannot possibly prosper unless they know at least as much as we do. This is, of course, our folly. Much of what we know and think should, and will, be forgotten as mankind moves forward. We only hope to show students some of the span of chemical science, from elementary reasoning to fairly sophisticated model building, and to give some impression of the way in which we see chemistry as one of the threads in the fabric of human experience.

Questions and Problems

1   Scientists are not the only people who think in terms of intellectual models. What kinds of models are involved in your thoughts about the following subjects: (1) the people of India, (2) the stars in the sky, (3) poverty, and (4) the spirit of man?

2   A discarded model of the shape of the earth pictured it as flat. See how well you can fit the facts about the earth, as you have experienced them personally, to this model.

3   Translate the following mathematical statements into English. Which of the statements are definitions and which are not?

$$3 + 4 = 7$$
$$2x + 4y = 27$$
$$a \times a = a^2$$
$$A = \pi r^2$$

4   Would chemists still exist if the molecular model were discarded?

5   Do you think that the ripening of a green apple involves one chemical change, or several? Note that we do not expect you to know *the* answer, just to think about the problem. What kinds of models did you use in your thinking?

6   Do you think that blue eyes contain different kinds of molecules than brown eyes contain?

7   The usual unit for thinking by chemists is the molecule. What would you consider to be useful thinking units for people in the following fields: biology, psychology, sociology, economics, and astronomy?

8   What is the relationship between models that exist in our minds and physical models, such as plastic model airplanes? Is the physical airplane model any help in thinking about a real airplane?

9   What makes the leaves on a tree move? Would tree leaves move in the same way if the tree were suddenly put down on the surface of the moon? Can the movement of tree leaves have anything to do with molecules?

10 Considerable criticism has been aimed at chemists because some products of chemical technology, such as unburned fuel in automobile exhausts and the effluent from chemical manufacturing processes, contribute to pollution of the environment. Do you think that chemists, therefore, should alter their ethical standards? If so, how?

ATOMS AND MOLECULES

Much of what is known about chemistry today can be explained by describing the way in which *molecules* are built from *atoms.* We all know the taste of sugar and salt, the smell of vanilla, and the appearance of water. We have learned to recognize these materials because they have certain characteristic properties shared by all samples of the substances. The constancy of the properties of familiar substances like these is something we take for granted and seldom think about. But this constancy has fundamental scientific significance, for it leads us to ask why.

Why does every grain of sugar taste the same as every other grain of sugar? A simple and reasonable explanation for this is that all grains of sugar contain the same fundamental building blocks that are organized into units in the same way. These units must be very small, because the taste of sugar, and many other of its properties, remain the same, even though we examine samples too small to be seen with the naked eye. It seems reasonable that, if a pure substance such as sugar is subdivided enough times, some very small unit must at last be obtained that cannot be divided further without changing the fundamental character of the material. There are many experiments that demonstrate that this is correct. *The smallest units of a pure substance are called molecules.*

In this chapter, we will see that the matter of the universe can be divided into pure substances that have definite properties. We will show how the property of *constant composition* led to the idea that smaller units called atoms combine to form molecules. Finally, we will use models to show how much we can learn about the nature of pure substances from the simple idea that molecules consist of atoms hooked together in a definite way.

1–1 What Is Matter?

Matter, such as water, a cube of sugar, or a shirt, has mass and volume. We can measure the mass of some quantity of matter by using the property that substances with the same mass exert the same force when acted on by gravity; that is, they have the same weight. We can measure the weight of a quantity of matter by comparing its weight with standard weights. We can measure the volume of a quantity of matter either by measuring its dimensions and calculating its volume or by measuring the volume it displaces. For instance, we could measure the volume of an irregularly shaped rock by immersing it in water and measuring the volume of water displaced.

The volume and mass of an object are related by a property called *density.* Density, which is defined as weight over volume, is usually expressed in grams per cubic centimeter (g cm^{-3}). If we know the weight of a sample and its density, we can calculate its volume

$$\text{volume} = \frac{\text{weight}}{\text{density}}$$

If we know the volume and the density, we can calculate the weight

$$\text{weight} = \text{density} \times \text{volume}$$

Example 1–1. An irregularly shaped piece of aluminum weighs 36.93 g and displaces a volume of water equal to 13.67 cm^3. Calculate the density of aluminum.

Solution.

$$\text{density} = \frac{\text{weight of substance}}{\text{volume of substance}}$$

$$\text{density} = \frac{36.93 \text{ g}}{13.67 \text{ cm}^3} = 2.702 \text{ g cm}^{-3}$$

Example 1–2. What volume would a piece of aluminum weighing 10.32 g occupy?

Solution.

$$\text{volume} = \frac{\text{weight of substance}}{\text{density of substance}} = \frac{\text{g}}{\text{g cm}^{-3}} = \text{cm}^3$$

$$\text{volume} = \frac{10.32 \text{ g}}{2.702 \text{ g cm}^{-3}} = 3.819 \text{ cm}^3$$

Weight and volume are properties of matter that chemists measure every day in their investigations. They are fundamental properties that provide a starting point for chemical investigations. In particular, we will see in this chapter how important weight measurements were in the development of ideas about the units of matter.

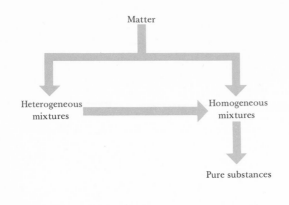

Figure 1–1. Classes of matter.

1–2   Mixtures and Pure Substances

Most samples of matter we see around us are mixtures. These mixtures can be divided into two broad classes—*heterogeneous mixtures* and *homogeneous mixtures.* Heterogeneous mixtures are not uniform, rather they have physically distinct portions that are separated from each other by definite boundaries.

An example of a heterogeneous mixture is navy bean soup. We can easily see that there are definite boundaries separating the beans from the rest of the soup. The mixture can easily be separated by straining out the beans. In this example, we have a mixture of solid (the beans) and liquid (the rest of the soup). Concrete is an example of a solid heterogeneous mixture. An example of a heterogeneous mixture that is of worldwide concern is the atmosphere around heavily industrialized areas. The air contains mixtures of gases, condensed vapors, and solid particles. The separation and removal of these pollutants from the air is often a complex and technologically difficult task. A very large effort is being made by industry and government to regulate and control these mixtures because of the adverse effect they have on plants and animals, including humans.

Heterogeneous mixtures usually can be separated into different homogeneous mixtures that are uniform throughout the sample. For example, smog can be passed through a filter that removes the solid particles and leaves a homogeneous mixture of gases.

Suppose that we prepare a homogeneous mixture by dissolving table salt (sodium chloride) in water. How could we separate the mixture to retrieve the original substances? In this case, it is easy. We would boil the mixture, catching the steam that then condenses to water when it is cooled. When all the water had boiled away we would have solid salt and pure water. This process of purification is known as *distillation.* This simple example illustrates how man might go about solving another of his most serious problems; that is, providing sufficient quantities of usable water from oceans to meet the increasing water need of an expanding population. However, the purification of sea water by distillation requires a large amount of energy, and energy is expensive. Today many scientists and engineers are working actively to improve the efficiency of methods for separating water from salt and to find more economical sources of energy to make the purification of sea water technologically and economically practical.

It is not always easy to determine whether a sample of material is a homogeneous mixture or a single substance. This is an important question to answer, for it is embarrassing for a chemist to find that he has been studying a mixture when he thought it was one substance. An example is the confusion among chemists before the nineteenth century when it was not recognized that the atmosphere is a mixture of several gases, not simply one gas.

Heterogeneous mixtures often can be separated by simple mechanical means, such as filtering. Homogeneous mixtures can be separated by physical changes such as distillation (the salt–water case), freezing, and melting. A homogeneous material is one pure substance, and not a mixture, if all possible methods fail to separate it into two or more substances. There are also some tests that help chemists decide if a substance is pure or not. For example, a single pure substance has a fixed boiling point that is characteristic of that substance, whereas the boiling point of a mixture depends on its composition.

The development of chemistry as a science and the systematic study of chemistry today depends on working with pure substances. In the following sections we will describe pure substances and the ideas used to explain their properties.

1–3   Compounds and Elements

Pure substances can be obtained from mixtures by physical separation. These methods of separation do not involve chemical change. Most pure substances can be decomposed into two or more substances by a treatment that causes a chemical change. For example, water can be decomposed into hydrogen and oxygen by passing an electric current through water (Figure 1–2). The salt calcium carbonate (limestone) can be decomposed into calcium oxide (lime) and carbon dioxide by heating

$$CaCO_3 \quad \rightarrow \quad CaO \quad + \quad CO_2$$
calcium carbonate   calcium oxide   carbon dioxide

Millions of pure substances are known to chemists. These substances either have been separated from mixtures occurring in nature or have been prepared in the laboratory. Most of

Figure 1–2.   Decomposition of water by an electric current.

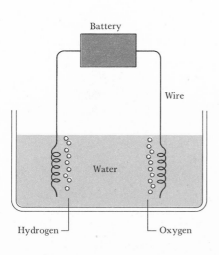

these pure substances are *compounds,* which can be decomposed by various chemical means to a relatively small number of simpler materials called *elements.* All of the millions of pure compounds can be made, directly or indirectly, by suitable combination of the 105 known elements. One of the most important steps in the development of chemistry was the discovery of the relationships between elements and compounds.

ELEMENTARY ATOMIC THEORY

During the sixteenth century, the idea developed gradually that all of the hundreds of pure substances then known were composed of atoms. The idea that there are fundamental, indestructible particles from which all matter is made is actually a very old idea that was discussed by the Greek philosopher Democritus in the fourth century B.C. He had no clear idea of what these particles were like, and he reached no satisfactory conclusion to the problem of what the difference is between atoms of two different elements.

However, the ideas of Democritus about atoms were valuable because they greatly stimulated the thought of men who wondered about the nature of matter in the first years of the growth of modern science. For example, the great physicist Newton believed that atoms must be the indestructible units of matter of which all substances are made.

The impact of the atomic theory on the development of chemistry did not occur until John Dalton (1776–1844) combined the idea of atoms with his knowledge of *chemical composition.* After we discuss chemical composition, we will return to Dalton's atomic theory, which provided the basis for the development of chemistry as a science.

1–4 Chemical Composition

We have discussed the notion that the properties of a pure substance would remain unchanged if a sample of the material were divided into smaller and smaller pieces. However, we were not specific about what properties were involved. The taste of sugar is not a good example because the taste would become too faint to be detectable long before the sample was reduced to the size of one molecule. Perhaps the best property to observe is chemical composition.

METHODS OF DETERMINING CHEMICAL COMPOSITION

Suppose that we had a sample of a compound and wished to determine its chemical composition; that is, what elements it contains, and how much of each is in the sample. We mentioned in Section 1–3 that water can be decomposed into hydrogen and oxygen by passing an electric current through it. If we decomposed 1.000 g of water this way and measured the weight of oxygen and hydrogen produced, we would find 0.888 g of oxygen and 0.112 g of hydrogen

$$2H_2O \quad \rightarrow \quad 2H_2 \quad + \quad O_2$$
$$\text{water} \qquad \text{hydrogen} \quad \text{oxygen}$$
$$1.000 \text{ g} \qquad 0.112 \text{ g} \qquad 0.888 \text{ g}$$

The percent of hydrogen in water is

$$\% \text{ hydrogen} = \frac{\text{weight of hydrogen}}{\text{weight of water}} \times 100 = 11.2\%$$

The percent of oxygen in water is

$$\% \text{ oxygen} = \frac{\text{weight of oxygen}}{\text{weight of water}} \times 100 = 88.8\%$$

Therefore, by this *elemental analysis* of water we find that its chemical composition by weight is

oxygen: 88.8%
hydrogen: 11.2%

We could find the same kind of information by making a substance from its elements, instead of decomposing a substance into its elements. The compound carbon dioxide can be made by burning carbon in oxygen. If we burn 1.000 g of carbon in an excess of oxygen (to make sure we had enough oxygen) we would find that 2.667 g of oxygen had been used and 3.667 g of carbon dioxide formed.

$$\begin{array}{ccccc} \text{C} & + & \text{O}_2 & \rightarrow & \text{CO}_2 \\ \text{carbon} & & \text{oxygen} & & \text{carbon dioxide} \\ 1.000\,\text{g} & & 2.667\,\text{g} & & 3.667\,\text{g} \end{array}$$

The percent composition of carbon dioxide by weight is

$$\frac{1.000 \text{ g carbon}}{3.667 \text{ g carbon dioxide}} \times 100 = 27.3\% \text{ carbon}$$

$$\frac{2.667 \text{ g oxygen}}{3.667 \text{ g carbon dioxide}} \times 100 = 72.7\% \text{ oxygen}$$

Let us return to our example of sugar. If we burn a sample of sugar by heating it in a stream of oxygen, we find that only two substances are formed, water and carbon dioxide

sugar + oxygen → water + carbon dioxide

This means that the only elements in sugar are carbon, hydrogen, and, perhaps, oxygen. Now let us try a quantitative experiment.

If we burn a 0.1000-g sample of sugar, we produce 0.0584 g of water and 0.1542 g of carbon dioxide. Since water is 11.2% hydrogen, the number of grams of hydrogen produced can be calculated by multiplying the weight of the water produced by the fraction of its weight that is hydrogen

weight of hydrogen = weight of water × 0.112
weight of hydrogen = 0.0584 g × 0.112 = 0.00654 g

Since carbon dioxide is 27.3% carbon

weight of carbon = weight of carbon dioxide × 0.273
weight of carbon = 0.1542 g × 0.273 = 0.0421 g

Therefore, the composition of sugar is

$$\frac{0.00654 \text{ g hydrogen}}{0.1000 \text{ g sugar}} \times 100\% = \frac{6.54 \times 10^{-3} \text{ g}}{1 \times 10^{-1} \text{ g}} \times 10^2$$

$$= 6.54\% \text{ hydrogen}$$

$$\frac{0.0421 \text{ g carbon}}{0.1000 \text{ g sugar}} \times 100\% = \frac{4.21 \times 10^{-3} \text{ g}}{1 \times 10^{-1} \text{ g}} \times 10^2$$

$$= 42.1\% \text{ carbon}$$

Since oxygen was the only element in addition to hydrogen and carbon in the combustion products, the rest of the weight of the sugar must have been oxygen. The weight of oxygen cannot be measured directly, since some of the oxygen in the products must have come from gaseous oxygen used to burn the sample. However, the amount can be calculated as the difference between the total sample weight and the weight of hydrogen and carbon

$$100.0\% - 6.5\% - 42.1\% = 51.4\% = \text{percent of oxygen in sugar}$$

These are some examples of how a compound can be broken up into elements, or converted into other compounds whose composition we know. Experiments like these were important in arriving at some explanation of the nature of compounds, and today they are necessary and routine operations performed by chemists studying new materials and new reactions.

CONSERVATION OF MASS

When we converted carbon and oxygen into carbon dioxide and used the weights of the elements involved to calculate the chemical composition of carbon dioxide, we made one very important assumption. We assumed that the weight of carbon used is equal to the weight of carbon in the product, carbon dioxide, and, similarly, that the weight of oxygen used is equal to the weight of oxygen in the carbon dioxide formed. This assumption is a very important principle that is true of all chemical reactions. It is called the *conservation of mass.*

Physicists often deal with nuclear reactions in which atoms of one element are converted to atoms of other elements. In these nuclear reactions, mass is often converted to energy, or energy converted to mass, according to Einstein's equation, $E = mc^2$ (E = energy, m = mass, c = speed of light). But in all *chemical reactions,* mass is conserved. The weight of the reactants (for instance, the weights of carbon and oxygen) is equal to the weight of the products (carbon dioxide in this example). Furthermore, the weight of each of the elements in the reactants is equal to its weight in the products.

Example 1–3. Metals may be heated in an excess of oxygen and converted completely to metal oxides. For example, when 6.078 g of magnesium is heated in the presence of excess oxygen, it is converted to 10.08 g of magnesium oxide. How many grams of oxygen are used, and what is the percentage of oxygen in the metal oxide?

Solution.

$$2Mg + O_2 \rightarrow 2MgO$$
magnesium   oxygen   magnesium oxide
6.078 g x g   10.08 g

Since the total weight of the product, magnesium oxide, must be equal to the combined weights of the reactants, magnesium and oxygen (according to the law of conservation of mass), it follows that

10.08 g magnesium oxide − 6.078 g magnesium = 4.00 g oxygen

The percent oxygen in magnesium oxide is

$$\frac{\text{g oxygen}}{\text{g metal oxide}} \times 100\% = \frac{4.00 \text{ g}}{10.08 \text{ g}} \times 100\% = 39.7\%$$

CONSTANT COMPOSITION

The results of the elemental analysis of sugar presented previously in this section are independent of the source of sugar or the size of the sample. This is the most important property of all compounds—all compounds have *constant chemical composition.* When we discussed mixtures we had some difficulty in describing the difference between a homogeneous mixture and a pure substance. But now the distinction is clear. Mixtures have variable composition. A homogeneous mixture of sugar and water at room temperature may contain any amount of sugar up to 68% by weight. (The water is saturated with sugar at this percent composition; that is, no more sugar will dissolve at this temperature.) But sugar and water, which are pure substances, have fixed compositions that never vary.

The property of constant composition allows us to divide all of the matter in the universe into two distinct groups: matter which does not have constant composition (mixtures) and matter which does (pure substances; that is, compounds and elements). This property also demands an explanation. After all, such a constant and widespread property of matter surely is not a coincidence.

During the sixteenth to nineteenth centuries, the principle of constant composition was apparently recognized, at least tacitly, because chemists published recipes that called for definite weights of the ingredients corresponding to the composition of the compound they wanted to make. During the nineteenth century, many systematic elemental analyses of compounds were performed, usually with the handicap of inadequate equipment and poor methods. The results of these experiments were often so inaccurate that the law of constant composition seemed in doubt. But many chemists realized that their experiments could contain errors, and, as more evidence was accumulated, the law was finally stated by Joseph Proust, in 1797. It was gradually accepted as a fundamental property of compounds that must be explained by some theory. The first successful attempt to describe the nature of compounds in a way that would explain the law of constant composition was made by the English chemist John Dalton, in 1802.

1–5 Dalton's Atomic Theory

Dalton accepted the law of constant composition and believed that compounds were composed of atoms, the indestructible particles of which matter is made. He wanted to describe the fundamental properties of atoms and compounds in a way that would explain the law of constant composition. Realizing that the important measurable property was weight, he gave the following explanation known as Dalton's atomic theory:

1 All atoms of each element are the same. In particular, they have the same weight, and atoms of different elements have different weights.

2 Compounds are made of atoms combined with each other in simple ratios.

These two simple assumptions neatly explain constant chemical composition. Each compound has a fixed (definite) ratio of atoms, and each atom has a fixed weight. Therefore, the composition of each compound, by weight, is fixed.

But more important, Dalton's theory implies that atoms are joined together to form molecules. Because the ratio of atoms in each molecule is fixed, the large samples we deal with, which contain molecules, necessarily have a fixed ratio of atoms.

The next question that occurs to us is: Why is the ratio of atoms in molecules of a compound fixed? It must be because atoms of each element can fit together with other atoms only in a few definite ways. Thus, molecules must contain atoms hooked together in a definite structure. In the next three sections of this chapter, we will explore this idea further.

Dalton's theory also showed chemists the direction they must take to find out more about the chemical behavior of matter. Dalton stated two problems that dominated chemical research for a century. First, what are the relative weights of the atoms of different elements (*atomic weights*); second, what are the combining ratios of atoms in compounds?

An example will show how these problems are connected and why something more than the composition by weight of compounds must be known. Water has the composition hydrogen, 11.2%; oxygen, 88.8%. The ratio of the weight of oxygen to the weight of hydrogen in water is

$$\frac{\text{weight of oxygen}}{\text{weight of hydrogen}} = \frac{88.8}{11.2} = 7.93$$

This is the ratio of the weights in each individual molecule of water. Suppose that we want to find the ratio of the weight of one oxygen atom to that of one hydrogen atom, that is, the ratio of their atomic weights. The ratio of the weights of oxygen and hydrogen in water is

$$\frac{\text{number of atoms of oxygen per molecule} \times \text{atomic weight of oxygen}}{\text{number of atoms of hydrogen per molecule} \times \text{atomic weight of hydrogen}} = 7.93$$

The two factors that we would like to know are the ratios

$$\frac{\text{number of atoms of oxygen per molecule}}{\text{number of atoms of hydrogen per molecule}}$$

and

$$\frac{\text{atomic weight of oxygen}}{\text{atomic weight of hydrogen}}$$

But to find either ratio from the chemical composition, we need to know the other.

We will see in Chapter 2 how this sort of problem was solved using additional information. In the meantime, because the relative weights of the atoms are so important, we will return to this subject at the end of this chapter and discuss atomic and molecular weights.

1–6   Molecular Structure

The fact that pure compounds have constant composition leads to the most important theory in chemistry—the theory of *definite molecular structure*. We believe that every compound has a definite composition because individual molecules of the compound have the same definite structure and because the molecules are made by the combination of atoms of the constituent elements in definite (usually predictable) ways.

At the present time, the theory of definite molecular structure is so familiar that appreciation of its importance is difficult. Most first-year college students have known that the formula for water is H_2O for so long that they have forgotten when they first learned the fact. But there are several ways in which the three atoms in the water molecule could be bound together. Two of the easiest to think of are represented by structures (a) and (b)

H—O—H   H—H—O
   (a)   (b)

Our ideas about molecular structure lead us to believe only one of these structures can be correct. We are convinced that structure (a) is the correct one because of direct evidence obtained from the study of the ways in which beams of x rays, neutrons, or electrons are scattered when passed through samples of water. These methods of studying molecular structure tell us not only the sequence of atoms, H—O—H, but also the shape of the molecule. The atoms are not arranged in a straight line, as is implied by (a); the molecule is bent. Thus, we can describe water molecules in more detail by giving the distance separating the oxygen and the hydrogen atoms and the angle formed by the three atoms

    The distance between two attached (bonded) atoms is called the *bond length.* These are very short distances compared with the dimensions of objects we can see. An angstrom unit (Å) is one hundred millionth of a centimeter (10^{-8} cm). For contrast, the diameter of the earth is about one hundred million centimeters (10^8 cm). Another way to think about molecular sizes is provided by the fact that one cubic centimeter (cm^3) of liquid water contains 3×10^{22} molecules.

1–7 Models

Models play an important role in a chemist's thinking about chemistry. A model can be either a real or an imaginary device that acts like the more complicated natural object the model represents. In a way, models are like toys. By studying the way a model behaves a scientist learns about nature, just as a child learns by playing with toys. A good model is useful because it is simpler than the natural object that it represents. But because models are oversimplified, they can never tell the scientist all he may want to know about the real physical system. The theory of molecular structure is a kind of intellectual model. Many of the conclusions of the theory can be illustrated, and new predictions can be made by using simple physical models that can be handled like a simple set of tinker toys.

    When a chemist thinks about water, he is likely to think "H_2O" rather than thinking about the colorless liquid that flows from the tap in the kitchen. The existence of different kinds of models is also demonstrated by a chemist's thoughts about water. He may simply think of the molecular formula, H_2O, or he may think about the more descriptive formula H—O , or he may think
                                             H

of a physical model that can be built and handled in a way that could never be possible with a real molecule.

Figure 1–3. Models of a water molecule.

MOLECULAR MODELS

The physical models constructed by scientists are made to help their thinking. There are several kinds of molecular models that are constructed to show different ideas about molecular structure. Figure 1–3 shows various kinds of models of water molecules. Each model attempts to show different things. All show that hydrogen atoms are bound to oxygen atoms, and all give some indication of the shape of the molecule. In addition, space-filling models give an indication of the relative sizes of the atoms and indicate that two molecules cannot actually become entangled, as might be guessed by looking at other kinds of models.

In these models, actual molecules are, in a sense, duplicated. Experience with model building shows that it is a very useful tool for predicting chemical structure and behavior; that is, model building works. When we wonder why models work so well, it is tempting to conclude that the models imitate molecules because molecules are like very small Tinker Toys. And this is to some extent true. But the ways in which molecules are different from Tinker Toys are very important, and these differences limit the utility of models. A striking example of the importance of model making may be found in the Nobel Prize-winning work of James Watson and Francis Crick, who used models as a primary tool for establishing the double helix structure of the biologically important compound deoxyribonucleic acid (DNA).[1]

Ball-and-stick models are worth studying in some detail. Balls of different colors represent atoms of different elements. Small sets, such as are usually used by students, contain only eight to ten kinds of balls of as many different colors. This may seem surprising in view of the fact that there are 105 elements. However, only a relatively small number of elements are found in the vast majority of all known compounds. Also, a considerable number of compounds cannot be well represented by molecular models because they do not have simple molecular structures. Sodium chloride, common table salt, is an example of such a compound.

---

[1] For an interesting account of the search for this important structure as well as a look into the lives of some well-known contemporary scientists, read J. D. Watson, *The Double Helix* (Atheneum, New York, 1968).

Sodium chloride is a solid, having a definite structure. However, the structure is an infinite, repeating pattern with sodium and chlorine atoms arranged so that one cannot pick out individual molecules. Dalton didn't know this, and for our purposes in this chapter we don't need to consider it. We will confine our discussion to compounds that have discrete molecules and defer substances such as sodium chloride to Chapter 8.

Although there is no special significance to the choice of colors for various atomic models, the colors used in different sets are often the same (e.g., hydrogen, white; carbon, black; nitrogen, blue; and oxygen, red). The model atoms have holes bored in them for the insertion of sticks, which represent bonds that hold the atoms together in molecules. Examination of the balls shows that the number of holes varies. All the white balls have a single hole. This is because a hydrogen atom is almost always bound to only one other atom. This property of hydrogen atoms is expressed by saying that the *valence* of hydrogen is one. The red balls all contain two holes, indicating that the usual valence of oxygen is two. In other words, an oxygen atom can form two bonds to other atoms, as in a water molecule. Similarly, since carbon usually has a valence of four, the black balls have four holes. Nitrogen commonly has a valence of three, thus the blue balls have three holes. Figure 1–4 shows ball-and-stick models of some compounds of hydrogen with the other elements.

THE USE OF MOLECULAR MODELS TO MAKE PREDICTIONS

Molecular models allow us to speculate about the possible existence of other, more complex compounds. For example, we can assemble molecular models in which two red balls are bonded together

The formula of the compound represented by this model would be H_2O_2. A compound having this composition and structure, hydrogen peroxide, is known. However, the fact that a model can be constructed is no guarantee that the compound will exist. We can also build a model of H_2O_3, but all attempts to prepare such a substance have been unsuccessful. Many compounds have been prepared recently after many years of fruitless attempts, so we dare not guarantee that H_2O_3 cannot exist.

Models also indicate that several kinds of atoms can be put together in the same molecule. The following are examples of well-known compounds for which molecular models can readily

Figure 1–4.   Ball-and-stick models of some hydrogen compounds.

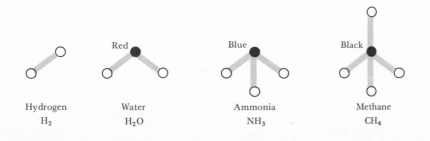

be assembled. Notice that we have written the formula of methyl alcohol as CH_3OH, rather than, for instance, CH_4O. By looking at the model for this molecule, you can see that this is a way of using the formula to tell something about the way the molecule is built. Three hydrogen atoms are attached to the carbon atom, so we write CH_3, which is called the methyl group. Then we write O, because an oxygen atom is attached to the carbon of the methyl group. Finally, we write H for the hydrogen atom that is attached to the oxygen atom. (It would be even more instructive to write the formula of methyl alcohol H_3COH.) There are no absolute rules to follow for doing this, but you see that the formula for hydroxylamine is written in the same sort of way to indicate something about how the atoms are joined together.

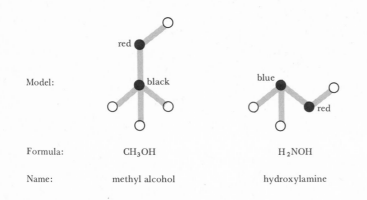

Model:

Formula: CH_3OH H_2NOH

Name: methyl alcohol   hydroxylamine

VALENCE

We have introduced the term valence in the preceding paragraphs. The valence of an element tells how many bonds each of its atoms forms with other atoms. Since most elements combine with hydrogen, the valence of an element is usually equal to the number of hydrogen atoms that can be bound by an atom of the element. Some elements have more than one valence in various compounds. Theories of structural chemistry, which will be introduced in Chapters 5, 6, and 7, will provide a theoretical basis for understanding valences. However, very useful rules of valence can be learned from working with atomic models.

1–8 Atomic and Molecular Weights

In the section on Dalton's atomic theory we mentioned the problem of determining the relative weights of atoms. In this section, we want to describe the scale of atomic weights that chemists use today.

Now we know the weights of individual atoms. But long before individual atomic weights were known, chemists developed a completely adequate system for expressing weight relationships in chemistry, using a scale of *relative weights*. A relative scale of measurement is one in which relationships between sizes, weights, and so forth, of the members of a group of things are known, but the absolute values are not necessarily known. For example, we could describe the members of a family by giving their weights relative to each other. We might do this by giving the baby a relative weight of one and then assigning the other relative weights accordingly

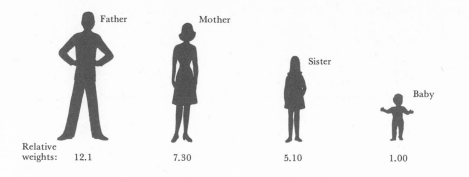

Relative weights: Father 12.1   Mother 7.30   Sister 5.10   Baby 1.00

Example 1–4. If the weight of the father is 90.0 kilograms (kg), what is the weight of the baby?

Solution. The problem states that the weight of the baby, relative to that of the father, expressed as a ratio, is

$$\frac{1.00}{12.1}$$

Whatever the weight of the father, the ratio of the baby's weight to his must be 1/12.1. Since the father's weight is 90.0 kg, we can let x be the baby's weight and solve the problem

$$\frac{1.00}{12.1} = \frac{x \text{ kg}}{90.0 \text{ kg}}$$

or

$$x \text{ kg} = \frac{90.0 \text{ kg} \times 1.00}{12.1}$$

which gives

$$x = 7.44 \text{ kg}$$

When a scientist obtains a result, he usually checks it against intuition based on experience. Is 7.44 kg a reasonable weight for a baby? Can you guess the approximate age of the baby?

It would have been reasonable to assign relative atomic weights on a scale with hydrogen, the lightest element, being given a relative atomic weight of one. Instead, a scale has been adopted that assigns a relative atomic weight of 1.008 to hydrogen. The reason for this choice will be discussed in Chapter 4. The inside back cover lists the relative atomic weights on this scale for all elements.

Example 1–5. A compound consists of 85.62% carbon and 14.38% hydrogen. Given that the relative weights of carbon and hydrogen are 12.01 and 1.008, respectively, what can you say about the relative numbers of carbon and hydrogen atoms present in the hydrocarbon compound?

Solution. If there were equal numbers of carbon and hydrogen atoms present, the ratio of the weights of these two elements would be

$$\frac{n_{\text{carbon}} \times 12.01}{n_{\text{hydrogen}} \times 1.008} = 11.90$$

The observed carbon-hydrogen weight ratio is

$$\frac{85.62}{14.38} = 5.95$$

We note that this is just one half the value for 1 carbon/1 hydrogen ratio. This suggests 1 carbon for 2 hydrogens

$$\frac{n_{\text{carbon}} \times 12.01}{2n_{\text{hydrogen}} \times 1.008} = 5.95$$

We can conclude, then, that whenever elements are present in quantities such that their weight ratios are the same as the ratios of their relative weights, there must be equal numbers of atoms of each element present. Now we define a new term called the *gram atomic weight,* or *gram-atom,* as that quantity of an element whose weight in grams is numerically equal to the atomic weight of the element. This is a useful definition because it allows us to work with measurable quantities of the elements and still be sure about the relative number of atoms present in the quantity of matter we are working with.

Example 1–6. How many gram-atoms are there in (a) 45.98 g sodium and (b) 0.0523 g carbon?
Solution.

(a) $45.98 \text{ g} \times \dfrac{1 \text{ g-atom}}{22.99 \text{ g}} = 2.000 \text{ g-atom}$

(b) $0.0523 \text{ g} \times \dfrac{1 \text{ g-atom}}{12.0 \text{ g}} = 4.36 \times 10^{-3} \text{ g-atom}$

Example 1–7. How many grams are there in 0.0342 g-atom of magnesium?
Solution.

$$0.0342 \text{ g-atom} \times \frac{24.31 \text{ g}}{1 \text{ g-atom}} = 8.31 \times 10^{-1} \text{ g}$$

RELATIVE MOLECULAR WEIGHTS

Relative molecular weights are assigned on the same scale as relative atomic weights. The relative molecular weight of a compound is the sum of the relative atomic weights of the atoms in one molecule. Thus, the relative molecular weight of water is twice the relative atomic weight of hydrogen plus the relative atomic weight of oxygen, or $2 \times 1.008 + 15.999 = 18.015$.

Chemists have defined a term similar to gram atomic weight for dealing with molecules. It is the gram molecular weight or, more briefly, the *mole.* One mole of a substance is the amount contained in one gram molecular weight. For example, one mole of water is 18.015 g of water, and one mole of carbon dioxide is 44.01 g of the substance.

Example 1–8. How many moles of substance are there in (a) 0.047 g nitrogen gas (N_2) and (b) 31.6 g $KMnO_4$ (potassium permanganate)?

Solution.

(a) moles of N_2 $= 0.047 \text{ g} \times \dfrac{1 \text{ mole}}{28.0 \text{ g}} = 1.7 \times 10^{-3} \text{ mole}$

(b) moles of $KMnO_4 = 31.6 \text{ g} \times \dfrac{1 \text{ mole}}{158 \text{ g}} = 0.200 \text{ mole}$

Example 1–9. How many grams are there in 0.540 mole HCl?
Solution.

$$\text{grams HCl} = 0.540 \text{ mole} \times \dfrac{36.4 \text{ g}}{\text{mole}} = 19.7 \text{ g}$$

The mole is a unit, like dozen or gross, which specifies a certain number of objects. In this sense of the definition, the mole is often used when referring to atoms and other particles, as well as molecules, almost to the exclusion of the other term gram-atom. For example, it is common to refer to 4.00 g of helium as a mole of helium, even though helium exists as uncombined atoms. Chemists can use this unit without knowing how many molecules are in one mole. The important aspect about the mole is that one molecular weight (one mole) of any compound or element contains the same number of molecules as one molecular weight of any other compound or element.

Now we know that one mole of molecules consists of 6.022×10^{23} molecules. This number is called Avogadro's number, N. There are several methods of determining the value of N, but these need not concern us at this time. What is important is that one mole of a substance is Avogadro's number of molecules of that substance. However, the value of N is not needed in many of the calculations of chemistry.

Example 1–10. How many moles of sugar are in 22.3 g of sugar? The molecular weight of sugar is 342.3 g.
Solution.

$$\text{weight of one mole} = 342.3 \text{ g}$$
$$\text{number of moles in 22.3 g} = \dfrac{22.3 \text{ g}}{342.3 \text{ g mole}^{-1}} = 0.0652 \text{ mole}$$

Example 1–11. What is the weight of 0.221 mole of water? The molecular weight of water is 18.02 g.
Solution.

$$\text{weight of one mole} = 18.02 \text{ g}$$
$$\text{weight of 0.221 mole} = 0.221 \text{ mole} \times 18.02 \text{ g mole}^{-1} = 3.98 \text{ g}$$

SIMPLEST FORMULA AND MOLECULAR FORMULA

Often it requires only a straightforward experiment to determine the composition of a pure substance, and from this information to calculate the relative numbers of atoms of the various elements in the substance. The relative number of atoms in the substance is known as the *simplest formula.* This should not be confused with the *molecular formula,* which gives the absolute number of atoms of a particular kind in each molecule of the substance. For example, the molecular formula of a certain hydrocarbon compound is C_6H_6. However, the simplest formula of this compound is C_1H_1, which gives the relative number of each kind of atom.

Example 1–12. Elemental analysis of a substance indicates it consists of 87.42% nitrogen and 12.58% hydrogen. Other studies indicate that its molecular weight is 32.04. What are the simplest formula and molecular formula for this compound?

Solution. Since we are given the percent composition by weight, we can assume any quantity of material we desire. It is convenient to start with 100 g, since 12.58% of that weight is 12.58 g of hydrogen, and 87.42% of the total weight is 87.42 g of nitrogen. We know that there are equal numbers of atoms in a gram-atom of any substance, and once we know the number of gram-atoms of nitrogen and hydrogen present, we will know the relative numbers of atoms of each

$$87.42 \text{ g} \times \frac{1 \text{ g-atom nitrogen}}{14.00 \text{ g}} = 6.244 \text{ g-atom nitrogen}$$

$$12.58 \text{ g} \times \frac{1 \text{ g-atom hydrogen}}{1.008 \text{ g}} = 12.48 \text{ g-atom hydrogen}$$

$$\frac{12.48 \text{ g-atom hydrogen}}{6.244 \text{ g-atom nitrogen}} = \frac{2}{1}$$

Therefore, the simplest formula is NH_2. If this were the molecular formula, the molecular weight would be 16.02. The experimental molecular weight is twice this, so the molecular formula must be N_2H_4.

Example 1–13. Calculate the simplest formula of a compound that consists of 40.28% potassium (K), 26.78% chromium (Cr), and 32.94% oxygen.

Solution. Calculate the number of gram-atoms of each element present, assuming a 100-g sample

$$40.28 \text{ g K} \times \frac{1 \text{ g-atom K}}{39.10 \text{ g K}} = 1.030 \text{ g-atom K}$$

$$26.78 \text{ g Cr} \times \frac{1 \text{ g-atom Cr}}{52.00 \text{ g Cr}} = 0.515 \text{ g-atom Cr}$$

$$32.94 \text{ g O} \times \frac{1 \text{ g-atom O}}{15.99 \text{ g O}} = 2.060 \text{ g-atom O}$$

To determine the relative number of gram-atoms of each element present, divide each by the smallest number of gram-atoms present

$$\frac{1.030 \text{ g-atom K}}{0.515 \text{ g-atom Cr}} = \frac{2.00 \text{ g-atom K}}{\text{g-atom Cr}}$$

$$\frac{2.060 \text{ g-atom O}}{0.515 \text{ g-atom Cr}} = \frac{4.00 \text{ g-atom O}}{\text{g-atom Cr}}$$

We can conclude that the simplest formula must be K_2CrO_4.

It is not possible to determine the molecular formula from the percent composition of a compound alone. Some additional information, usually the molecular weight, is required. The molecular weight can be measured in a variety of ways, some of which will be discussed in later chapters.

1–9 Summary

1 The matter of the universe is classified in the following way:

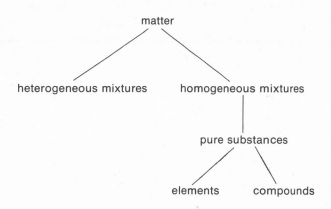

2 Compounds have constant chemical composition. This is explained by Dalton's atomic theory, which states that:

   (a) All atoms of each element are the same. In particular, they have the same weight, and atoms of different elements have different weights.

   (b) Compounds are made of atoms combined with each other in simple ratios.

3 The atomic theory leads to the idea of definite molecular structure. Molecules can be described, and predictions can be made about what kinds of molecules might exist, by using simple molecular models.

4 The weights of molecules are expressed in terms of atomic weights. One mole of a substance is one molecular weight expressed in grams.

5 The simplest formula and the molecular formula are not always the same. The simplest formula of a compound may be calculated from the percent composition of the substance; additional information, usually the compound's molecular weight, is required to obtain the molecular formula.

1–10 Postscript: The Relevance of Molecular Structure

There are a number of reasons for thinking that a sample of a pure material consists of many molecules, each having the same atomic structure. What do we gain from this model? We could, for example, state the law of constant composition without any reference to molecules.

   The value of the model comes from the ways scientists use the molecular theory in their thinking. Comparison of the properties of different substances is especially easy to discuss in terms of molecular structures. When we think about liquid water and wonder why it is a liquid rather than a vapor or a solid, we think about the ways in which individual water molecules interact with neighboring water molecules (Section 8–7). The picture that we imagine places emphasis on the properties of O—H bonds. By comparison, we should be able to make some predictions about other molecules containing O—H bonds. For example, methyl alcohol (CH_3OH) has properties that resemble those of water.

Much of the reasoning in chemistry is based upon complex analogy, and these analogies are usually based upon comparisons of the structural formulas of molecules. The way in which the strands of DNA, the basic genetic material of living things, are bound together (Section 1–7) is closely related to the way in which molecules of liquid water are bound to each other. Chemical formulas are symbols, so a chemist who thinks about his work by thinking about chemical formulas is using a mental process called symbolic logic. The symbol used to think about water provides a ready-made system for thinking about most complicated materials involved in the chemistry of life.

The beginning student must accept on faith the assertion that ideas about water, formulated in the chemist's terms, will pay rich dividends in formulation of ideas about many other substances. There is no way in which we can illustrate the real growth potential for the thought system in a chapter, or even in a book. However, we believe that the full potential for formulation of ideas in the symbolic language of chemistry is probably inexhaustible, as is the case in other forms of expression, such as natural languages, mathematics, music, and painting.

New Terms

Atom: The smallest chemically indivisible particle of an element.

Atomic weights: The relative weights of the atoms of the elements.

Avogadro's number (N):   The number of molecules in one mole, 6.022×10^{23}.

Bond length: The distance between the centers of two atoms bound together.

Compound: A pure substance composed of elements in fixed proportions.

Conservation of mass: The principle that the total mass of each element present does not change in a chemical reaction.

Constant chemical composition: The property that characterizes a pure substance. A substance has constant chemical composition if the relative amounts by weight of all the elements present do not change.

Density: The weight of an object divided by its volume, or the weight per unit volume.

Distillation: The process of slowly boiling a liquid and condensing the vapor given off; a separation technique for homogeneous mixtures.

Element: The basic chemical unit of matter. All chemical substances are made from atoms of the elements.

Elemental analysis: The process of breaking down a chemical substance into its elements and finding their relative proportions.

Gram atomic weight (gram-atom):   The quantity of an element whose weight is equal to its atomic weight expressed in grams.

Heterogeneous mixture: A nonuniform mixture with different phases; usually separable by mechanical means.

Homogeneous mixture: A uniform mixture that usually can be separated by some physical change.

Mole (gram molecular weight):   The amount of a substance whose weight is equal to its molecular weight expressed in grams. Avogadro's number of particles.

Molecular formula: The formula that gives the actual number of atoms of each element in one molecule of a substance.

Molecular weight: The relative weight of a molecule on the atomic weight scale; the sum of the atomic weights of the atoms in the molecule.

Molecule: The smallest unit of a pure substance. Molecules are composed of atoms hooked together in definite ways.

Simplest formula: The formula that gives the relative number of atoms of the various elements in a substance.

Valence: The number of bonds an element forms in a molecule.

Questions and Problems

1 What is matter and what are some of its properties that we can measure?

2 What do we call the ratio of the weight of an object to its volume?

3 What would be the length of a cylindrical column of liquid mercury weighing 515 g, if the cross-sectional area is 0.500 cm^2? The density of mercury is 13.59 g cm^{-3}.

4 What is the difference between a homogeneous and a heterogeneous mixture?

5 To which civilization do we trace the notion that matter is made of fundamentally indestructible particles?

6 Describe a chemical experiment that would illustrate the principle of conservation of mass.

7 Sodium chloride can be separated into pure sodium and chlorine by passing an electric current through the melted salt. Calculate the percent composition of sodium chloride if a 5.64-g sample produces 3.42 g of chlorine.

8 Describe Dalton's view of the atom and tell what impact this had on chemical research.

9 When we express the atomic weight of an element, is it an absolute weight or a relative weight?

10 Why do we know that there will always be the same number of sodium atoms in 23 g of sodium, but cannot be certain about the number of oranges in 10 pounds of oranges without counting them?

11 If 0.87 mole of a substance weighs 5.30 g, what is the molecular weight of the substance?

12 (a) Suppose that we were preparing one of the new microencapsulated medicines commonly found in "timed" capsules used as remedies for the common cold. The remedy is prepared for distribution by mixing two preparations supplied in tiny capsules of different sizes. The proper mixture contains twice as many large capsules as small capsules, but the capsule sizes are actually too small to be counted and mixed by numbers. The ratio of the capsule weights is 22.99/15.99, or one is 1.438 times as heavy as the other. What weight of the heavier capsules would be needed to mix with 750 g of the lighter ones? Do you need to know the actual numbers of capsules to mix them in proper proportions?

   (b) Consider combining the two elements sodium (atomic weight 22.99) and oxygen (atomic weight 15.99). Experiments show that two sodium atoms combine with one oxygen atom to give Na_2O. What weight of sodium would be required to combine with 64.0 g of oxygen? State the relationship between this problem and 12(a).

13 Calculate the molecular weight of each of the following compounds:

   (a) CH_4 (d) $C_7H_6O_2$

   (b) H_2SO_4 (e) B_2H_6

   (c) $Cu(NO_3)_2 \cdot 4H_2O$ (f) $XeO_2(OH)_4$

14 How many moles are there in the indicated weights of each of the following substances?

   (a) 0.8 g O_2 (d) 4.3×10^{-2} g SO_2

   (b) 421 g gold (Au)   (e) 62.4 g NH_3

   (c) 0.00245 g N_2 (f) 6.84×10^3 g uranium (U)

15 In 1962, the first of a series of compounds of xenon, Xe, with fluorine and oxygen was discovered. What is the molecular weight of potassium xenate, K_6XeO_6? What is its xenon
composition in percent by weight?

16 How many grams of matter are there in each of the following?
(a) 0.00387 mole O_2 (d) 721 moles H_2
(b) 55.55 moles H_2O (e) 32×10^{-5} mole CO_2
(c) 7.2×10^{-4} mole K_2PtCl_4 (f) 6.48×10^3 moles He

17 How many molecules are there in each of the following?
(a) 2.3 g sodium
(b) 0.002 mole O_2
(c) 1.28×10^{-3} g S

18 How many grams are there in each of the following?
(a) 3.011×10^{22} molecules O_2
(b) 1.66×10^{20} atoms sodium, Na
(c) 2.01×10^{24} molecules NO_2

19 The atomic weight of thulium is 169. What is the weight of one average atom of thulium?

20 A compound consists of 5.40 g of aluminum and 21.3 g of chlorine. Calculate its simplest

formula. This substance has a molecular weight of 266.66. What is its molecular formula?

21 How many grams and gram-atoms of oxygen are in 0.0100 mole of ascorbic acid (vitamin C), $C_6H_8O_6$?

22 A compound has the formula CaC_2. What weights of calcium and of carbon are present if the total weight of a sample of the compound is 8.00 g?

23 A balloon is filled with 0.0570 g of helium. How many atoms of helium are in the balloon?

24 Two oxides of carbon are known. One contains 57.14% by weight oxygen, and the other, 72.73% oxygen. Starting with two 1.000-g samples of carbon, how many grams of each oxide can be produced? Is there any simple weight relationship between the weights of oxygen used in each oxide preparation?

25 Phosphorus pentachloride, PCl_5, decomposes to produce two new products, phosphorus trichloride, PCl_3, and chlorine, Cl_2, according to the equation

$$PCl_5 \rightarrow PCl_3 + Cl_2$$

If a 5.00-g sample of PCl_5 decomposes completely to the products PCl_3 and Cl_2, how many grams of chlorine will be produced? Phosphorus pentachloride is 85.13% chlorine.

GASES AND AVOGADRO'S HYPOTHESIS

2

Matter exists in three phases—gaseous, liquid, and solid. Solids such as ice and steel have definite shape and definite volume. Liquids such as water have definite volume, but are fluid; that is, they adapt to the shape of their containers. A gas, however, not only adopts the shape, but also the volume of its container.

Solids have definite shape and volume because they have definite structures. They contain atoms or molecules that are packed closely together in a systematic way. Each atom or molecule has a fixed position in the solid and is not free to move. Liquids are somewhat like solids because the molecules in a liquid also are close to each other. This is why liquids have definite volumes. Liquids and solids are called *condensed phases.* But liquids do not have regular structures as solids do, thus the molecules in a liquid are free to move past one another. This is why liquids are fluid.

Gases have neither a fixed volume nor a fixed shape. A sample of gas fills any container of any size or shape. In this chapter, we will discuss the properties of gases and the laws that describe quantitatively how gases behave. Knowing these laws makes it possible to study the volume changes that occur when gases react with each other. This information, in turn, can be used to determine the combining ratios of atoms in molecules in the gas phase. This provides part of the solution to Dalton's problem (Chapter 1) of determining both the relative atomic weights of the elements and their combining ratios in compounds. Finally, we will discuss the kinetic molecular theory, which explains the way gases behave. We will come back to liquids and solids in Chapter 8.

2–1 Properties of Gases

Although gases were the last substances to be understood chemically, they were the first substances whose physical properties could be explained in terms of simple laws. What are the common properties of gases?

MASS AND VOLUME

The first obvious statement is that gases are matter. Today, none of us would question the idea that gases occupy space, but this was in dispute among Greek philosophers. The oldest recorded demonstration that gases occupy volume is the famous water clock experiment of Empedocles, a Sicilian–Greek philosopher of the fifth century B.C. A water clock is a cone-shaped object with holes at the base and the apex. When placed in water, base first, it takes a definite amount of time for the cone to sink, providing a rough measurement of time. Empedocles immersed the cone in water with his finger over the hole in the apex and observed that the water did not fill the cone completely. When he removed his finger, air rushed out of the opening. This demonstrated that air occupies space.

It is also easy to demonstrate that gases have weight. For example, we can weigh a glass bulb filled with air. Then a vacuum pump can be attached to it and the air pumped out. When we reweigh the bulb, we find that it weighs less.

PRESSURE

The second obvious property of gases is that they exert pressure. The most famous demonstration of this is the experiment of Otto von Guericke with his "Magdeburg hemispheres." He took two large metal hemispheres, placed them together, and evacuated the resulting sphere. He then attached a team of horses to each hemisphere and had them pull in opposite directions; they were unable to pull the hemispheres apart. The only force holding the two hemispheres together was the pressure of the atmosphere.

Another demonstration that gases exert pressure is provided by the barometer, invented in 1643 by Evangelista Torricelli (1608–1647). A simple barometer can be made by filling a glass tube open at one end with mercury and inverting the tube in a dish filled with mercury (Figure 2–1). When the tube is placed in the dish, a small amount of mercury flows out of the tube. The rest of the mercury is prevented from running out of the tube by the pressure that the atmosphere exerts against the surface of the mercury in the dish.

Gas pressure is often measured in scientific work with a barometer. Some different pressure conditions are shown in Figure 2–2. The height of a column of mercury supported by the gas pressure is a measure of the pressure. The pressure is expressed in millimeters of mercury. One millimeter of mercury is called one *torr*, in honor of Torricelli. The normal pressure of the atmosphere at sea level supports a mercury column 760 mm high, so a pressure of *1 atmosphere* (atm) is 760 *torr*.

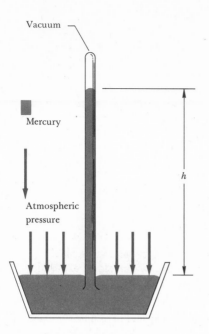

Figure 2–1. Torricellian barometer.

Example 2–1. The weather bureau usually reports the atmospheric pressure in inches of mercury. What would be the pressure expressed in torr and atmospheres if the bureau reported a pressure of 30.75 in.?

Solution. There are 25.40 mm per inch, so to calculate the pressure in torr

$$25.40 \text{ mm in.}^{-1} \times 30.75 \text{ in.} = 781.1 \text{ mm} = 781.1 \text{ torr}$$

One atmosphere of pressure equals 760.0 torr. To convert torr to atmospheres

$$781.1 \text{ torr} \times \frac{1 \text{ atm}}{760.0 \text{ torr}} = 1.026 \text{ atm}$$

Figure 2–2. Barometers showing different gas pressures.

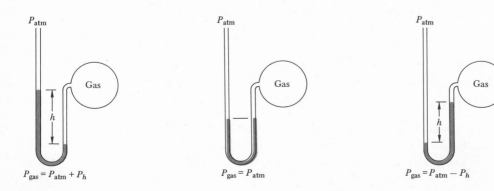

COMPRESSIBILITY

A third property of gases is that they can be compressed. We observe this property every day. Bicycle and automobile tires are "filled" with compressed air, and when sufficiently "full" the air pressure inside the tire is greater than the air pressure outside. Gases commonly used in chemical research, such as nitrogen, oxygen, and hydrogen, are stored in metal cylinders at a pressure of 1.1×10^5 torr, which is about 150 atm.

TEMPERATURE

A fourth common property of gases is that they expand when heated and contract when cooled. These properties give rise to large-scale convections in the atmosphere and small-scale convections around any source of heat. A fire warms the air around it, causing the air to expand and therefore have lower density. Being less dense than the surrounding cool air, the hot air rises, sucking more cool air into the vicinity of the fire.

Because gases expand when heated, they exert more pressure when heated. Anyone who has ever had the unfortunate experience of heating an aerosol spray container until it exploded will confirm that this is true. A more common example, and a very important one to highway safety, is the fact that air pressure in automobile tires increases during high-speed driving, because the air in the tires is heated.

DIFFUSION

A fifth common property of gases is that they diffuse. This is demonstrated readily by the way in which the odor of a piece of Limburger cheese, for instance, can pervade an entire room (or house) in which the cheese is kept. The molecules of gas responsible for the odor diffuse through the air, thus spreading the odor. This can occur even if the cheese is in a plastic wrapper, because gases not only diffuse through each other, but also through many solid substances.

2–2 The Quantitative Relationships of Volume, Temperature, and Pressure

In experiments involving gases, there are four variables to consider. These variables are pressure (P), volume (V), temperature (T), and the mass of the gas. To study the behavior of gases, scientists generally hold two of these four variables constant and study the relationship between the other two.

BOYLE'S LAW

The first quantitative study of gases was done by the English chemist Robert Boyle (1627–1691), who investigated the relationship between volume and pressure in a given mass of gas at constant temperature.

The first experiment that Boyle did is illustrated in Figure 2–3. He filled the bottom of a U-tube closed at one end with mercury, and adjusted the volume of air trapped in the closed end so that the levels of mercury in the two sides of the tube were equal. This meant that the air trapped in the closed end was at atmospheric pressure. He then added successive amounts of mercury to the open end of the tube, forcing the mercury level up in the closed end, thus compressing

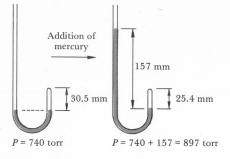

Figure 2–3.   Schematic of Boyle's gas experiment.

the trapped air. The difference in the height of mercury in the two sides of the tube is equal to the difference between the gas pressures in the closed end of the tube and in the open end of the tube. The pressure at the open end is the pressure of the atmosphere, which was 740 torr during the experiment. The total pressure in the closed end is the pressure difference, P, plus the atmospheric pressure, 740 torr. The results of this experiment are given in Table 2–1.

The numbers in Table 2–1 are plotted in Figure 2–4(a) to show more clearly the relationship between volume and pressure. The shape of the curve suggests that P is proportional to $1/V$. To test this we find the reciprocals of the volume in Table 2–1 and plot P vs. $1/V$, shown in Figure 2–4(b). We obtain a straight line that passes through the origin (i.e., when the pressure is zero the volume is infinite). The line is described by the equation $P = a/V$, where a is a proportionality

Table 2–1.   Boyle's Data Relating Pressure and Height of Columns of Air

Height of trapped air column (mm)[a]	Height of mercury column plus height of mercury supported by the atmospheric pressure ($P + 740$ torr)	$P_{total} \times h$
30.5	740	22,600
25.4	897	22,800
21.6	1058	22,800
17.8	1279	22,800
15.2	1497	22,800
13.3	1705	22,700
11.4	1980	22,600
10.2	2235	22,800
9.5	2365	22,500
8.9	2555	22,700
8.2	2740	22,500
7.6	2990	22,700

[a]*The tube has constant cross-sectional area, so the height of the column of trapped air is proportional to its volume.*

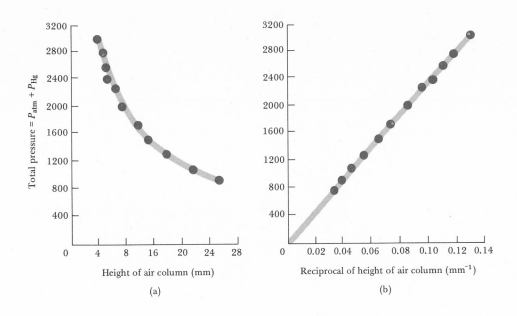

Figure 2–4.   Plot showing relationship between volume and pressure of a gas.

constant. This relationship is known as *Boyle's law.* The last column in Table 2–1 shows that the length of the column of trapped gas times the pressure is essentially equal to a constant.

Example 2–2. One liter of hydrogen gas is at a pressure of 0.77 atm. What would be the pressure if the gas were expanded to 3.0 liters?

Solution. Initially, we have $P_1 = 0.77$ atm and $V_1 = 1.0$ liter. When the gas is expanded to 3.0 liters, we have $P_2 = ?$, $V_2 = 3.0$ liters. According to Boyle's law, $P_1V_1 = a$ and $P_2V_2 = a$. But since a is a constant, $P_1V_1 = P_2V_2$ or $P_2 = P_1(V_1/V_2)$. Therefore, we can find the new pressure, P_2, by multiplying the old pressure, P_1, by the ratio of the volumes

$$P_2 = 0.77 \text{ atm} \times \frac{1.0 \text{ liter}}{3.0 \text{ liters}} = 0.26 \text{ atm}$$

In a problem of this kind, the first step toward a solution is always to decide whether the new pressure would be larger or smaller than the original pressure. Here, the gas was expanded, so the pressure must decrease. Therefore, to get the answer we multiply by a volume ratio less than one (1/3, not 3/1).

Boyle's law tells us that a measurement of gas volume has little meaning unless the pressure is specified. For convenience in comparing different experiments with gases, a *standard pressure* has been decided on. This standard pressure is 760 torr (1 atm). In an experiment, a gas volume may be measured at any convenient pressure and then converted, using Boyle's law, to the volume at 760 torr.

Example 2–3. A gas occupies a volume of 2.40 liters at a pressure of 740 torr. What volume would it occupy at the standard pressure?

Solution. The standard pressure is 760 torr (1.00 atm). The pressure is greater, so the volume must be less

$$V = 2.40 \text{ liters} \times \frac{740 \text{ torr}}{760 \text{ torr}} = 2.34 \text{ liters}$$

VOLUME AND TEMPERATURE

We have mentioned already that the pressure and volume of a gas depend on temperature. The first quantitative measurements of the dependence of gas volume on temperature were made, about 1787, by the French physicist J. A. C. Charles (1746–1823), but his results were never published. His work was repeated, in 1802, by Joseph L. Gay-Lussac (1778–1850).

The temperature scale commonly used in scientific work is the centigrade scale, introduced in 1742 by Anders Celsius, a Swedish astronomer. It arbitrarily defines the freezing point of pure water as 0°C (centigrade) and the boiling point as 100°C. The Fahrenheit scale fixes these two points at 32°F and 212°F, respectively. The temperature difference between these two points is 100 centigrade degrees, or 180 Fahrenheit degrees. Thus, one centigrade degree is 180/100, or 9/5, as large as one Fahrenheit degree. Fahrenheit temperatures are converted readily to centigrade temperatures by subtracting 32 and multiplying by 5/9. Conversely, converting from centigrade to Fahrenheit temperatures is done by multiplying by 9/5 and adding 32. We can express this algebraically by the formula °F = 9/5°C + 32, or °C = 5/9(°F − 32).

Example 2–4. Normal body temperature is 98.7°F. What is this temperature on the centigrade scale?

Solution. We first subtract 32° from the Fahrenheit temperature to standardize the freezing point and then multiply by 5/9 to convert the Fahrenheit degrees to the centigrade degrees

$$°C = \frac{5}{9}(98.7 - 32.0)$$

$$°C = \frac{5}{9}(66.7) = 37.1°C$$

An example of the type of results obtained by Gay-Lussac is shown in Figure 2–5. The first point to notice about this graph is that it is a straight line. The second point is that the line does not go through the origin. When the volume equals zero, the temperature equals −273°C. Suppose that we added 273 to all the values of temperature. Then we would obtain the dashed line, which passes through the origin. The equation for this line is very simple

$$V = b(t + 273)$$

(2–1)

where b is a constant.

Remember that the centigrade temperature scale was *arbitrarily* chosen with zero being the melting point of ice. Suppose that we define another temperature scale, using centigrade degrees, *but with zero at −273°C*. In other words, we arrive at the new temperature scale by adding 273° to the centigrade temperature ($T = t + 273°$). Using this scale, the relationship between volume and temperature is more simple

$$V = bT$$

(2–2)

where T is measured on the scale with zero at −273°C, shown on the upper abscissa in Figure

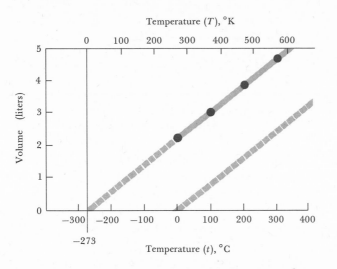

Figure 2–5. Plot showing relationship between volume and temperature of a gas.

2–5. This relationship is known as Charles' law. We see that V and T are in simple proportion to each other. For example, if T is doubled at constant pressure, V also doubles. This new temperature scale is called the Kelvin scale, after William Thomson, Lord Kelvin (1824–1907), who first suggested it. Temperature measured on the Kelvin scale is given the symbol T. Kelvin temperature must be used in all gas law calculations. Zero degrees on the Kelvin scale is called *absolute zero*. We will learn the significance of this temperature later in the chapter, when we discuss the kinetic theory of gases. A comparison of the three common temperature scales is shown in Table 2–2.

Example 2–5. Suppose 3.00 liters of gas at 30°C is heated to 370°C at constant pressure. What is the final volume of the gas?

Solution. From Charles' law we have $V_1 = bT_1$, where $V_1 = 3.00$ liters and $T_1 = (30 + 273)$°K; and $V_2 = bT_2$, where $V_2 = ?$ and $T_2 = (370 + 273)$°K. Since b is constant

$$\frac{V_1}{T_1} = b = \frac{V_2}{T_2}$$

or

$$V_2 = V_1\left(\frac{T_2}{T_1}\right)$$

Table 2–2.   Comparison of Temperature Scales

	Fahrenheit	Centigrade	Kelvin
Boiling point of water	212°	100°	373°
Freezing point of water	32°	0°	273°
Absolute zero	−460°	−273°	0°

Therefore,

$$V_2 = 3.00 \text{ liters} \times \frac{(370 + 273)°\text{K}}{(30 + 273)°\text{K}}$$

$$= 3.00 \text{ liters} \times \left(\frac{643°\text{K}}{303°\text{K}}\right) = 6.37 \text{ liters}$$

Since the gas is heated, it expands, and the new volume is larger than the initial volume. This requires multiplying the old volume by a temperature ratio that is greater than one.

In an experiment, gas volumes can be measured at any convenient temperature and then converted, using Charles' law, to the volume at a *standard temperature,* which is chosen to be 0°C (273°K).

Example 2–6. A gas sample occupies 17.2 liters at 32°C. What volume would it occupy at standard temperature?

Solution. The standard temperature is 0°C (273°K). Since this is less than 32°C, the gas will occupy a smaller volume, and we must multiply the initial volume by a temperature ratio less than one

$$V = 17.2 \text{ liters} \times \frac{273°\text{K}}{305°\text{K}} = 15.4 \text{ liters}$$

COMBINED GAS LAW

Boyle's law, $P = a/V$ is true for any given temperature; Charles' law, $V = bT$, is true for any given pressure. What if the pressure, volume, and temperature of a gas sample all change? For example, suppose that we take a 1.000-liter sample of gas at 0.836 atm and 24°C and compress it to 0.963 liter at 47°C. What is the final pressure? We can work this problem in two parts. First, suppose that we change the temperature from 24°C to 47°C at constant pressure. The new volume, from Charles' law, is

$$V = 1.000 \text{ liter} \times \frac{(47 + 273)}{(24 + 273)} = 1.000 \text{ liter} \times \frac{320°\text{K}}{297°\text{K}}$$

$$= 1.077 \text{ liters}$$

Now let us change the volume from 1.077 liters to 0.963 liter at constant temperature and find the new pressure. From Boyle's law

$$P = 0.836 \text{ atm} \frac{1.077 \text{ liter}}{0.963 \text{ liter}} = 0.935 \text{ atm}$$

Now notice that we could have made the entire calculation in one step

$$P = 0.836 \text{ atm} \times \frac{320°\text{K}}{297°\text{K}} \times \frac{1.000 \text{ liter}}{0.963 \text{ liter}} = 0.935 \text{ atm}$$

In the last equation, we have combined Boyle's law ($PV = a$) with that of Charles and Gay-Lussac ($V/T = b$) to produce the *combined* or *universal gas law,* which in general terms has the form

$$\frac{P_1 V_1}{T_1} = \frac{P_2 V_2}{T_2} = \cdots \frac{P_n V_n}{T_n} = c \tag{2–3}$$

or

$$PV = cT \qquad\qquad\qquad (2-4)$$

Clearly, if temperature is held constant (cT = constant = a), the equation reduces to Boyle's law; if pressure is constant (c/P = constant = b), Charles' law results.

2–3 Volume Changes in Gas Reactions

Once the gas laws were known it became possible to study the meaning of changes in volume when two gases react with each other. Gas reactions can be studied at constant temperature and pressure, so that any volume change will be due to the nature of the reactions. Alternatively, gas reactions can be performed at constant temperature and volume. Then the pressure of the reaction mixture can be measured, and, from the change in pressure, we can calculate, using Boyle's law, what the change in volume would be if the reaction were performed at constant pressure.

In 1805, Gay-Lussac established the fact, first observed by the English chemist Henry Cavendish (1731–1810), that two volumes of hydrogen react with one volume of oxygen to form two volumes of steam

Gay-Lussac and other chemists studied many other reactions of gases, and obtained similar results

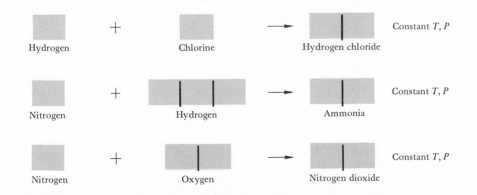

From these investigations, Gay-Lussac was able to state, in 1808, the *law of combining volumes:* The ratios of volumes of reacting gases are small whole numbers. This law suggests that there might be a very simple relationship between the number of molecules in a gas sample and its volume.

2–4 A Model for Gaseous Matter

In this chapter, we have described how Boyle found that the pressure of a gas varies inversely with volume at constant temperature, have stated that Charles and Gay-Lussac found that the volume of a gas varies directly with temperature at constant pressure, and have combined these two laws to obtain the combined gas law. We also have stated the law of combining volumes, deduced by Gay-Lussac: Gases combine in simple, whole-number volume ratios (at constant temperature and pressure). Now our main objective is to try to understand why these laws describe how gases behave. The way we do this is construct a simple model of a gas that obeys these laws. Then we can learn more about the behavior of gases by seeing how the model behaves.

AVOGADRO'S HYPOTHESIS

The first bold step in the formulation of a successful model for gaseous matter was taken by Amadeo Avogadro (1776–1856), an Italian physicist. He knew of Gay-Lussac's work and reasoned that, if gases combine in simple volume ratios and compounds contain atoms in simple ratios, "it must be admitted that very simple relationships also exist between the volumes of gaseous substances and the number of . . . molecules which form them."

In 1811, Avogadro published a hypothesis to explain this relationship:

1   Gases consist of discrete molecules.

2   Each molecule of a gas "occupies" a given amount of space at the same temperature and pressure.

3   Therefore, *equal volumes of gases contain equal numbers of molecules at the same temperature and pressure.*

The great importance of Avogadro's hypothesis is that it gives a method of obtaining the relative weights of molecules.

According to Dalton's atomic theory, atoms and molecules have definite weights. But Dalton had no method to determine those weights. Avogadro's hypothesis states that *relative weights of molecules* of a gas can be determined simply by weighing samples of the gases under known conditions. The weight of a gas sample is the number of molecules in the sample times the weight of each molecule

$$\text{weight} = \text{number of molecules} \times \text{weight of one molecule}$$

To determine the relative weights of molecules, the quantity we want to measure is, for instance,

$$\frac{\text{weight of one hydrogen molecule}}{\text{weight of one oxygen molecule}}$$

If samples of hydrogen and oxygen having the same volume, temperature, and pressure are weighed, the sample weights have the ratio

$$\left(\frac{\text{weight of hydrogen}}{\text{weight of oxygen}}\right)_{\substack{\text{same} \\ \text{V,T,P}}} = \frac{\begin{array}{c}\text{number of molecules of hydrogen} \\ \times \text{ weight of one hydrogen molecule}\end{array}}{\begin{array}{c}\text{number of molecules of oxygen} \\ \times \text{ weight of one oxygen molecule}\end{array}}$$

But since the number of hydrogen molecules is the same as the number of oxygen molecules, by Avogadro's hypothesis, the number terms cancel, and we are left with

$$\left(\frac{\text{weight of hydrogen}}{\text{weight of oxygen}}\right)_{\substack{\text{same} \\ \text{V,T,P}}} = \frac{\text{weight of one hydrogen molecule}}{\text{weight of one oxygen molecule}}$$

Avogadro's hypothesis also helps us to deduce molecular formulas. As an example, consider the reaction of hydrogen and oxygen to form steam (Section 2–3). Gay-Lussac found that two volumes of hydrogen combine with one volume of oxygen to form two volumes of water vapor (steam). According to Avogadro's hypothesis, this means that two *molecules* of hydrogen combine with one *molecule* of oxygen to form two *molecules* of water vapor

2 molecules of hydrogen + 1 molecule of oxygen → 2 molecules of water vapor

Each molecule of water vapor contains at least one atom of oxygen. But one molecule of oxygen is sufficient to make two molecules of water vapor. Therefore, each molecule of oxygen must contain *at least two* oxygen atoms. Though we cannot be sure from this evidence that the oxygen molecule does not contain four, six, or even more atoms, let us assume it contains only two, because this is the simplest formula. This assumption is justified, in the absence of any direct evidence, if it explains all of the observed reactions of oxygen molecules.

Now let us consider a second reaction, that of hydrogen and chlorine to give hydrogen chloride (Section 2–3). One volume of hydrogen reacts with one volume of chlorine to produce two volumes of hydrogen chloride. These volume ratios prove that hydrogen molecules must contain *at least two* hydrogen atoms.

Suppose that the hydrogen molecule is H_2 and the oxygen molecule is O_2. Then the reaction to form water (steam) is

$$2H_2 + O_2 \rightarrow 2 \text{ water}$$

To have the same number of atoms of hydrogen and oxygen on each side of the equation, the formula of water must be H_2O

$$2H_2 + O_2 \rightarrow 2H_2O$$

THE MOLE

We have defined one mole of a compound as that amount containing one molecular weight, expressed in grams. For instance, the molecular weight of a hydrogen molecule (H_2) is 2.016. Therefore, one mole of hydrogen molecules weighs 2.016 g. Another way of saying this is that one mole of hydrogen molecules is the number of molecules that weighs 2.016 g.

What volume would one mole of hydrogen gas have? Since we know that gas volume varies with temperature and pressure, we have to specify what the temperature and pressure are. We have defined standard temperature (0°C or 273°K) and standard pressure (760 torr or 1 atm). Measurements of gas volume are usually converted to these standard temperature and pressure (STP) conditions. We know how to calculate the volume of a gas sample at STP from a volume measurement at any other conditions. Thus, we can measure the gas volume under any convenient conditions and calculate what the volume would be under standard conditions.

Experiments have shown that at STP the volume of 2.016 g of hydrogen gas is 22.4 liters. However, 2.016 g of NH_3 under these same conditions occupies only 2.7 liters. Upon closer examination of these two numbers, we see that 2.016 g of NH_3 is 0.118 mole of NH_3, and that 2.7 liters is 0.118 of the volume occupied by 2.016 g (1 mole) of hydrogen. There seems to be a direct relationship between the molar amount of gas present and the volume it occupies. On the basis of these observations, we would predict that one mole of NH_3 (17.03 g) would occupy 22.4 liters at STP; this is confirmed experimentally. We may conclude, then, that equal volumes of gases at the same temperature and pressure contain the same number of molecules. Also, one mole of all compounds contains the same number of molecules. Therefore, under STP conditions, one mole of any gas has a volume of 22.4 liters, which is called the *molar volume* of a gas.

For instance, oxygen has a density of 1.429 g liter^{-1} at STP. The atomic weight of oxygen is 16.0, and the molecular weight of O_2 is 32.0; therefore, one mole of oxygen gas weighs 32.0 g. The volume that one mole of the gas would occupy is

$$V_{STP} = \frac{1.00 \text{ liter}}{1.429 \text{ g}} \times 32.0 \text{ g} = 22.4 \text{ liters}$$

In addition to knowing that 32.0 g of oxygen at STP occupies 22.4 liters, we also know that there are 6.022×10^{23} (Avogadro's number, Section 1–8) molecules of oxygen present. (It is very difficult to imagine such a large number. If every man, woman, and child on earth—3 billion people—counted objects at the rate of one per second, it would take more than 6 million years to count Avogadro's number of objects.)

The observed direct relationship between the volume of a gas and the number of moles (or molecules) present allows us to write the combined gas law as

$$PV = nRT$$

where n is the number of *moles* of gas; R is called the gas constant. Since one mole of a gas occupies 22.4 liters at STP,

$$R = \frac{1 \text{ atm} \times 22.4 \text{ liters}}{1 \text{ mole} \times 273°K} = 8.20 \times 10^{-2} \text{ liter atm mole}^{-1} \text{ °K}^{-1}$$

Example 2–7. Calculate the volume occupied by 21.0 g of nitrogen gas at 734 torr and 48°C.
Solution. Calculate the number of moles of nitrogen present

$$\text{moles of } N_2 = 21.0 \text{ g} \times \frac{1 \text{ mole}}{28.0 \text{ g}} = 0.750 \text{ mole}$$

Before using the combined gas law, convert to absolute temperature

$$T = 48°C + 273° = 321°K$$

and convert pressure in torr to pressure in atmospheres

$$P_{atm} = 734 \text{ torr} \times \frac{1 \text{ atm}}{760 \text{ torr}}$$

$$P_{atm} = 0.966 \text{ atm}$$

Then substituting into the equation

$$V = \frac{nRT}{P}$$

we obtain

$$V = \frac{0.750 \text{ mole} \times 0.0820 \text{ liter atm mole}^{-1} \text{ deg}^{-1} \times 321°K}{0.966 \text{ atm}}$$

$$V = 20.4 \text{ liters}$$

The combined, or universal, gas law expressed in the form $PV = nRT$ is very versatile. For example, we can substitute for the number of moles, n, the grams of gas present divided by the molecular weight of the gas since

$$n = \frac{g}{\text{mol wt}} \tag{2–5}$$

Rearrangement gives

$$\text{mol wt} = \frac{gRT}{PV} \tag{2–6}$$

which provides a simple way to calculate the molecular weights of various gases.

THE KINETIC MOLECULAR THEORY

Avogadro's hypothesis is the basis for a general theory of gaseous matter called the *kinetic molecular theory*. This theory begins with a model of a gas to which we can apply the laws of mechanics to determine how the model behaves. We will not go through a detailed derivation, but we will point out the main features of the theory and show qualitatively that the model agrees with the observed properties of real gases.

The main features of the model for an ideal gas are

1   All gases consist of separate and distinct particles, which we call molecules.

2   Individual molecules are very small. We can get a rough idea of the size of molecules by calculating the volume available to each molecule in a liquid. For instance, since water has a density of about 1 g cm^{-3}, one mole of water occupies about

$$\frac{18 \text{ g mole}^{-1}}{1 \text{ g cm}^{-3}} = 18 \text{ cm}^3 \text{ mole}^{-1}$$

One mole contains Avogadro's number of molecules (6.022×10^{23}), so the volume for each molecule is about

$$18 \text{ cm}^3 \text{ mole}^{-1} \times \frac{1 \text{ mole}}{6 \times 10^{23} \text{ molecules}} = 3 \times 10^{-23} \text{ cm}^3 \text{ per molecule}$$

Therefore, a molecule occupies a cube about

$$\sqrt[3]{30 \times 10^{-24} \text{ cm}^3} = 3 \times 10^{-8} \text{ cm} = 3 \text{ Å}$$

on an edge. Since the molecules in a liquid are close together, this is roughly the size of the water molecule.

3   There is a large empty space between molecules in a gas. The size of one molecule is about 3×10^{-23} cm^3. At STP, 6.022×10^{23} molecules occupy a volume of 22.4 liters. The volume per molecule is

$$\frac{22.4 \text{ liters}}{6.022 \times 10^{23} \text{ molecules}} = 3.72 \times 10^{-23} \frac{\text{liter}}{\text{molecule}} = 3.72 \times 10^{-20} \text{ cm}^3 \text{ per molecule}$$

Therefore, the volume per molecule is about one thousand times the size of the molecule.

4   Molecules are in random, straight-line motion.

5   Molecular collisions are elastic. This is another way of saying that there are no repulsive or

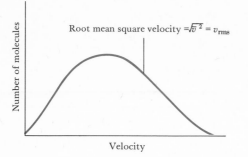

Figure 2–6. Distribution of molecular velocities at a given temperature.

attractive forces between molecules. When they collide with each other they neither tend to spring apart nor to stick to each other. They simply hit and bounce apart as billiard balls do when they collide.

6   The molecular kinetic energy, $\frac{1}{2}m\overline{v^2}$, is constant for constant temperature, T, and is proportional to T

$$\tfrac{1}{2}m\overline{v^2} = \text{constant} \times T \tag{2–7}$$

(The $\overline{v^2}$ term is a type of average value of the squares of the molecular velocities, and is called the mean square velocity.) The molecules in a gas do not all have the same velocity. Figure 2–6 shows the distribution of molecular velocities at a given temperature. A useful, typical value for molecular velocity is $\sqrt{\overline{v^2}}$, the square root of the mean square velocity, or v_{rms} (rms stands for root mean square). Equation 2–7 says that temperature is a measure of kinetic energy, and that absolute zero (0°K) is the temperature where all motion stops. This suggests that absolute zero is the lowest possible temperature.

7   We can get a rough idea of the velocities of molecules in a gas by using the relationship between average kinetic energy and temperature to calculate v_{rms}. The exact relation between $\frac{1}{2}m\overline{v^2}$ and T is

$$\tfrac{1}{2}m\overline{v^2} = \frac{3RT}{2N} \tag{2–8}$$

where N is Avogadro's number.

   If we express the mass in grams and velocity in cm sec^{-1}, the kinetic energy, $\frac{1}{2}m\overline{v^2}$, is in g cm^2 sec^{-2}, or ergs. Therefore, we must express the gas constant in ergs mole^{-1} °K^{-1}

   $R = 8.20 \times 10^{-2}$ liter atm mole^{-1} °K^{-1} = 8.31 $\times 10^7$ ergs mole^{-1} °K^{-1}

The v_{rms} can be calculated from this equation as follows:

$$v_{\text{rms}} = \sqrt{\overline{v^2}} = \sqrt{\frac{3RT}{mN}} = \sqrt{\frac{3RT}{\text{mol wt}}} \tag{2–9}$$

The molecular weight is equal to the weight of one molecule, m, times Avogadro's number,

N. At 27°C (300°K) for helium gas (mol wt = 4.00)

$$v_{\mathrm{rms}}(\mathrm{He}) = \sqrt{\frac{3 \times 8.31 \times 10^7 \text{ g cm}^2 \text{ sec}^{-2} \text{ mole}^{-1} \text{ }^\circ\mathrm{K}^{-1} \times 300^\circ\mathrm{K}}{4.00 \text{ g mole}^{-1}}} = 1.37 \times 10^5 \text{ cm sec}^{-1}$$

The high average speed of the helium atom at 300°K is perhaps more striking when expressed as 3060 miles per hour.

The velocity of a gas molecule at any given temperature depends clearly on its mass. This is an important conclusion that we can make from the kinetic model, since it can be verified experimentally. For example, experimental rates of leakage of gases through an orifice, a process which we call *effusion,* can be measured. The leakage, or the effusion rate, of a particular gas should be proportional to the velocity of the gas molecules. For oxygen gas (mol wt = 32) and sulfur dioxide gas (mol wt = 64), experiments show that the lighter gas, O_2, effuses at a rate that is the $\sqrt{2}$ times faster than the heavier SO_2. This is exactly what would be predicted from the kinetic model, since

$$R_{O_2} = \text{rate of effusion } O_2 \propto \text{ velocity } O_2 \propto \sqrt{\frac{3RT}{\text{mol wt}_{O_2}}}$$

$$R_{SO_2} = \text{rate of effusion of } SO_2 \propto \text{ velocity } SO_2 \propto \sqrt{\frac{3RT}{\text{mol wt}_{SO_2}}}$$

Therefore, since the gases are at the same temperature

$$\frac{R_{O_2}}{R_{SO_2}} = \sqrt{\frac{\text{mol wt}_{SO_2}}{\text{mol wt}_{O_2}}} = \sqrt{2}$$

Now let us summarize what we have been discussing to see if the properties of gases and the kinetic molecular theory model are compatible:

1   Gases are compressible, because there are large empty spaces between molecules.
2   Gases exert pressure, because the gas molecules collide with the walls of their container with a force determined by their mass and velocity.
3   Gases fill any vessel, because the molecules are in random motion.
4   Lighter gas molecules effuse faster than heavier gas molecules; kinetic molecular theory predicts an effusion ratio to be inversely proportional to the square roots of molecular weights.
5   Gases diffuse, because the individual molecules are in rapid, random motion.

Pressure is due to collisions of molecules with the wall of the container. So if we double the number of molecules in a container, we double the number of collisions, hence we double the pressure. But the pressure does not depend on the size or shape of the container. So if we double the number of molecules per liter by halving the volume of the container, we also double the pressure. This agrees with Boyle's law ($P = a/V$). If a gas is heated, each molecule has a higher velocity, which increases the number of collisions with the container walls, and also gives each collision a greater impact. Therefore, the pressure increases. This agrees with Charles' law.

Now we see that the kinetic molecular theory—the model—accounts for all of the properties of gases that we have discussed so far. But remember that real gases are not exactly the same as the model, so we would expect to find some conditions for which gases would not behave as predicted by this simple theory. We will discuss some examples of this behavior in Section 2–6.

2–5 Partial Pressures

Because there are large spaces between gas molecules, we might expect a gas in a mixture of gases to act independently of the other gases present. This is, in fact, very nearly the case. The oxygen gas in the atmosphere (air is about 20% oxygen) behaves as if it were pure oxygen at 0.2 atm. The fact that gases in a mixture act essentially independently of each other was discovered by John Dalton (1766–1844); his *law of partial pressures* states that each component in a gas mixture exerts its own pressure (its partial pressure) independent of the others, and that the total pressure is the sum of the partial pressures.

If we know the partial pressures of gases in a mixture, it is easy to find the total pressure. But what if we know only the amount of each gas present, and we want to find the partial pressure of each? Suppose that we have a mixture of one mole of oxygen and four moles of nitrogen at total pressure P. According to the law of partial pressures, we can write the gas law for each gas; we will use p for partial pressure

$$p_{O_2}V = n_{O_2}RT$$

$$p_{N_2}V = n_{N_2}RT$$

or

$$p_{O_2} = n_{O_2}\frac{RT}{V} \qquad p_{N_2} = n_{N_2}\frac{RT}{V}$$

We also know that

$$p_{O_2} + p_{N_2} = P$$

or

$$P = (n_{O_2} + n_{N_2})\frac{RT}{V}$$

then

$$\frac{p_{O_2}}{P} = \frac{n_{O_2}(RT/V)}{(n_{O_2} + n_{N_2})(RT/V)} = \frac{n_{O_2}}{n_{O_2} + n_{N_2}}$$

or

$$p_{O_2} = \left(\frac{n_{O_2}}{n_{O_2} + n_{N_2}}\right)P$$

The quantity $\dfrac{n_{O_2}}{n_{O_2} + n_{N_2}}$ is called the mole fraction of oxygen, X_{O_2}. The mole fraction of any gas in a mixture of gases is the number of moles of that gas divided by the total number of moles of gas present. The partial pressure of any gas in a mixture is

$$p = XP \tag{2–10}$$

where X is the mole fraction of that gas. The sum of the mole fractions of gases in a mixture always equals one, just as the sum of the partial pressures equals the total pressure.

There is one practical application of these relationships that occurs frequently in studying gas reactions. Gas formed in a reaction is often trapped over water, as illustrated in Figure 2–7.

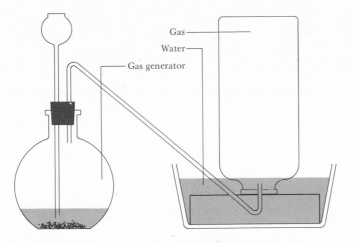

Figure 2–7.   Gas collection apparatus.

But water vapor is present and mixes with the gas that has been trapped. To find the amount of gas formed, the partial pressure of the water vapor must be subtracted from the total pressure.

Example 2–8. A 3.20-liter sample of nitrogen collected over water had a total pressure of 742 torr at 27°C. The water vapor pressure at 27°C is 27 torr. How many moles of nitrogen were formed?

Solution. We can work this problem either of two ways. First, we could say that the total pressure is 742 torr or 0.976 atm, and using $PV = nRT$

$$n = \frac{PV}{RT} = \frac{0.976 \text{ atm} \times 3.20 \text{ liters}}{8.20 \times 10^{-2} \text{ liter atm mole}^{-1} \text{ °K}^{-1} \times 300 \text{ °K}} = 0.127 \text{ mole (of nitrogen}$$
$$+ \text{ water vapor)}$$

Since the partial pressure of nitrogen is $742 - 27 = 715$ torr, the mole fraction of nitrogen is

$$X = \frac{p}{P} = \frac{715 \text{ torr}}{742 \text{ torr}} = 0.964$$

Therefore, the number of moles of nitrogen is

$$n_{N_2} = X_{N_2}n = 0.964 \times 0.127 \text{ mole} = 0.122 \text{ mole}$$

Alternatively, we could calculate the number of moles of nitrogen from the gas equation, using the partial pressure of nitrogen (715 torr = 0.941 atm)

$$n_{N_2} = \frac{p_{N_2}V}{RT} = \frac{0.941 \text{ atm} \times 3.20 \text{ liters}}{8.20 \times 10^{-2} \text{ liter atm mole}^{-1} \text{ °K}^{-1} \times 300 \text{ °K}} = 0.122 \text{ mole}$$

2–6   Real Gases

We have indicated that the gas laws can be derived from the kinetic molecular theory and have discussed how this theory can be used to predict the behavior of gases. However, the kinetic molecular *model* of a gas is not the same as a *real* gas. As a practical approach to describing the behavior of real gases, we consider the model of the kinetic molecular theory to be an *ideal gas* whose behavior is exactly described by the combined, or ideal, gas law

$$PV = nRT \tag{2-11}$$

Then we describe the behavior of real gases by describing how they deviate from ideal behavior. An example of how some gases deviate from ideality is shown in Figure 2–8. The ideal gas law equation may be written as follows

$$\frac{PV}{nRT} = 1 \tag{2-12}$$

Notice that if we plot *PV/nRT* versus pressure, we should obtain a straight line with zero slope. A gas whose behavior deviates from that predicted by the ideal gas law will clearly display itself in the graph. Notice that at low pressures the gases come closer to ideal behavior.

At sufficiently low temperature, all gases are more compressible at lower pressures than predicted by Boyle's law. In Figure 2–8 we see that at 0°C methane (CH_4) is more compressible than Boyle's law predicts up to about 170 atm. This is because there are relatively long-range forces

Figure 2–8.   Deviation of real gases from ideal behavior.

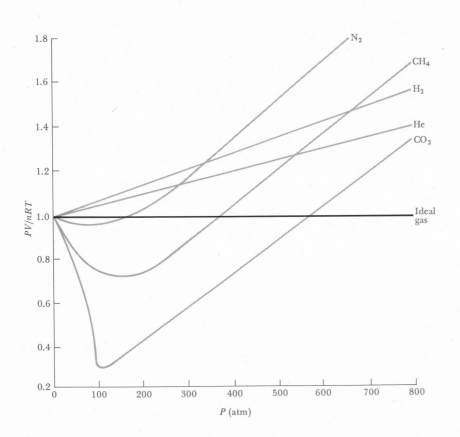

between molecules that attract them to one another. These attractive forces (called van der Waals forces) are more important at low temperatures, where the molecules are moving more slowly. This tends to make the volume of a gas smaller than the Boyle's law volume.

At sufficiently high temperature, all gases are less compressible than predicted by Boyle's law. This is because gas molecules do occupy some volume, and there are very strong short-range forces that repel molecules from one another. This effect is more pronounced the higher the temperature, because then the molecules collide with each other more often.

Most gases obey the gas laws at ordinary pressure, and the behavior of any gas becomes more nearly ideal the lower the pressure.

The powerful role of models in the development of science is illustrated by the ideal gas laws. If Avogadro and Dalton had known the detailed behavior of real gases, they would have realized that their simple models did not correspond exactly to reality. We can speculate whether they would have been led to endless attempts to improve the models to account for the properties of real gases before enunciating the powerful theories such as Avogadro's hypothesis and Dalton's law of partial pressures. Now we regard the molecular theory of gases as close to absolute truth, and we patch up discrepancies between simple theory and experiment by postulating phenomena such as "sticky" collisions between molecules.

A sticky collision is an encounter in which molecules do not rebound with perfect elasticity. The idea might also be expressed by saying that, in some collisions, molecules seem soft. The encounter between two billiard balls comes fairly close to being an elastic collision between hard bodies, whereas the collision between two soft pillows would be a "super sticky" encounter. The simple kinetic molecular theory treats colliding molecules as very hard spheres. We know that molecules can be compressed and deformed, and that most molecules are not spherical. Therefore, the sticky collision model provides a reasonable way for thinking about deviations from the ideal gas law.

We can imagine an entirely different history of the theory of gases that would have led to a different model. If Avogadro's hypothesis had not been firmly established as a useful concept, we might count the molecules in a gas differently. Instead of thinking of sticky collisions, we might say that every real gas contains a few molecules of twice the size of the usual ones. These double molecules would be thought of as unstable and short-lived. If we think about it, there probably is little difference between the idea of short-lived, double molecules and two single molecules in a long encounter. Consequently, either model could be made to serve the desired purpose. However, the language created by the double molecule model might be quite different from that now used in discussing nonideal gases.

2–7 Summary

1 The combined gas law is $PV = nRT$, where P is pressure; V, volume; n, the number of moles of gas; T, absolute temperature (Kelvin scale); and R, a constant. If P is expressed in atmospheres and V in liters, the value of R is 0.0820 liter atm mole^{-1} °K^{-1}.

2 The volume of one mole of an ideal gas at STP is 22.4 liters.

3 The law of combining volumes states that gases combine with each other in simple volume ratios, at constant temperature and pressure.

4 Avogadro's hypothesis states that equal volumes of gases contain equal numbers of molecules, at the same temperature and pressure. This hypothesis can be used to determine the

formulas and relative molecular weights of gas molecules. The number of molecules in one mole of any substance, Avogadro's number, is 6.022×10^{23}.

5   The partial pressure of a gas in a mixture of gases is given by $p = XP$, where P is the total pressure of the mixture and X is the mole fraction of that gas in the mixture.

6   The combined gas law can be derived from the kinetic theory of gases, which has the following assumptions:

   (a) All gases are made of distinct particles, which we call molecules.

   (b) Individual molecules are very small.

   (c) There is a large space between molecules.

   (d) Molecules are in random, straight-line motion.

   (e) Molecular collisions are elastic.

   (f) The quantity $\frac{1}{2}m\overline{v^2}$ (the average kinetic energy of the molecules) is a constant at constant temperature, T, and is proportional to T. Absolute zero, $0°K$, is the temperature where all molecular motion stops.

   (g) Molecular velocities are very high, except at very low temperatures.

7   Under conditions where any of the assumptions of the kinetic theory are not true (for instance, at very high pressure, where there is not a large space between molecules) gases do not obey the combined gas law. At sufficiently low temperatures, all gases are more compressible at lower pressure than predicted. At sufficiently high temperatures, all gases are less compressible. Most gases approximately obey the gas law at ordinary pressures, and the behavior becomes more nearly ideal the lower the pressure.

2–8   Postscript: The Use of the Gas Laws

We have shown how the gas laws can be inferred reasonably from the molecular model of matter, and we also have shown how the study of gases can shed light on chemical reactions. The behavior of gases is important in many kinds of problems. The combustion of gasoline in an automobile engine, the formation of smog in a polluted urban atmosphere, and the sonic boom generated when a jet plane exceeds the speed of sound, all depend on the physical and chemical properties of gases.

   In an internal combustion engine, the mixture of fuel and air is sucked into the cylinders, compressed, and ignited. Burning of the fuel produces carbon dioxide and water. In the combustion, the number of molecules is increased, and the gases are heated by the energy of the reaction. This causes rapid gas expansion, which drives the piston back to the recessed position and, at the same time, turns the wheels of the vehicle by way of the crankshaft and gear system. Analysis of the dynamics of gas flow and the delivery of energy released to run the automobile is done using essentially the same kinetic molecular theory described in this chapter.

   Air pollution chemistry involves complicated chemical reactions of oxygen with gaseous pollutants, mostly hydrocarbons released as unburned fuel from automobile exhausts. The oxidation reactions leading to smog formation are slow, but they are triggered by the action of sunlight on oxides of nitrogen, which are also present in very small amounts in polluted atmospheres. The nitrogen oxides also are produced in automobile engines by reactions of oxygen and nitrogen at the high temperatures of the combustion. Unfortunately, nitrogen oxides are produced from other sources, such as power plants, where natural gas is burned to produce electrical power. The understanding of the basic chemistry of smog formation has proceeded a long way. But even after the chemical reactions are known, we must turn our attention to the problem of understanding the mixing and movement of the slowly reacting gas mixture in large atmospheric pockets. One of the most crucial chemical engineering problems of our time is analysis of the huge chemical "reactors" found in places such as the Los Angeles basin. We are not yet out of the woods in finding a solution to the problem of air pollution, but we would be nowhere at all without the molecular theory of vapors—clean and unclean.

Sound is caused by the compression of a gas. When we clap our hands, the gases in the atmosphere are compressed, and the compression travels out from the original source as a wave, something like the water waves caused by throwing a rock into a pond. When the compression wave reaches a human ear, the pattern of pressure on the ear drum is transmitted to the brain and registered as sound. The speed of sound depends upon the temperature, pressure, and the molecular weight of the gases in ways predicted directly by the kinetic molecular theory.

A disturbance in a gas is caused by anything passing through it. At ordinary speeds, a moving object pushes the gas out of the way by compression. However, an object moving faster than the speed of sound cuts through the gas without compressing it, because the motion is too fast to allow the gas to respond. When an airplane passes through the sound barrier, a tremendous shock wave is created, because of the sudden change in pattern of air flow around the plane. Engineers designing supersonic planes make use of gas laws and kinetic theory of gases in their effort to eliminate or attenuate sonic booms. Attempts so far have been unsuccessful, and the task may prove to be impossible.

New Terms

Absolute zero: The temperature at which all motion stops; the lowest possible temperature. Absolute zero is 0°K or −273°C.

Avogadro's hypothesis: Equal volumes of gases contain equal numbers of molecules at the same temperature and pressure.

Boyle's law: At constant temperature, the volume of a gas sample varies inversely with the pressure, or $PV = a$.

Charles' law: At constant pressure, the volume of a gas sample is directly proportional to its temperature, or $V/T = b$.

Combined gas law: $PV = nRT$.

Condensed phase: A phase in which molecules are closely packed together; a liquid or solid, for example.

Gas constant (R):   The proportionality constant in the combined gas law, $PV = nRT$.

$$R = 8.20 \times 10^{-2} \text{ liter atm mole}^{-1} \text{ °K}^{-1}$$

Ideal gas: The imaginary gas whose properties are predicted exactly by the kinetic molecular theory.

Kinetic molecular theory: A model based on tiny particles in rapid motion, which can be used to predict the properties of gases.

Law of combining volumes: The ratios of volumes of reacting gases are small whole numbers (Gay-Lussac).

Law of partial pressures: Each component of a gas mixture exerts its own partial pressure independent of the others, and the total pressure is the sum of the partial pressures of the gases present.

Molar volume: The volume occupied by one mole of a gas at STP; 22.4 liters.

Mole fraction: The mole fraction of A in a mixture is the number of moles of A divided by the total number of moles of all substances present.

Standard temperature and pressure (STP):   Standard conditions for reporting the properties of gases and other substances. The standard temperature is $0°C = 273°K = 32°F$, and the standard pressure is 760 torr $= 1$ atm.

Questions and Problems

1   Why should gases obey simpler laws than liquids or solids?

2   Give some simple examples that illustrate that gases exert pressure.

3   To what volume will a 1.50-liter balloon expand if the pressure is reduced from 0.98 atm to 0.79 atm, while the temperature is held constant at 64°F?

4   Why is a plot of experimental data that produces a straight line useful or desirable?

5   How is an absolute scale of temperature defined in terms of gas behavior?

6   The 1.50-liter balloon of Problem 3 experiences a temperature drop to −31°C, while the pressure is maintained at 0.98 atm. What will the volume of the balloon be at the lower temperature?

7   Under what conditions does Boyle's law apply? When is Charles' law applicable? How are these laws derived from the complete ideal gas law?

8   A gas at an initial pressure of 700 torr is allowed to expand until the pressure is 150 torr. What is the ratio of the final volume to the initial volume?

9   What does STP signify, and why is it useful?

10   What does Figure 2–5 suggest happens to the volume of a gas when it is cooled to absolute zero? Why would it be impossible to confirm this experimentally?

11   A gas occupies a volume of 375 ml at 75°C and 734 torr. What volume will this gas occupy at standard temperature and pressure?

12   Industrial chemical processes are developed on a small scale in reaction systems called pilot plants. What volume would be required for a reaction vessel in which 64.6 g of gaseous propylene, C_3H_6, is contained at 65.4°F and 3.40 atm?

13   One of the fuels used by a Saturn V rocket is liquid oxygen. Assuming the ideal gas law, what pressure would result if 3200 g of liquid oxygen were placed in a heavy-gauge metal fuel tank with a capacity of 100 liters and allowed to warm to 300°K?

14   The temperature of a 0.0100-g sample of chlorine gas (Cl_2) in a 10-ml sealed glass container is raised in an oven from 20°C to 250°C. What is the initial pressure at 20°C? What is the pressure at 250°C?

15   Why is the gas volume of 22.4 liters significant?

16 What pressure will be exerted by 5.0×10^{13} molecules of an ideal gas in 1.000 ml at 0°C?

17 Three gases in a container exert a total pressure of 1.5 atm. The partial pressures of two of the gases are 0.75 atm and 0.30 atm. Calculate the mole fractions of all three gases.

18 How many atoms of helium are in a balloon that has been filled to a volume of 2.34 liters at 1147 torr and 35°C?

19 What does Dalton's law of partial pressures indicate about the behavior of gases in a mixture?

20 What is the density of XeO_4 gas at STP in g liter^{-1}?

21 At 300°K and 730 torr, 7.0 g of a gas occupies a volume of 5.63 liters. What is the molecular weight of the gas?

22 A mixture of gases contains 0.5 mole of oxygen, 0.1 mole of hydrogen, and 0.8 mole of nitrogen. The total pressure is 0.8 atm. What is the partial pressure of each gas?

23 If the molecules in a liter of hydrogen gas and a liter of oxygen gas are moving with the same root mean square speed, which gas is hotter?

24 The average speed of O_2 molecules at STP is 1030 miles per hour. What is the average speed of H_2 molecules under the same conditions? Which gas would be more likely to escape from the gravitational attraction of the earth?

25 What volume will 0.50 ml of water occupy, if it is completely converted into steam at 100°C and 800 torr? Assume the density of water is 1.00 g ml^{-1}.

26 A sample of NO_2 gas occupies a volume of 5750 ml at STP. How many grams of NO_2 are there in the sample?

27 A container is filled with 27.4 g of O_2 gas and 17.5 g of H_2 gas at a total pressure of 3.70 atm. What is the partial pressure of each gas?

28 Calcium carbide, CaC_2, reacts with water to produce acetylene according to the reaction

$$CaC_2(s) + 2H_2O(l) \rightarrow Ca(OH)_2(s) + C_2H_2(g)$$

The acetylene gas is collected over water at a pressure of 754 torr at 22.0°C and the gas occupies 874 ml. How many grams of acetylene are present? How many moles of CaC_2 are required for this reaction?

29 A 250-ml sample of a compound with the empirical formula CH_2 weighs 0.395 g at 700 torr and 27°C. What are the molecular weight and molecular formula of the compound?

30 During World War II, it was necessary to separate two different forms of the element uranium, one having an atomic weight of 235 and the other of 238. The separation was effected by making gaseous UF_6 and then utilizing the different effusion rates of the two substances of different molecular weights. How much faster will the 235 form of UF_6 diffuse than the 238 form? Does this seem to be an effective way to separate these two isotopes of uranium?

PERIODICITY OF
CHEMICAL COMBINATION

3

Previously, we have established how the laws of chemical composition paved the way for Dalton's atomic theory of the structure of matter. Probably the most fundamental chemical property of an element is the characteristic ratios in which it combines with other elements. If we probe more deeply and examine these ratios for several different elements, we find certain striking similarities. This discovery allows us to group elements with similar chemical properties into a single family. It is to our advantage to make these groupings, because it allows us to store our knowledge of important chemical properties in substantially larger chunks. If the properties of one element in a family are known, those of the other members of the family can be predicted with some confidence. Recognition of *families of elements* is the first important subject to be explored in this chapter.

Another fundamental property of elements becomes apparent when we arrange the elements in a tabular form according to an index called the *atomic number.* The atomic numbers of the elements approximately, but not exactly, increase with the atomic weights of the elements. When the elements are arranged according to increasing atomic number, we find that the families of elements lie in vertical columns of a table. This demonstrates a definite repeating, or *periodic,* characteristic of the elements. This compact table of elements, arranged emphasizing their periodic behavior, is called a *periodic table.* Since the first 103 elements can be grouped into eight "main" families, three "transition" series, and two "innertransition" series, we have a systematic way of discussing the properties of millions of known chemical compounds. You can imagine how hopeless our task would be without such a classification!

In this chapter, we will use facts concerning the composition of some simple compounds to show the periodic pattern of behavior that leads to the grouping of elements in families. For example, the knowledge that the metal lithium forms an oxide having the formula Li_2O leads to the prediction that there will be a family of metallic elements that will all form oxides having the general formula M_2O. At the same time, we will show why chemists need knowledge concerning the structures of atoms. The fact that elements show periodic chemical properties, rather than a regular series of changes as the atoms become larger, is very difficult to explain by Dalton's hypothesis that atoms are hard, structureless particles.

3–1   The Simple Hydrides

We will begin our discussion of periodic properties by examining the composition of compounds formed between hydrogen and other elements. Compounds in which hydrogen atoms are bound to other elements are often called *hydrides.* Thus, water could be called oxygen dihydride. Conversely, the name dihydrogen oxide also could be used. The ending "ide" in a name implies that the compound is a *binary compound;* that is, it contains only two elements. The name of a compound also has other implications (to be discussed later) that make dihydrogen oxide, for example, more appropriate than oxygen dihydride. Nonetheless, the entire group of binary compounds containing hydrogen are often referred to as hydrides.

The names, structural formulas, and models of the hydrides of the lightest elements are shown in Figure 3–1. The atomic numbers shown were originally chosen so the elements were arranged in order of their increasing atomic weights. In Chapter 4, we will discuss a more fundamental basis for the assignment of atomic numbers. Note carefully that atomic weights are not the same as atomic numbers.

Some clarification is needed concerning the molecular models used to illustrate the hydrides. Lithium hydride and beryllium hydride are crystalline solids with very high melting points. The actual structures of these solids are like that of potassium chloride (Section 8–9), in that individual molecules, LiH and BeH_2, cannot be found in the crystals. Boron hydride is a gas under ordinary conditions, but it consists exclusively of double molecules (dimers) having the composition B_2H_6. The simple formulas and models in Figure 3–1 for the hydrides of lithium, beryllium, and boron represent molecular forms that can be observed only under special conditions, such as very high temperatures. In this chapter, we are interested primarily in composition, rather than molecular structure and will, therefore, use simple models and formulas.

A key factor in the organization of the elements in Figure 3–1 is the fact that neither helium nor neon forms any compounds with hydrogen, or with any other element. Other similarly inert elements are argon (atomic number 18), krypton (atomic number 36), xenon (atomic number 54), and radon (atomic number 86). These *noble gases* constitute a family of elements. They are

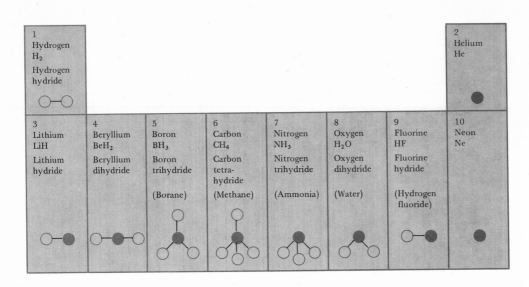

Figure 3–1.   A table of the first ten elements and their binary hydrogen compounds. Each entry in this figure contains (1) the atomic number of the element, (2) the name of the element, (3) the formula of the hydride, (4) the name of the hydride, (5) the common name of the hydride, and (6) a model of the hydride.

called noble gases because they are all much less reactive than other elements. Other families of elements also show many similarities in their chemical behavior. A tabular form, known as the periodic table, has proved to be one of the most useful devices for presenting relationships among the elements. The organization is called periodic because study of the properties of elements shows that, as the atomic numbers increase, characteristic properties recur periodically. The inert character of the noble gases is an example.

Figure 3–2.   An arrangement of the first 18 elements in order of increasing atomic number. The hydride formulas are similar for the vertically grouped elements.

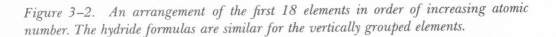

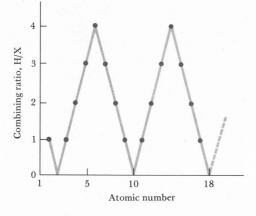

Figure 3–3. Periodicity of the combining ratios of the lightest elements in their compounds with hydrogen.

If we arrange the elements having atomic numbers between neon (10) and argon (18) in tabular form, placing the members of the series below the series helium to neon, we obtain the pattern shown in Figure 3–2. The formulas of the simplest hydrides are also given. Note that, in every case, the simplest hydrides of elements separated by eight atomic number units have the same general formula XH_n, where X may be any element of a family. Note also that the combining ratios, H/X, form a symmetrical pattern (see Figure 3–3)

number of H atoms in the hydride: 1 2 3 4 3 2 1

Some family relationships are suggested in Table 3–1. It seems that eight is not a magic number, since the first row in the table has only two members, hydrogen and helium. The fact that the noble gas krypton (atomic number 36) has an atomic number 18 units higher than that of argon suggests that the next row in the table may be even longer.

Table 3–1.   Family Groups of Elements Based on Their Abilities to Form Hydrides

Family 0	He, Ne, Ar
Family 1	H, Li, Na
Family 2	Be, Mg
Family 3	B, Al
Family 4	C, Si
Family 5	N, P
Family 6	O, S
Family 7	F, Cl

1 H HF Hydrogen fluoride							2 He
3 Li LiF Lithium fluoride	4 Be BeF_2 Beryllium fluoride	5 B BF_3 Boron tri-fluoride	6 C CF_4 Carbon tetra-fluoride	7 N NF_3 Nitrogen tri-fluoride	8 O OF_2 Oxygen difluoride	9 F F_2 Fluorine	10 Ne
11 Na NaF Sodium fluoride	12 Mg MgF_2 Magnesium fluoride	13 Al AlF_3 Aluminum tri-fluoride	14 Si SiF_4 Silicon tetra-fluoride	15 P PF_3 Phosphorus tri-fluoride	16 S SF_2 Sulfur difluoride	17 Cl ClF Chlorine mono-fluoride	18 Ar

Figure 3–4. A further indication of the periodic chemical behavior of the first 18 elements, based on the formulas of their compounds with fluorine.

3–2   The Composition of Fluorides

It is interesting to examine the composition of the simplest fluorides of the first 18 elements. Figure 3–4 shows such a summary in the same form as the hydride table, Figure 3–2. It is a striking fact that the periodicity of chemical combination for these fluorides is identical to that of the hydrides. The same family relationships are observed.

3–3   Classification of the Elements

The key place of the noble gases in the periodic table is established from observations of chemical combination. However, if composition of the simple compounds were the only guideline, we would be inclined to define fewer families, each with more members. For example, beryllium, magnesium, oxygen, and sulfur might be placed in the same family, since they all form dihydrides and difluorides. However, in many other ways these elements are so different that we would feel uncomfortable placing them in the same family. For example, the elements beryllium and magnesium are metals having high melting points and being good conductors of electricity. In contrast, oxygen is a gas, and sulfur is a solid with a relatively low melting point. Furthermore, sulfur does not have the shiny appearance of a metal and is not a good electrical conductor. Many other differences convince us that the relationships suggested by the formulas of the hydrides and fluorides alone are not sufficient to allow us to arrange the elements in families.

All of the elements are divided roughly into two groups, *metals* and *nonmetals*. The classification is useful, since most people have some notion of the common, characteristic properties of metals. However, the separation is not precise, and some unexpected classifications are made. For example, boron is usually classified as a nonmetal, whereas aluminum, a member of the same chemical family, is a familiar metal.

Some characteristic properties of metals are lustrous appearance, ability to conduct electricity and heat, *ductility* (ability to be drawn into thin wires), and *malleability* (ability to be shaped into thin sheets). Metals are solids at room temperature with the exception of mercury, which melts at −38.9°C. In contrast to these properties, the nonmetals generally are not lustrous, are poor conductors of electricity and heat, and are not ductile or malleable. Nonmetals often have much lower melting points than metals have.

3–4   Some Families of Elements

Some of the families of elements show such strong similarities among their members that they have been associated as a group since the work of the German chemist Johann Döbereiner, in 1829. We will summarize briefly some of these characteristics, emphasizing the similarities in the composition of simple hydrides and fluorides. However, there are many other grounds for the group associations.

THE ALKALI METALS

This group of metals forms monohydrides (MH) and monofluorides (MF). They are also highly reactive toward many chemicals. For example, they all react violently with water, forming compounds known as *alkalis.* For example,

*s*olid sodium + *l*iquid water = *aq*ueous solution of sodium ion

$$+ \ aqueous \ solution \ of \ hydroxide \ ion$$
$$+ \ hydrogen \ gas$$
$$2Na(s) + 2H_2O(l) = 2Na^+(aq) + 2HO^-(aq) + H_2(g)$$

This property is responsible for the name of the group. The *alkali metals* are listed in Table 3–2, along with the formulas of the simple hydrides and fluorides.

Table 3–2.   The Alkali Metals and the Formulas of Their Hydrides and Fluorides

Element[a]	Symbol	Atomic number	Formula of hydride	Formula of fluoride
Lithium	Li	3	LiH	LiF
Sodium	Na	11	NaH	NaF
Potassium	K	19	KH	KF
Rubidium	Rb	37	RbH	RbF
Cesium	Cs	55	CsH	CsF
Francium	Fr	87	FrH[b]	FrF[b]

[a]*The more important members of the series are shown in boldface type.*

[b]*The compound is not known, because the metal is only available in trace amounts. However, any chemist probably would expect the compound to exist and have the indicated composition.*

The relationship between the names of elements and their symbols is worth comment. Usually, the symbol is taken directly from the name, and it may be only the first letter of the name (e.g., H for hydrogen). The first letter of the symbol is always capitalized. In some cases, the symbol bears no obvious relationship to the name. Examples are sodium (Na) and potassium (K). In these cases, the symbols are taken from the Latin names (natrium = sodium, and kalium = potassium). To English-speaking students, the system may seem senseless. However, the chemical symbols are used almost universally throughout the world, and the common names of the elements are different in different languages. In French, the names of sodium and potassium are natron and kalium; in German they are natrium and kalium. Consequently, French and German students see no problem in this choice of symbols.

THE ALKALINE EARTHS

The next family, suggested by the relationship between beryllium and magnesium, is the *alkaline earth* metals. The members of the family are listed in Table 3–3, along with the formulas for the simple hydrides and fluorides of the elements. There is a regular progression from the alkali metals to the alkaline earth metals. The alkalis have a combining ratio with H of one, the alkaline earths of two. Each alkali metal is followed by an alkaline earth metal, having an atomic number one higher. The alkaline earth metals react with water, though less violently than the alkalis

$$Ca(s) + 2H_2O(l) = Ca^{2+}(aq) + 2HO^-(aq) + H_2(g)$$

They are hard metals, beryllium being hard enough to scratch glass.

GROUP III

The third family is Group III, whose members are listed in Table 3–4. The name Group III is an extension of the unsystematic nomenclature used previously. The alkali metals form Group I, while Group II comprises the alkaline earth metals. The first noticeable characteristic of the

Table 3–3.   The Alkaline Earth Metals and the Formulas of their Hydrides and Fluorides

Element[a]	Symbol	Atomic number	Formula of hydride	Formula of fluoride
Beryllium	Be	4	BeH_2[b]	BeF_2
Magnesium	Mg	12	MgH_2[b]	MgF_2
Calcium	Ca	20	CaH_2	CaF_2
Strontium	Sr	38	SrH_2	SrF_2
Barium	Ba	56	BaH_2	BaF_2
Radium	Ra	88	RaH_2[b]	RaF_2

[a] *The more important members of the family are given in boldface type.*
[b] *Although a compound of the indicated composition probably can be made, it has not been identified definitely.*

Table 3–4.   Group III Metals and the Formulas of Their Hydrides and Fluorides

Element[a]	Symbol	Atomic number	Formula of hydride[b]	Formula of fluoride
Boron	B	5	(BH_3)	BF_3
Aluminum	Al	13	(AlH_3)	AlF_3
Gallium	Ga	31	(GaH_3)	GaF_3
Indium	In	49	$InH_3{}^c$	InF_3
Thallium	Tl	81	$TlH_3{}^c$	TlF_3

[a] *The more important members of the family are listed in boldface type.*

[b] *The compounds in parentheses are known, but they do not form simple molecules of* MH_3. *Under conditions where* BH_3 *is formed, it reacts with itself immediately to form* B_2H_6. *The aluminum and gallium hydrides are not well characterized, but it is suspected that they exist as* Al_2H_6 *and* Ga_2H_6, *respectively.*

[c] *Although a compound of the indicated composition probably can be made, it has not been identified definitely.*

third family is that, whereas each alkaline earth differs from a corresponding alkali metal by one in its atomic number, no such easy correspondence exists between the alkaline earth and Group III elements. Thus, an anticipated regularity in the periodic arrangement of the elements must be more carefully considered.

In Group III we meet the first nonmetal, boron, whose main source is the compound borax, found in the dry lake beds of California and Nevada. Aluminum, the next element in the group, is a well-known metal.

GROUP IV

The elements of Group IV are given in Table 3–5. The symbols for tin and lead (Sn and Pb) stand for the Latin names stannum and plumbum, respectively. Tin and lead are the names of Anglo-Saxon origin. These two metals have been known since antiquity.

Table 3–5.   Group IV Elements and the Formulas of Their Hydrides and Fluorides

Element	Symbol	Atomic number	Formula of hydride[b]	Formula of fluoride
Carbon	C	6	CH_4	CF_4
Silicon	Si	14.	SiH_4	SiF_4
Germanium	Ge	32	GeH_4	GeF_4
Tin	Sn	50	SnH_4	SnF_4
Lead	Pb	82	PbH_4	$PbF_4{}^a$

[a] *Although a compound of the indicated composition probably can be made, it has not been identified definitely.*

Table 3–6. Group V Elements and the Formulas of Their Hydrides and Fluorides

Element[a]	Symbol	Atomic number	Formula of hydride	Formula of fluoride
Nitrogen	N	7	NH_3	NF_3
Phosphorus	P	15	PH_3	PF_3
Arsenic	As	33	AsH_3	AsF_3
Antimony	Sb	51	SbH_3	SbF_3
Bismuth	Bi	83	BiH_3	BiF_3

[a]*The more important members of the family are given in boldface type.*

With Group IV we see a heartening return to regularity. Each element in Group IV differs from the corresponding element in Group III by one in the atomic number. Group IV also contains a mixture of metals and nonmetals. Carbon, silicon, and germanium are nonmetals, whereas tin and lead are metals.

GROUP V

The Group V elements are given in Table 3–6. The symbol Sb for antimony comes from the Latin name for the element, stibium.

Once again, the regular progression in atomic number from one family to the next is preserved. The trend toward nonmetallic characteristics and away from metallic characteristics continues in this family. The first three elements are nonmetals, but antimony and bismuth are metals.

Table 3–7. Group VI Elements and the Formulas of Their Hydrides and Fluorides

Element[a]	Symbol	Atomic number	Formula of hydride	Formula of fluoride
Oxygen	O	8	H_2O	OF_2
Sulfur	S	16	H_2S	SF_2
Selenium	Se	34	H_2Se	SeF_2[b]
Tellurium	Te	52	H_2Te	TeF_2[c]
Polonium	Po	84	PoH_2	PoF_2[d]

[a]*The more important elements of the family are given in boldface type.*
[b]*The existence of the compound is suspected, but positive identification has not been achieved.*
[c]*The compound is unknown, but probably can be prepared.*
[d]*Scarcity of the element has prevented extensive investigation of its compounds. The difluoride has not been made, although its existence is predicted.*

Table 3–8.   Group VII Elements and the Formulas of Their Hydrides and Fluorides

Element[a]	Symbol	Atomic number	Formula of hydride	Formula of fluoride
Fluorine	F	9	HF	F_2
Chlorine	Cl	17	HCl	ClF
Bromine	Br	35	HBr	BrF
Iodine	I	53	HI	IF[b]
Astatine	At	85	HAt[c]	AtF[c]

[a] *The more important members of the family are given in boldface type.*
[b] *This compound reacts immediately with itself to form higher fluorides (e.g., IF_3), so it has not been identified positively.*
[c] *Astatine is so rare, being produced artificially, that its chemistry is little known. However, almost all chemists would predict the existence of these two compounds.*

GROUP VI

The elements of Group VI are given in Table 3–7. The regular progression of atomic numbers from group to group continues to Group VI. The trend to nonmetallic character also continues. Polonium is the only member of the group that is a metal. Polonium is also the first element that was discovered because of its radioactivity. It was found mixed with bismuth (Group V), and was isolated from the ore pitchblende by Marie and Pierre Curie in 1898. (For their work on radioactive elements, they received the Nobel prize in physics, in 1903.)

GROUP VII, THE HALOGENS

The name halogen comes from the Greek root halo (meaning of the sea) and refers to the fact that salts of the halogens, such as sodium chloride, can be obtained by evaporating sea water. The halogens are listed in Table 3–8. The most prominent characteristic of the halogens is that they react vigorously (often explosively, in the case of fluorine) with metals. The elements of this group are the most nonmetallic of the nonmetals. And of the members of the group, fluorine reacts most vigorously with metals and is itself the most nonmetallic. An example of a metal–fluorine reaction is

$$2Na(s) + F_2(g) = 2NaF(s)$$

Notice that in Groups III to VI in the transition from metals to nonmetals, the heavier elements (higher atomic number) of each group tend to have metallic characteristics and the lighter ones, nonmetallic characteristics. This same trend within a given group is maintained in the halogens; accordingly, fluorine is the most nonmetallic element. Also, not surprisingly, francium (or, for practical purposes, cesium, since francium is so rare) is the most metallic element (see Group I).

The regular procession of atomic numbers from group to group continues from Group VI to Group VII and, finally, to the noble gas group, which ends the rows, or periods, of elements.

Table 3–9.   Families of Elements and Their Most Common Valences

Family	Common valence
I (alkali metals)	1
II (alkaline earths)	2
III	3
IV	4
V	3
VI	2
VII (halogens)	1

3–5 Valence

Useful rules for the common binding capacity of elements, called *valence,* can be derived from the study of the composition of simple hydrides and fluorides. If we assume that hydrogen and fluorine have a valence of one, then the alkali metals also must have a common valence of one. As a matter of fact, they are not known to form compounds in which they show any other valences, although some elements do show more than one valence. Table 3–9 is a compact tabulation of the common valences of members of the families of elements discussed above. The valences in this table are the same as the number of holes that would appear in a model atom of the element (Section 1–7). Part of the utility of model sets is that so many elements exhibit the same common valence. For this reason, one kind of model atom can be used to represent many different elements.

Including the noble gases, there are eight families of elements. Notice that the common valence is the same as the family number for the first four families, and eight minus the family number for the remaining families. This is either numerology, or striking evidence for an underlying structure that determines the behavior of atoms.

Some elements show more than one valence when they form different compounds with a given element. Sulfur, for example, forms the following fluorides:

Formula	Name	Valence of sulfur
SF_2	sulfur difluoride	2
SF_4	sulfur tetrafluoride	4
SF_6	sulfur hexafluoride	6

Other variations in the valence of an element will be encountered later. However, the principal valences shown in Table 3–9 are useful enough to warrant their being committed to memory. With this table and the periodic table, the existence of an enormous number of compounds can be predicted accurately.

3–6   Self-Bonding of Elements

The composition of some compounds does not appear to fit simple valence rules. For example, two compounds of hydrogen and oxygen are known. One, water, has two hydrogen atoms for each oxygen atom; the other, hydrogen peroxide, contains atoms of the two elements in 1:1 ratio. This might imply that oxygen can have a valence of one. However, the molecular weight of hydrogen peroxide is 34. Therefore, each molecule contains two atoms each of oxygen and hydrogen. As was shown in the discussion of molecular models (Section 1–7), we can assemble a model in which the common valence rules are followed by bonding oxygen atoms to each other: H—O—O—H.

When the atoms of an element form strong bonds to each other, a large number of compounds may be formed with any given element. Spectacular examples are found in carbon compounds. Indeed, the most outstanding property of carbon atoms is their ability to form strong bonds with each other. This ability, combined with its valence of four, permits carbon to form large numbers of compounds containing many carbon atoms linked together. In addition, carbon forms strong bonds with nitrogen, oxygen, hydrogen, and many other elements. The occurrence of these elements in carbon compounds helps give them the variety that provides the chemical basis for all living things. The principles that apply in the chemistry of carbon compounds are the same as those that apply in all of chemistry. Yet the large number and diversity of carbon compounds, and their special role in our understanding how living systems work, give carbon a special place among the elements. Figure 3–5 shows molecular models of some simple examples of carbon–hydrogen compounds.

The preceding discussion indicates that there would be some advantage in replacing the term valence with the term *bond number* (i.e., the number of bonds atoms of an element can form with atoms of other elements). However, this would introduce other systematic problems. For present purposes, valence can be considered to mean bond number.

Bonding of like atoms of elements is most important with boron, carbon, nitrogen, oxygen, silicon, and sulfur. However, many other elements show the same behavior. Hydrogen and the halogens are *diatomic* molecules in their elementary states. The H–H bond in hydrogen mole-

Figure 3–5. Models of several carbon–hydrogen compounds, showing their three-dimensional features.

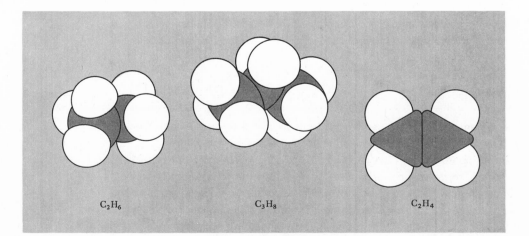

C_2H_6 C_3H_8 C_2H_4

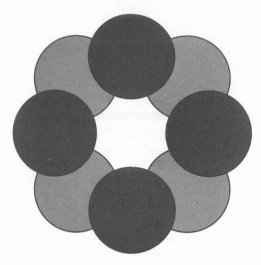

Figure 3–6. A molecular model of S_8*, which is the most common form of elemental sulfur. In these molecules, sulfur is bound to itself in such a way that the valence of each sulfur atom is two.*

cules is very strong. If hydrogen had a valence higher than one, it would provide the basis for another large family of compounds like the compounds of carbon.

It is important to distinguish carefully between variable bond number and self-bonding. The structures of the sulfur fluorides involve variation in the number of fluorine atoms bound to sulfur

$$SF_2 \qquad\qquad SF_4 \qquad\qquad SF_6$$

In these compounds, sulfur shows variable bond number (2, 4, and 6). Solid elemental sulfur consists of molecules having the formula S_8 (Figure 3–6). This is an example of self-bonding, in which the valence of sulfur is two.

3–7   Transition Metals

There is a large group of elements that has been omitted from our discussion of the family groupings. These are the elements that are found in the periodic table between Group II and Group III, beginning with the fourth period (row). (The complete periodic table is shown in Figure 3–7.) These elements, known as the *transition metals,* have been set apart in the periodic table as a collection of subgroups with the ''B'' label, because their electronic structures (Chapter 5) give rise to characteristic chemical and physical properties. These metallic elements are all high-melting solids of relatively high density. They exhibit variable combining abilities, or valences,

IA	IIA	IIIB	IVB	VB	VIB	VIIB	├──────VIIIB──────┤			IB	IIB	IIIA	IVA	VA	VIA	VIIA	Noble gases
1 H 1.008																		2 He 4.003
3 Li 6.939	4 Be 9.012												5 B 10.811	6 C 12.011	7 N 14.007	8 O 15.999	9 F 18.998	10 Ne 20.183
11 Na 22.990	12 Mg 24.312												13 Al 26.982	14 Si 28.086	15 P 30.974	16 S 32.064	17 Cl 35.453	18 Ar 39.948
19 K 39.102	20 Ca 40.08	21 Sc 44.956	22 Ti 47.90	23 V 50.942	24 Cr 51.996	25 Mn 54.938	26 Fe 55.847	27 Co 58.933	28 Ni 58.71	29 Cu 63.54	30 Zn 65.37		31 Ga 69.72	32 Ge 72.59	33 As 74.922	34 Se 78.96	35 Br 79.909	36 Kr 83.80
37 Rb 85.47	38 Sr 87.62	39 Y 88.905	40 Zr 91.22	41 Nb 92.906	42 Mo 95.94	43 Tc (99)	44 Ru 101.07	45 Rh 102.91	46 Pd 106.4	47 Ag 107.87	48 Cd 112.40		49 In 114.82	50 Sn 118.69	51 Sb 121.75	52 Te 127.60	53 I 126.90	54 Xe 131.30
55 Cs 132.91	56 Ba 137.34	57 La 138.91	72 Hf 178.49	73 Ta 180.95	74 W 183.85	75 Re 186.2	76 Os 190.2	77 Ir 192.2	78 Pt 195.09	79 Au 196.97	80 Hg 200.59		81 Tl 204.37	82 Pb 207.19	83 Bi 208.98	84 Po (210)	85 At (210)	86 Rn (222)
87 Fr (223)	88 Ra (226)	89 Ac (227)	104	105														

58 Ce 140.12	59 Pr 140.91	60 Nd 144.24	61 Pm (145)	62 Sm 150.35	63 Eu 151.96	64 Gd 157.25	65 Tb 158.92	66 Dy 162.50	67 Ho 164.93	68 Er 167.26	69 Tm 168.93	70 Yb 173.04	71 Lu 174.97
90 Th 232.04	91 Pa (231)	92 U 238.03	93 Np (237)	94 Pu (242)	95 Am (243)	96 Cm (247)	97 Bk (249)	98 Cf (251)	99 Es (254)	100 Fm (253)	101 Md (256)	102 No (253)	103 Lr (257)

Figure 3–7.   A complete periodic table. Atomic weights in parentheses are for one isotope of the element. Elements 104 and 105 have not been named.

illustrated by the compounds of iron and palladium $FeCl_2$, $FeCl_3$, PdO, and PdO_2. Generally, the elements grouped under one "B" heading have similar chemical properties.

Another group of elements, called the *lanthanides* (rare earths), are distinguished by setting them below the body of the periodic table. This is mainly a convenience, since placing those elements from cerium (58) through lutetium (71) between lanthanum (57) and hafnium (72) would make the table much wider and lopsided. The chemical properties of all the lanthanides are very similar. The elements starting with actinium (the *actinides*) are placed just below the lanthanides, thereby completing the periodic table.

3–8   The Periodic Table

In the full periodic table (Figure 3–7) the elements are grouped in families, for the reasons discussed already. It is clear that the chemical properties of the elements have a definite periodic character. Thus, it would seem desirable to explore *the structure of the atom* in an attempt to uncover some basis for the chemical periodicity of elements. If this is a basis, then we can go on to a full and more meaningful consideration of the chemistry of all the families in the periodic table.

3–9 Summary

1 Elements may be grouped into families having chemical similarities.
2 Families of elements may be arranged in tabular form, such that there is a regular or periodic repetition of chemical properties. This arrangement of elements is known as the periodic table.
3 In a broad sense, the elements may be divided into three groups—noble gases, metals, and nonmetals.
4 The term valence describes the capacity of an atom to bind other atoms to itself.

3–10 Postscript: The Periodic Table—Law, Theory, or Philosophy?

A Russian, Dmitri Mendelyeev, generally is credited with working out the periodic relationships, and he announced his "periodic law" in 1869. Lothar Meyer, a German, developed essentially the same system independently and almost simultaneously. Unknown to either man was the work of John Newlands, an English chemist, who had worked for many years trying to establish a kind of periodic law. Even in his own country, Newlands received little sympathy, because of his insistence that there was an analogy between the eightfold pattern of chemical behavior and the octaves of a musical scale. Use of the musical scale as a model caused him trouble, because of the existence of long periods and the differences between the members of the A and B families of elements. We wonder what differences would have occurred in Newlands' approach to modeling if he had been familiar with the Chinese musical scales, which consist of five notes rather than eight?

The periodic table has been of great value in shaping chemical thought. First, the existence of family relationships allows remarkably accurate predictions to be made with great ease. A chemist who learns much about the properties of methyl chloride (H_3C—Cl) can confidently make a number of predictions about methyl bromide (H_3C—Br), because both chlorine and bromine belong to the halogen family. Mendelyeev provided several brilliant predictions when he first formulated the periodic law. For example, he believed that there should be an element in Group IV having properties intermediate between those of silicon and tin. He referred to the *then unknown* element as ekasilicon, and he predicted its atomic weight, color, melting point, and many chemical properties. The predictions were all remarkably close to facts reported some years later by Winkler for the element now called germanium.

The attempts by Mendelyeev and others to see order among the properties of the various elements reflected a point of view now almost universally accepted. We are thoroughly convinced that chemical and physical properties of matter can be related in systematic ways to chemical structure. This view was not always popular, and early searchers for systematic relationships were often scorned as irresponsible dreamers. Acceptance of the periodic table was a giant step in establishing the notion that chemistry must have a coherence that would be understandable to rational man. At the present time, our understanding of chemistry is far from completely coherent, but we cherish the conviction that a system exists in nature that can be learned by man.

As far as we know, Mendelyeev had no model in mind to explain atomic properties when he did his work. However, the empirical establishment of periodic relationships turned out to be invaluable to scientists of a later generation. In Chapters 4, 5, and 6 we will show how our present models for atomic and molecular structure were developed. Many of the observations that led to modern theories, with all of their intricacies, were fitted together systematically only because the periodic table was available to serve as a guide.

Much has been written about the interplay between theory and experiment in science. The history of the periodic table tells us something about interplay. Theories, or models, are intended to provide sensible correlation of experimental facts. Useful theories usually do not spring from the consideration of a single fact, but are inspired by patterns seen in a number of facts. Experimental data gathered in a random manner are not of much general value until someone is able to find some kind of system in the results.

New Terms

Actinides: The group of elements beginning with thorium and ending with lawrencium.

Alkali metals: The elements that constitute Group I in the periodic table.

Alkaline earth metals: The elements that constitute Group II in the periodic table.

Atomic number: The characteristic number of an element that places it in correct order in the periodic table.

Binary compound: A compound composed of two different elements, such as $NaCl$, Fe_2O_3, and BF_3.

Chemical family: A group of elements that have very similar chemical properties and predictable trends in their physical properties. Another name for any of the groups shown in the periodic table.

Diatomic: Having two atoms, in reference to molecules.

Ductility: The ability to be drawn into thin wires.

Halogens: The family name given the elements of Group VII (F, Cl, Br, I, and At).

Hydride: The term applied to compounds containing hydrogen and one other element. Most often used for metal–hydrogen compounds, such as NaH and CaH_2.

Lanthanides: A group of heavy, similar elements beginning with cerium and ending with lutetium.

Malleability: The ability to be hammered into thin sheets.

Metals: A general term used to describe elements that are good conductors of heat and electricity, and are ductile, malleable, and have a lustrous appearance.

Noble gases: The name of the members of the group of elements after Group VII. So called because they are chemically unreactive.

Nonmetals: A general term used to describe those elements that do not have metallic properties. Nonmetals are nonconducting, nonductile, nonmalleable, and usually dull in appearance.

Period: A row in the periodic table; distinguished from a family, which occupies a column.

Periodic table: A tabular arrangement of the elements in order of increasing atomic numbers; the columns represent families with similar chemical properties.

Rare earths: Common name for the lanthanides.

Transition metals: The group of metals in the center of the periodic table; the "B" group elements.

Questions and Problems

1   Explain and give examples of the concept "families of elements."

2   Consult a periodic table, and, using the information regarding the combining properties of the elements given in this chapter, predict the formulas of the hydrides of the elements Rb, Ba, Ga, Ge, Sb, Te, and I.

3   What unusual chemical characteristics do we associate with the noble gases?

4   List the elements that constitute the alkali metal group. Write an equation that describes the reaction of a typical alkali metal with water.

5   Give the names of the elements that have the following symbols: K, Na, P, Ba, Sb, Te.

6   One tenth of a mole of the alkali metal sodium reacts with 4.1 liters of H_2 gas at 1000°K and 1 atm. One tenth of a mole of the alkaline earth calcium reacts with 8.2 liters of H_2 under the same conditions. What can be said about the relative combining power of these two elements?

7   Describe what is meant by the term valence. What common valences would you expect for the elements in Groups II, IV, and VII?

8   Write the proper formulas for the following binary compounds: BaCl, MgO, BeS, CsTe, and SiH.

9 How do metals and nonmetals differ? Why are the elements called noble gases grouped together?

10 In addition to the periodicity of chemical properties of the elements, there are certain regular trends in physical properties. Assuming that there is a regular trend in the change of atomic weights of the alkali metals, calculate the atomic weight of the alkali metal rubidium, given that the atomic weights of potassium and cesium are 39.10 and 132.91, respectively. How does this calculated value compare with that listed for rubidium in the periodic table?

11 The atomic radii of several of the elements of the third period are Na (1.90 Å), Mg (1.60 Å), Al (1.43 Å), Si (1.32 Å), P (1.28 Å), S (1.27 Å). Is the apparent trend in the atomic radii what you would expect it to be? Can you suggest an explanation for this trend?

12 Using the valences listed in Table 3–9, predict a general formula for the combinations of elements from the following groups: (a) Groups I and VI, (b) Groups IV and VII, and (c) Groups III and VI.

13 Suppose that a new element was discovered that had an atomic number of 119. Which group would it fit into? To which element would it be similar in chemical properties?

THE PARTICLES OF MATTER

4

Atoms are composed of three kinds of smaller particles. Two of the particles, *electrons* and *protons,* carry electrical charges; the third particle, the *neutron,* is uncharged. In this chapter, we will introduce several important ideas concerning the behavior of particles of matter and the fundamental nature of energy. Probably the most important of these concepts is the idea that energy comes in small, discrete packages called *quanta.* Various kinds of radiation (light, x rays, γ rays, β rays, and α rays) are described. The significance of two important properties of atoms, *atomic number* and *atomic weight,* will be explained. The atomic number turns out to be of principal importance in determining the chemical behavior of atoms. The atoms of an element all have the same atomic number, although they may have different atomic weights. Atoms of an element having different atomic weights are called *isotopes.* Finally, we will consider briefly the absorption of light by matter and the emission of light from energy-rich atoms and molecules.

4–1   Atomic Theory

In Chapter 1, we mentioned briefly Dalton's atomic theory, and Avogadro's hypothesis was discussed in Chapter 2. These are two examples of a marvelous inductive logic that led to the formulation of a rather definite theory of molecular structure by the end of the nineteenth century. The main features of the theory have been introduced earlier but should be restated:
1   Atoms are fundamental building blocks of matter.
2   All atoms of a given element are identical.
3   Atoms in molecules are found together in perfectly reproducible patterns. Even the geometric shapes of molecules are fixed in three-dimensional space.

Each of the three hypotheses is now known to be incorrect in some ways. But even with modification, all three remain fundamental to theories of structural chemistry. For example, consider the second hypothesis. Now we know that there are isotopic forms of atoms of many elements. These isotopes have different atomic weights. However, the chemical properties of the isotopic atoms are almost identical. Therefore, a structural theory based upon the notion that all atoms of an element are identical works very well.

4–2 Faraday's Electrolysis Experiments

During the early part of the nineteenth century, Michael Faraday performed careful experiments with the *electrolysis* (decomposition by an electric current) of water solutions of several compounds. For example, the electrolysis of hydrogen chloride in water (hydrochloric acid) produces hydrogen and chlorine

$$\underset{\substack{\text{hydrochloric} \\ \text{acid}}}{2\text{HCl}} \quad \xrightarrow[\text{current}]{\text{electric}} \quad \underset{\text{hydrogen}}{\text{H}_2} \quad + \quad \underset{\text{chlorine}}{\text{Cl}_2}$$

Hydrogen is evolved at the *cathode* (the negatively charged electrode) and chlorine is produced at the *anode* (positive electrode). A simple laboratory electrolytic setup is shown in Figure 4–1. The gases produced at the two electrodes must be kept separated, because chlorine and hydrogen react with each other explosively.

Faraday discovered two fundamental relationships by making careful measurements of the amount of electric current required to produce a given chemical change:

1 The chemical action of an electric current is directly proportional to the total amount of electricity that passes through the solution. For example, if the amount of current passed through a solution of hydrochloric acid is doubled, the amounts of hydrogen and chlorine produced are doubled.

2 The weights of products formed are proportional to their atomic weights, or some fraction thereof. It was known that 35.46 g of chlorine combines with 1.008 g of hydrogen, and Faraday found that in the electrolysis of hydrochloric acid the two products are formed in exactly that ratio.

We can illustrate these two relationships best by considering the following information. The amount of electricity flowing through a system is expressed in terms of the current, and current is defined as the rate at which electrical charge flows past a point. The basic unit of electrical charge is called the *coulomb,* and the basic unit of current is the *ampere* (amp). One ampere of current corresponds to a charge flow rate of one coulomb per second. Now consider the elec-

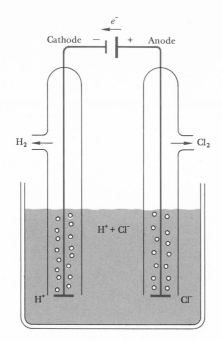

Figure 4–1. Laboratory apparatus for the electrolysis of hydrogen chloride.

trolysis of hydrochloric acid again, and imagine that a current of 1.00 amp has passed through the electrolytic cell for 2.68 hours (h). The total charge passing through the system is

$$2.68 \text{ h} \times \frac{60 \text{ min}}{\text{h}} \times \frac{60 \text{ sec}}{\text{min}} \times \frac{1.00 \text{ coulomb}}{\text{sec}} = 96{,}500 \text{ coulombs}$$

where 1.00 amp is replaced with 1.00 coulomb sec^{-1}. Under these conditions, 35.5 g of chlorine would be produced. If the current were doubled to 2.00 amp and allowed to flow for 2.68 h, 71.0 g of chlorine would be produced. If the original current (1 amp) flowed for 5.36 h, 71.0 g of chlorine again would be produced.

   Notice that the quantity of charge, 96,500 coulombs, is sufficient to produce one gram atomic weight of chlorine. The question to be answered now is: Will this same quantity of electrical charge, when passed through some other chemical system, produce the same mass of product as in the HCl cell, or will the quantity of the product be related to the atomic weight of the substance being electrolyzed? The following evidence provides the answer. When 96,500 coulombs is passed through cells in which various elements are produced, definite quantities of the products are observed, as illustrated in Table 4–1.

   A direct relationship between the quantity of electricity, 96,500 coulombs (called the *faraday*), and the atomic weights of the elements is suggested by these observations. It is convenient to introduce a new term at this point. Notice in Table 4–1 that one faraday of electricity does not always produce one mole of a substance. But when one faraday produces less than one mole, it always produces some simple fraction of a mole, such as $\frac{1}{2}$ or $\frac{1}{3}$. It is significant that the quantity of a substance produced by one faraday is the exact amount of that substance required to react with one gram-atom of hydrogen (1.0 g). We call the quantity of a substance that reacts with one gram-atom of hydrogen, or that is produced by one faraday of electrical charge, the *gram equivalent weight*. Remember that the gram equivalent weight of a substance is always equal to, or less than, its atomic weight; this fact is demonstrated in Table 4–1.

Example 4–1. A solution of nickel chloride ($NiCl_2$) is electrolyzed, and metallic nickel is produced by a current of 3.70 amp flowing through an electrolysis cell for 470 min. If 31.6 g of nickel is produced, what is the equivalent weight of nickel? How does the equivalent weight compare with the atomic weight?

Solution. We learned from Faraday's experiments that to produce one gram equivalent weight of any element requires 96,500 coulombs. The charge that passed through the system was

$$470 \text{ min} \times 60 \text{ sec min}^{-1} \times 3.70 \text{ coulombs sec}^{-1} = 104,000 \text{ coulombs}$$

This is slightly more than one faraday of charge, and we can conclude that the 31.6 g of nickel it produced is slightly more than one gram equivalent weight. A simple proportion will give us the equivalent weight of nickel

$$\frac{31.6 \text{ g}}{104,000 \text{ coulombs}} = \frac{\text{g equiv wt}}{96,500 \text{ coulombs}}$$

$$31.6 \text{ g} \times \frac{96,500 \text{ coulombs}}{104,000 \text{ coulombs}} = 29.3 \text{ g (g equiv wt)}^{-1}$$

Thus, we see that the equivalent weight of nickel is about half its atomic weight.

The relationship between the faraday and the equivalent (or atomic) weights of elements was confirmed by electrolysis of many substances. Faraday reasoned that the current must be carried through the solution by electrically charged particles, which he called *ions*. He supposed that in a solution of hydrochloric acid, hydrogen ions (H^+) and chloride ions (Cl^-) carry

Table 4–1. Electrolysis Data

Element	Atomic weight	Grams of element produced by 96,500 coulombs (1 gram equivalent weight)
Hydrogen	1.0	1.0
Oxygen	16.0	8.0
Sodium	23.0	23.0
Potassium	39.1	39.1
Magnesium	24.3	12.2
Chlorine	35.5	35.5
Calcium	40.1	20.0
Aluminum	27.0	9.0
Copper	63.5	31.7
Bromine	79.9	79.9

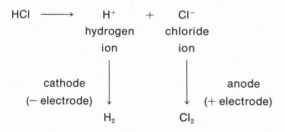

A = Anode
B = Battery
C = Cathode
D = Glow area

Figure 4–2.   A cathode ray tube.

the current. The signs of the electrical charges on the ions were assigned on the assumption
that an ion should be attracted to the electrode of opposite charge

$$HCl \longrightarrow \quad H^+ \quad + \quad Cl^-$$

hydrogen   chloride
ion   ion

cathode | | anode
(– electrode)   ↓   ↓   (+ electrode)
H_2 Cl_2

Faraday showed that electricity consists of discrete units, just as matter consists of discrete
units (atoms and molecules). But it remained for G. Johnstone Stoney, in 1891, to state clearly
that there must be a "natural unit of electricity," which he called the *electron*. However, he had
no way of knowing whether the electrons were present in matter before the electric current was
turned on. The fact that matter contains electrons was first demonstrated near the end of the
nineteenth century by experiments with cathode ray tubes .

Figure 4–3.   Crookes' cathode ray tube with two aligned pin holes.

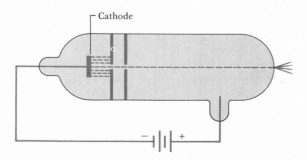

4–3 Cathode Ray Tubes

A *cathode ray tube* (Figure 4–2) is a glass tube into which two electrodes are sealed. The tube is evacuated by pumping, and sealed under a very low gas pressure (less than 0.01 torr). A large electrical potential difference can be induced on the electrodes by connecting them to the positive and negative terminals of a powerful set of storage batteries. Gases are normally poor conductors of electricity, but when a sufficiently high voltage is applied between the two electrodes, a current flows between them. When current is flowing in the tube, a distinct glow develops at the end of the tube near the anode.

If a metal shield is placed in front of the cathode, the area behind the shield does not glow. The shield casts a shadow, showing that something comes from the cathode and travels across the tube. The shadow cast by the shield is similar to shadows cast when an object is placed in the path of a beam of light. Consequently, early workers reasoned that some kind of radiation must come from the cathode, and they called the radiation *cathode rays.*

In 1874, William Crookes showed that cathode rays consist of some kind of matter. He placed two small plugs with carefully aligned pinholes in front of the cathode, shown in Figure 4–3. When the tube was operating, a bright spot appeared at the back of the tube. Apparently, the cathode rays that passed through the small hole in the first plug must be traveling in a straight line, since the rays also passed through the hole in the second plug.

Crookes then placed a small, delicately mounted paddle wheel behind the two pinholes in the path of the cathode rays. When the current was turned on in the tube, the paddle wheel turned, showing that the cathode rays consist of particles having mass, since they obviously struck the wheel and transferred momentum to it. (Remember that momentum is mass times velocity, so the rays would push the wheel only if they had mass.)

In 1897, J. J. Thomson determined more about the nature of cathode rays. He placed two more electrodes inside the tube, parallel to the line of flight. A schematic drawing of Thomson's apparatus is shown in Figure 4–4. A small voltage difference could then be induced on the electrodes. When this was done, the position of the spot at the back of the tube moved. The direction of motion showed that the cathode rays are attracted toward the positive electrode (the anode).

Figure 4–4.   Schematic of Thomson's apparatus for measuring e/m of the electron. Position a *represents the deflection of the electron beam with only the electrical field on; position* b *shows the beam deflected back to its original position by the addition of a magnetic field.*

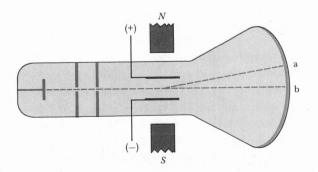

Thus, Thomson showed that the particles in cathode rays are negatively charged, and he concluded that they were identical to the electrons suggested by Stoney. Thomson also showed that the rays were deflected if an ordinary magnet were brought up to the cathode ray tube. By combining the results of the deflections by electrostatic and magnetic fields, Thomson was able to calculate the ratio of the charge of the electron to its mass

$$\frac{e^-}{m} = 1.76 \times 10^8 \text{ coulombs g}^{-1}$$

The value of the ratio was the same for all cathode materials (aluminum, iron, or platinum). Consequently, he reasoned that all materials contain electrons and that electrons must be parts of all atoms. Clearly, if negative electrons could be torn out of neutral atoms, some kind of positively charged fragments, or positive ions, must be left behind. Thomson reasoned that these positive ions are similar to those that carry current in electrolysis experiments.

Analysis of the relationship between the amounts of electric current required to produce equivalent weights of various elements in electrolysis had allowed Faraday to calculate the charge-to-mass ratio for ions believed to be current-carriers in electrolysis. The result showed that, for the hydrogen ion, the value was about a thousand times smaller than for electrons. Thomson assumed that the charges carried by an electron and by a hydrogen ion were equal, but opposite in sign. This assumption led to the conclusion that the mass of a hydrogen ion (a proton) is a thousand times larger than the mass of an electron. More accurate measurements have since shown that the ratio is closer to 1835

mass of a proton $= 1835$ mass of the electron
$$m_{H^+} = 1835 \, m_{e^-}$$

Another indication of the existence of positive ions had been noticed, in 1886, by E. Goldstein. He detected particles that moved toward the cathode of a specially designed cathode ray tube. These particles were called *canal rays.* In 1897, Wilhelm Wein proved that canal rays consist of positively charged particles, having about the same mass as atoms of the gas in the tube. The lightest ion found was the proton, H^+, which was detected when hydrogen was the gas in the tube.

To Thomson goes the credit for putting all of the facts together and suggesting that neutral hydrogen atoms must be made of two constituent parts—a proton, having almost all of the mass of the atom, and an electron, a very light, negatively charged particle. He extended the theory by suggesting that heavier atoms consist of larger, positively charged spheres with enough electrons embedded in the spheres to make the atoms neutral. This "raisin cake" or "plum pudding" model of the atom is shown in Figure 4–5.

4–4   Other High-Energy Radiation

The discovery of cathode rays was followed by the discovery of a number of other kinds of radiation. In 1895, Wilhelm Roentgen discovered x rays, which are produced when cathode rays strike a solid target. These rays are not deflected by electrical or magnetic fields, so they are not charged particles. Somewhat later, x rays were found to be reflected and scattered like rays of ordinary light, except that they are much more energetic. x Rays have many properties characteristic of high-energy radiation. For example, they penetrate and pass through many materials. This is the reason that thick lead sheets are used to protect technicians who use x-ray machines,

Figure 4–5.   A distribution of negative and positive charges throughout the volume of the atom, as described by J. J. Thomson.

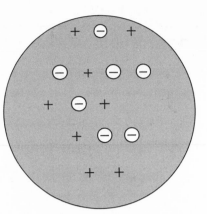

such as those used in medical and dental laboratories. When x rays pass through gases, ions are produced, as shown by an increase in the electrical conductivity of the gas. Like cathode rays and other forms of high-energy radiation, x rays cause darkening of photographic plates.

RADIOACTIVITY

In 1896, Henri Becquerel found that compounds of uranium emit radiation that resembles x rays in that it darkens photographic plates, induces chemical conductivity in gases, and can pass through objects such as thin metal films. Two other radioactive elements, polonium and radium, were discovered in 1898 by Pierre and Marie Curie, France's famous husband and wife scientific team.[1]

Study of the behavior of radioactive rays showed that they are a mixture of three kinds of radiation. Figure 4–6 shows the scheme used to separate the rays by an electrical field. The radioactive sample, a small amount of uranium sulfate, for example, is placed at the bottom of a narrow hole drilled in a lead block. Some of the rays emerging from the top of the hole are not deflected by the electrical fields; some, which are negative, are attracted to the positive electrode; others, which must be positive, are attracted to the negative electrode.

---

[1] Marie Curie introduced the term "radioactive" to describe the substances which emitted these unusual rays.

Figure 4–6.   Apparatus for the separation of α, β, and γ rays.

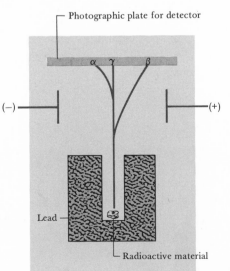

The negative particles are called *beta* (β) rays and are the same as electrons. The positive particles are called *alpha* (α) *rays*. They are deflected much less than β rays and must, therefore, be much heavier than β particles. Alpha rays turn out to be nuclei of helium atoms. They bear two positive charges and have an atomic mass number of four. Note that the atomic weight of helium is four, compared with an atomic weight of one for hydrogen. The electrically neutral *gamma* (γ) *rays* emitted from radioactive materials are high-energy x rays.

All these kinds of radiation, α rays, β rays, and γ rays (or x rays), are *high-energy radiation*. The term high energy does not refer to power in a beam of rays, but to the high energies carried by the individual particles, or packets of energy, in the beam. High-energy radiation of all kinds causes important chemical changes when it strikes almost any kind of material. The effects on living organisms are much discussed in this atomic age. We all know that large doses of radiation can be lethal to any living thing, and that small, controlled doses are used with some success to arrest the development of cancer. Study of radiation chemistry is now an active field of chemical research. For this book, however, our interest in radiation lies in the implications about the structures of atoms that are drawn from the study of high-energy radiation.

4-5 Characteristic Units of Radiation

Radiation of all types—radio waves, light, high-energy radiation, and so forth—has two kinds of energy characteristics. The first is the total amount of energy in a radiation beam. The second is the amount of energy associated with a single unit of the radiation. For our present purposes, the latter measure of energy is more important. Since radiations such as β rays (electrons) and α rays (helium nuclei) are streams of particles having mass, the concept of energy per particle is simple; it is the amount of kinetic or potential energy carried by the particle. For example, the energy of an α particle moving in the absence of any strong field (such as an electrical field) is approximately equal to the kinetic energy of the particle

$$\text{kinetic energy} = \tfrac{1}{2} \text{ mass} \times (\text{velocity})^2$$
$$KE = \tfrac{1}{2} mv^2 \tag{4-1}$$

Alpha particles released in the radioactive decay of one of the isotopes of radium have velocities of 1.70×10^9 centimeters per second (1.7 billion cm sec^{-1}). The mass of an α particle is 4.0028 atomic mass units.[2] To calculate the kinetic energy, in ergs, of the α particle under consideration, we need to express its mass in grams. Since we know that 4.0028 g of helium is equal to one mole (or 6.022×10^{23} atoms), the weight of one helium atom (one α particle) is

$$\frac{4.0028 \text{ g mole}^{-1}}{6.022 \times 10^{23} \text{ atoms mole}^{-1}} = 6.65 \times 10^{-24} \text{ g atom}^{-1}$$

The kinetic energy per particle is

$$KE = \tfrac{1}{2} (6.65 \times 10^{-24})(1.70 \times 10^9)^2 \text{ g cm}^2 \text{ sec}^{-2}$$
$$= 9.61 \times 10^{-6} \text{ g cm}^2 \text{ sec}^{-2} = 9.61 \times 10^{-6} \text{ erg}$$

The kinetic energy can be expressed in other units by using standard conversion formulas. For example, 1 erg $= 0.625 \times 10^{12}$ electron volts (eV) or 2.39×10^{-8} cal.

---

[2] One atomic mass unit (amu) is defined as exactly one twelfth the mass of one atom of carbon-12 and is 1.66024×10^{-24} g.

Example 4–2. Calculate the total energy in calories of one mole of α particles that have a velocity of 1.70×10^9 cm sec^{-1}

Solution. The preceding paragraph showed that an α particle traveling at this speed has an energy of 9.61×10^{-6} erg. We can convert this into calories by the following multiplication

$$9.61 \times 10^{-6} \text{ erg particle}^{-1} \times 2.39 \times 10^{-8} \text{ cal erg}^{-1}$$
$$= 2.30 \times 10^{-13} \text{ cal particle}^{-1}$$

Since there are 6.02×10^{23} α particles per mole, the total energy is

$$6.02 \times 10^{23} \text{ particles mole}^{-1} \times 2.30 \times 10^{-13} \text{ cal particle}^{-1}$$
$$= 1.38 \times 10^{11} \text{ cal mole}^{-1}$$

The *erg* is the common unit of energy used in classical mechanics. An *electron volt* is the amount of energy acquired by a charged particle carrying one unit of electrical charge when it moves from one electrode to another, and when the electrical potential difference between the electrodes is one volt. An electron volt is a tiny amount of energy (3.38×10^{-20} cal), but it is very useful for discussion of "happenings" with individual atoms, since such events often involve energy changes of a few electron volts. The *calorie* is the most familiar energy unit to the chemist, and it is defined as the amount of heat required to raise the temperature of one gram of water one degree centigrade (actually, from 14.5°C to 15.5°C). Since the normal human diet involves the intake of nutrients equivalent to about 2,000,000 cal (or 2000 kcal) per day, it is obvious that the energy of one α particle (2.3×10^{-13} cal) would not make much of a meal. However, Avogadro's number of α particles from radium A would have a very high energy content.

4–6 Photons, the Units of Light

Radiation such as light, x rays, and γ rays, also comes in discrete units, called *photons*. All such radiation travels with the same speed, the velocity of light, which is usually represented by the symbol c

$$\text{velocity of light} = c = 3.00 \times 10^{10} \text{ cm sec}^{-1}$$

In some of its properties, a beam of light acts as though it were a packet of particles traveling in straight lines. However, careful study of light shows that this description is not entirely correct. The progress of a beam of light is suggestive of the motion of a ripple on the surface of water. Light is said to travel as a wave.

We can make use of the analogy of the ripple on the surface of water to define the wavelength (usually represented by the Greek letter λ). The *wavelength* is the distance between neighboring peaks of the ripples, or between any two equivalent points along the wave (Figure 4–7). The *frequency* of the wave (usually represented by ν) is the number of peaks that pass a given point in unit time. For example, if a small stone tossed into a pond creates ripples, or waves, with their peaks separated by 2.0 ft (the wavelength), and if the peaks arrive at the edge of the pond at a rate of one every 5 sec, then the frequency is 12 peaks per minute. The velocity at which the wave advances is the product of the wavelength, λ, and the frequency, ν

$$2.0 \text{ ft} \times 12 \text{ min}^{-1} = 24 \text{ ft min}^{-1}$$

This same relationship applies to light, which has a much shorter wavelength and a correspondingly higher frequency. The velocity of light, c, which is constant in a vacuum, may be

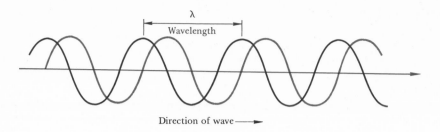

λ
Wavelength

Direction of wave ⟶

Figure 4–7. Propagation of a wave through space.

defined in terms of its frequency and wavelength

$$c = \lambda\nu = 3 \times 10^{10} \text{ cm sec}^{-1} \tag{4–2}$$

Example 4–3. The wavelength of a blue light is 4500 Å. Calculate the frequency of this light.
Solution. The wavelength 4500 Å is the same as 4500×10^{-8} cm, since 1 Å $= 1 \times 10^{-8}$ cm. Convert this to frequency, ν, by using the relationship

$$\nu = \frac{c}{\lambda} = \frac{3.00 \times 10^{10} \text{ cm sec}^{-1}}{4.500 \times 10^{-5} \text{ cm}} = 6.67 \times 10^{14} \text{ sec}^{-1}$$

This large number means that light waves with $\lambda = 4500$ Å move by a given point at an exceedingly rapid rate.

In 1900, Max Planck resolved a very difficult problem concerning the energy of the radiation emitted by hot objects. The solution to this problem came after Planck made a drastic departure from classical physics. He proposed that the energy emitted by heated, glowing elements is not emitted continuously, but in small packets of energy, which he called *quanta*. He postulated that the energy of the radiation is proportional to its frequency, and suggested the following relationship to calculate the energy of such radiation

$$E = h\nu = h\frac{c}{\lambda} \tag{4–3}$$

The proportionality constant, h, is known as Planck's constant, which equals 6.63×10^{-27} erg sec.

Example 4–4. Calculate the energy per quantum of light having a wavelength of 5000 Å. (Such light is in the visible range.)
Solution.

$$1 \text{ Å} = 10^{-8} \text{ cm}$$

$$\nu = \frac{c}{\lambda} = \frac{3.00 \times 10^{10} \text{ cm}}{5.00 \times 10^{-5} \text{ cm sec}} = 6.00 \times 10^{14} \text{ sec}^{-1}$$

$$E = 6.00 \times 10^{14} \times 6.63 \times 10^{-27} \text{ erg sec sec}^{-1}$$
$$= 3.98 \times 10^{-12} \text{ erg}$$
$$= 2.48 \text{ eV (electron volts)}$$

Table 4–2.   The Electromagnetic Spectrum

Region	Approximate range of wavelength, λ (cm)	Wave number, $\bar{\nu}$, for higher limit of wavelength (cycles cm^{-1})[a]
Radio waves	10^6–10	10^{-6}
Microwaves	10–10^{-1}	10^{-1}
Infrared	10^{-2}–10^{-4}	10^2
Visible light	8×10^{-5}–4×10^{-5}	10^4
Ultraviolet	10^{-5}–10^{-6}	10^5
x Rays	10^{-6}–10^{-8}	10^6
γ Rays	10^{-9}–10^{-12}	10^9
Cosmic rays	10^{-12}–	10^{12}

[a]*The wave number, $\bar{\nu}$, is the reciprocal of the wavelength (i.e., $1/\lambda$).*

Light and all other similar forms of radiation are known as *electromagnetic radiation*. Visible light comprises only a tiny fraction of the total spectrum of known electromagnetic radiation. Even visible light is a mixture of radiation having wavelengths ranging from 4000 Å to 8000 Å (4×10^{-5} cm to 8×10^{-5} cm). Table 4–2 is a summary of the various kinds of electromagnetic radiation.

4–7 The Rutherford Atom

Ernest Rutherford and his co-workers, in Manchester, England, used α rays to discover some enormously important facts about the structure of matter. Rutherford bombarded thin sheets of metal foil with α particles and placed a detector behind the target to "see" the helium ions that passed through the metal, as illustrated in Figure 4–8. The detector was a plate coated with zinc sulfide, which gives off light (i.e., it fluoresces) when struck by α rays. With no metal foil shield, the glowing area was a sharply defined circle. When metal foils were put in place, the area of the glowing circle increased, and the edges became less well defined. Rutherford reasoned that the spreading of the radiation was due to scattering of the α particles by the negative electrons and positive particles in the atoms of the metal foil.

Figure 4–8.   Schematic diagram of Rutherford's experiment to bombard thin metal foils with α particles.

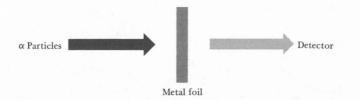

α Particles   Detector

Metal foil

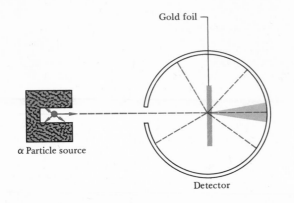

Gold foil

α Particle source

Detector

Figure 4–9.   Diagram of α-scattering apparatus with dashed lines showing wide angle scattering of α particles.

However, quantitative analysis of the results exposed some problems. For example, a large number of particles passed directly through thin films with almost no deflection. This indicated that the metal was mostly space, rather than being fully packed with Thomson's raisin-cake atoms. Hans Geiger and Ernest Marsden, two of Rutherford's students, then placed the detector at a right angle to the foil, instead of behind it. To their great surprise, they found that the flight paths of some α particles had been changed by as much as 90°. (Their apparatus is described in Figure 4–9.) Some α particles even bounced back in the general direction of the source. Rutherford said that it was "as if you had fired a 15-in. shell at a piece of tissue paper and it came back and hit you." He concluded that a few of the α particles had encountered a very dense charged particle. Putting the observations together, he concluded that most of the mass of atoms is concentrated in a very small, positively charged *nucleus.* He further postulated that the rest of the atom is nearly empty space, containing enough electrons to make the whole atom electrically neutral (see Figure 4–10).

Figure 4–10.   The Rutherford model of the atom.

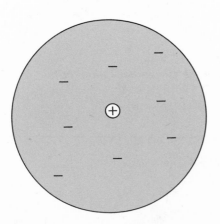

It is important to realize just how small the nucleus is in comparison with the total volume occupied by an atom. Rutherford's scattering experiments showed that atomic nuclei have radii of about 10^{-13} cm, whereas the radii of atoms are known to be about 10^{-8} cm. This means that the volume of the nucleus is only about 10^{-15} as large as the volume of the atom. If an atom were the size of the earth, the nucleus would be a sphere with a radius of about 20 ft. If an atom were the size of the solar system, the nucleus would be 1000 times smaller than the sun.

An atomic nucleus is extraordinarily dense. The radius (r) of a proton (hydrogen nucleus) is about 10^{-13} cm, and its weight is 1.7×10^{-24} g. The density may be calculated as follows:

$$\text{volume}_{\text{nucleus}} = \frac{4}{3}\pi r^3 = \frac{4}{3} \times 3.14 \times (1 \times 10^{-13}\ \text{cm})^3 = 4 \times 10^{-39}\ \text{cm}^3 \tag{4-4}$$

$$\text{density}_{\text{nucleus}} = \frac{1.7 \times 10^{-24}\ \text{g}}{4 \times 10^{-39}\ \text{cm}^3} = 4 \times 10^{14}\ \text{g cm}^{-3} \tag{4-5}$$

The density of about 10^{14} g cm^{-3} is staggeringly large compared to densities of common materials. For instance, lead, considered by most people to be a "heavy" metal, has a density of about 11 g cm^{-3}.

Rutherford's model of the atom, in which there is a small positive nucleus surrounded by electrons, was a significant advancement toward understanding the atom, and it provided a basis for a more sophisticated theory of atomic structure, which will be presented in the next chapter.

4–8   The Composition of Atomic Nuclei

As a consequence of Rutherford's theory of atomic structure, there followed an avalanche of exciting work designed to discover whether atomic nuclei were themselves made up of smaller parts. Shortly after Rutherford's scattering experiments, another of his students, Henry Gwyn Jeffreys Moseley,[3] began to study the emission of x rays from various elements. Moseley made the important discovery that elements bombarded with cathode rays (electrons) emit x rays of a characteristic frequency, and that the square root of the frequency of the x rays is proportional to the *atomic number* of the element (i.e., $\sqrt{\nu} = cZ$, where ν is frequency of the x rays; c is a constant; and Z is the atomic number of the target element). This linear relationship between Z and $\sqrt{\nu}$ is illustrated in Figure 4–11.

By the time of Moseley's experiment, the atomic weights of many elements were well known, largely due to the efforts of Theodore Richards (1868–1928) of Harvard University. Richards felt that atomic weights held the secret of the fundamental nature of matter, and devoted much of his life to their precise determination. Exact determinations of atomic weights led to some puzzling inconsistencies in the organization of the periodic table. For example, iodine is a member of the halogen family (Group VIIA), according to its chemical properties, and tellurium clearly belongs in the sulfur family (Group VIA). Yet the atomic weight of iodine is less than that of tellurium, suggesting that iodine should precede tellurium in the periodic table, rather than follow it, as chemical evidence indicates it should. A more striking example is the case of argon and potassium. Potassium must follow argon on the basis of chemical evidence, yet the atomic weight of potassium is less than that of argon. The work of Rutherford and others on atomic

---

[3] Moseley was killed in 1915 during the British landing at Suvla Bay, in the Dardanelles. He was 27.

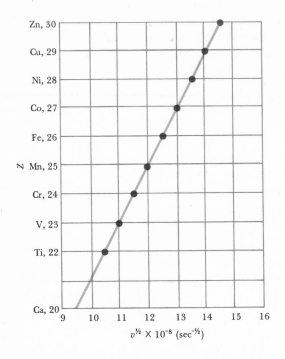

Figure 4–11. Moseley's relationship between atomic number, Z, and $\nu^{1/2}$.

structure, combined with Moseley's x-ray work, strongly suggested that the one important characteristic that defines each element and distinguishes it from all other elements is its *atomic number, Z.* The atomic weight must be less fundamental than atomic number.

4–9 Isotopes

In recent years, research in the field known as particle physics has shown that atomic nuclei contain two kinds of particles, protons and neutrons. A proton carries a positive charge, and the number of protons in the nucleus is equal to the atomic number of the element. The rest of the nuclear mass is contributed by neutrons, which are neutral particles having masses slightly larger than that of a proton. The neutron/proton mass ratio is 1.0014, so for most chemical purposes, we can consider the two to be equal. Furthermore, electrons are much lighter than either nuclear particle, thus the atomic weight of an atom of an element is approximately equal to the sum of the numbers of protons and neutrons

atomic weight $\cong$ no. of protons + no. of neutrons

A glance at a table of atomic weights of the elements (inside back cover) shows that most of the atomic weights are not whole numbers. We might have expected all atomic weights to be very close to integers, if all nuclei were composed of the same kinds of units. If this were true, it is difficult to see, for example, how chlorine can have an atomic weight of 35.45.

Actually, the answer to the problem was known before the development of our present theory of nuclear structure. Study of radioactive decay showed that products were often produced having the same chemical properties as known elements, but having different atomic weights. For example, a product of the decay of uranium known as "ionium" was found to be chemically indistinguishable from thorium, a known element. However, the atomic weight of ionium was 230, whereas that of thorium is 232.0. Several other element pairs were also known by 1913, when Soddy suggested that some elements have more than one kind of nucleus.

The different nuclear forms of the same element are called *isotopes*. Isotopes have the same number of protons, but have different numbers of neutrons. The isotope of carbon that has six protons and six neutrons in its nucleus has been assigned an atomic weight of exactly 12. All atomic weights are referred to this standard carbon isotope, which has the symbol ^{12}C. Superscripts are used in this way to designate the total number of protons and neutrons in the nucleus. Since chemical properties are determined largely by the electrons in an atom, isotopes have nearly identical chemical properties.

The atomic weights of naturally occurring elements are averages of the weights contributed by the various kinds of isotopic nuclei. Natural carbon is predominantly ^{12}C, with very small amounts of ^{13}C (six protons and seven neutrons). This explains the average relative atomic weight of 12.01 for carbon. Another example is chlorine, which occurs in nature as a mixture of 75.4% ^{35}Cl and 24.6% ^{37}Cl. We can calculate the average atomic weight of chlorine from this information

$$
\begin{aligned}
0.754 \times 35.00 \quad &= 26.39 \\
0.246 \times 37.00 \quad &= \underline{9.10} \\
\text{calculated average} \\
\text{atomic weight} \quad &= 35.49
\end{aligned}
$$

The calculated value is close, but not identical, to the observed value. The difference arises because the atomic weights of the pure isotopes are not exactly equal to the sum of the masses of the protons and neutrons. This startling fact is explained by the theory of relativity, which holds that mass can be converted to energy ($E = mc^2$). In the formation of complex nuclei, a tiny fraction of the mass of the neutrons and protons is converted into energy used to bind the nuclear particles together. Nuclear binding forces are called *short-range forces* to distinguish them from more familiar forces such as electrical, gravitational, and magnetic. Short-range forces are the subject of intensive study in the field of high-energy physics.

Example 4–5. The problem of calculating the average atomic weight for a mixture of isotopes of different weights is no different from any other problem dealing with an average value. For example, what would be the average age for a class of students having the following composition?

> 10%   17-yr olds
> 34%   18-yr olds
> 42%   19-yr olds
> 14%   20-yr olds

Solution. The average age is

$$
\begin{aligned}
\text{average age} &= (0.10)(17) + (0.34)(18) + (0.42)(19) + (0.14)(20) \\
&= 1.7 + 6.1 + 8.0 + 2.8 \\
&= 18.6 \text{ yr}
\end{aligned}
$$

Methods now are available for separating the isotopic forms of elements, and isotopes not occurring in nature can sometimes be made by nuclear reactions. Isotopes are often used as *tracers* in chemistry. A compound containing an unnatural proportion of some isotope is used to trace the fate of a material in some complex process. For example, compounds containing unusually large amounts of ^{13}C are often fed to animals. Detection of metabolic products that contain excess ^{13}C gives useful information concerning the chemistry of metabolism. The tracking of isotopes in various chemical and biological systems can be accomplished in different ways. One of the more common techniques takes advantage of the mass differences of the isotopes. Another very useful method involves detecting radiation (α, β, or γ) that is emitted by the radioactive isotope.

The most important concept to remember from this discussion is that the *atomic number* of an element gives:

1   The number of protons in the nucleus.
2   The number of electrons around the nucleus in the neutral atom.
3   Its position in the periodic table.

We have described several of the fundamental particles of which matter is composed. There are many other particles that are known to exist. However, those described here are sufficient for a working knowledge of atomic structure. Table 4–3 summarizes these particles and some of their properties.

4–10   The Interaction of Light with Matter

When light hits a material object many things can occur. The light may be absorbed, with the energy of the photon being taken up by the absorbing material. Absorption of visible light is responsible for the colors of materials. A solution of potassium permanganate in water appears purple because it absorbs light of wavelengths near 5250 Å, which is in the green region of the spectrum. The visible light of shorter wavelengths passes through the solution and registers in the eye, where chemical changes occur, resulting in transmission of a signal to the brain. The signal works so that the brain "sees" purple, the color that is complementary to that of the absorbed light. Table 4–4 lists the colors of light of various wavelengths and the complement of each color.

Light may be transmitted entirely by a material that does not contain any substance that absorbs the wavelengths in the light beam. An example is the transmission of sunlight by window

Table 4–3.   Fundamental Particles and Their Properties

Particle	Charge	Mass (amu)[a]	Mass (g)	Description
Electron	−1	0.000549	9.11×10^{-28}	Fundamental particle of (−) charge
Proton	+1	1.0073	1.673×10^{-24}	Fundamental particle of (+) charge
Neutron	0	1.0087	1.675×10^{-24}	Fundamental particle of no charge

[a] *amu = atomic mass units. One amu = 1.66024×10^{-24} g.*

Table 4–4.   Wavelengths of Several Colors and Their Complements

Wavelength (Å)	Color	Complementary color
4000	Violet	Yellow-green
4250	Indigo	Yellow
4500	Blue	Orange
4600	Blue-green	Red
5250	Green	Purple
5500	Yellow-green	Violet
5900	Yellow	Indigo
6200	Orange	Blue
7200	Red	Blue-green

glass. However, we should not conclude that no light is absorbed just because a material appears transparent to the eye. Light not detected by the eye—light in either the ultraviolet or infrared spectral regions (Table 4–2)—may be absorbed. Such absorption may be detected by devices known as *photocells,* which convert light energy into electrical signals. Infrared and ultraviolet absorption characteristics often provide useful information concerning the chemical nature of visibly transparent materials.

Study of the chemical changes caused by the absorption of light is called *photochemistry.* This field is intensely interesting, because light causes many remarkable chemical reactions. Probably the most important is the reaction of carbon dioxide with water to make carbohydrates, the *photosynthetic* reaction in green plants. Light energy absorbed by the green pigment chlorophyll causes the reaction

$$6CO_2 \ + \ 6H_2O \ \xrightarrow[\text{chlorophyll}]{\text{sunlight}} \ C_6H_{12}O_6 \ + \ 6O_2$$
$$\text{carbon} \qquad \text{water} \qquad\qquad\qquad \text{glucose} \qquad \text{oxygen}$$
$$\text{dioxide}$$

In 1887, Heinrich Hertz discovered that when a beam of light hits a metal surface, the metal emits electrons. The phenomenon is known as the *photoelectric effect.* In 1905, Albert Einstein explained this effect by proposing that photons are absorbed by atoms of the metal, and that, if the absorbed energy is sufficiently high, the *excited* atom ejects an electron. The detailed observations leading to Einstein's conclusion were the following:

1   The number of electrons emitted depends on the intensity of the light.

2   The energy of the emitted electrons depends on the frequency of the light, not on the intensity.

3   If the frequency of the light is below a definite threshold value, no electrons are emitted. Moreover, the threshold value is different for different metals.

The threshold value is the point at which the energy of the photon is equal to the minimum energy required to remove an electron from the atom. The requirement for photoionization is

energy of the photon $\geqslant$ binding energy of the electron

$$h\nu \geqslant E_B$$

(4–6)

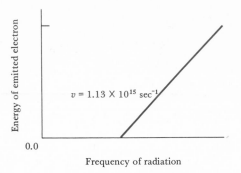

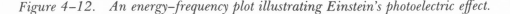

Figure 4–12.   An energy–frequency plot illustrating Einstein's photoelectric effect.

When the threshold energy is exceeded, the "extra" energy appears as kinetic energy of the ejected electrons. For example, to remove an electron from metallic silver requires radiation with a frequency of 1.13×10^{15} sec^{-1}. This radiation is in the ultraviolet region of the spectrum. As shown in Figure 4–12, there are no electrons emitted from the metal surface until the frequency reaches 1.13×10^{15} sec^{-1}. As the frequency passes this threshold value and exceeds the energy required to remove the electron, the excess energy is converted into kinetic energy of the electron. The intensity of the light is a measure of the number of photons, so it is reasonable that increasing the intensity should increase the number of electrons emitted, if the threshold energy has been exceeded.

Example 4–6. The threshold energy for photoelectric emission of electrons from potassium metal is 3.59×10^{-12} erg. Would light of 3000-Å wavelength cause electrons to be emitted from the metal?

Solution. First, let us calculate the threshold frequency, ν_0

$$E_B = h\nu_0$$
$$3.59 \times 10^{-12} \text{ erg} = (6.63 \times 10^{-27} \text{ erg sec}) \times \nu_0$$

$$\nu_0 = \frac{3.59 \times 10^{-12} \text{ erg}}{6.63 \times 10^{-27} \text{ erg sec}} = 5.43 \times 10^{14} \text{ sec}^{-1}$$

For the light to eject electrons from the metal's surface, it must have a frequency equal to or greater than $\nu_0 = 5.43 \times 10^{14}$ sec^{-1}. Now let us calculate the frequency of 3000-Å light. Since 1 Å $= 10^{-8}$ cm, 3000 Å $= 3.00 \times 10^{-5}$ cm.

$$\nu = \frac{c}{\lambda}$$

$$= \frac{3.00 \times 10^{10} \text{ cm sec}^{-1}}{3.00 \times 10^{-5} \text{ cm}} = 1.00 \times 10^{15} \text{ sec}^{-1}$$

Since $\nu > \nu_0$, light of 3000 Å *will* cause electrons to be emitted from the metal.

How much excess energy will there be to appear as kinetic energy of the electrons? The threshold energy is 3.59×10^{-12} erg. The energy of each photon is

$E = h\nu$

$E = (6.63 \times 10^{-27} \text{ erg sec}) \times (1.00 \times 10^{15} \text{ sec}^{-1})$

$E = 6.63 \times 10^{-12} \text{ erg}$

The extra energy is

$E - E_B = 6.63 \times 10^{-12} - 3.59 \times 10^{-12} = 3.04 \times 10^{-12} \text{ erg electron}^{-1}$

At the time Einstein explained the photoelectric effect, the greatest importance of the theory was in establishing the *quantum theory* of energy. Einstein's theory states that energy comes in discrete units called *quanta*. If the energy in a beam of light were not delivered in packages, the wavelength threshold for photoionization should not exist. Our interest in the phenomenon is largely in the information that it gives concerning the way in which electrons are bound in an atom. The quantum theory of light states that light of a given wavelength consists of photons, all having the same energy. If we accept this hypothesis, the photoelectric effect tells us that electrons in atoms are bound with definite binding energies. This suggests that the energy of the electrons may also be quantized.

4–11 Emission Spectra

There are many ways to produce highly energetic, or *excited*, atoms. Excited atoms are present in flames, and are formed when an electric spark is discharged in a gas. These energy-rich atoms can lose energy by emitting light. The light from flames, or from a flash of lightning, is emitted from excited atoms, molecules, or ions.

Instead of being continuous (i.e., consisting of light of all wavelengths), the light emitted by excited atoms consists of sharp lines having definite frequencies. The emitted light can be analyzed by passing it into a *spectroscope* (Figure 4–13), an instrument containing a prism that sorts out the components of a beam of light according to their frequencies. Figure 4–14 shows a portion of the spectrum of light emitted from excited hydrogen atoms when a spark is passed through gaseous hydrogen in a hydrogen discharge tube. The light emitted from each element consists of a unique assortment of lines of different wavelengths; these lines can be used to identify that element.

Figure 4–13. Prism spectroscope for analyzing light.

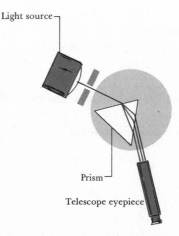

Light source

Prism

Telescope eyepiece

Figure 4–14.   Spectral lines of light emitted from excited hydrogen atoms. Wavelengths are in angstroms.

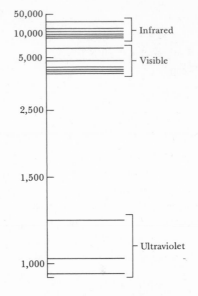

One particularly striking feature of these emission spectra is that there seems to be some relationship between the wavelengths of the many lines that appear. For example, the lines in the infrared region seem to be together as a group, and, similarly, the sets of lines in the visible and ultraviolet regions each seem to group together. The mathematical relationships that describe these sets of lines were not easily revealed. But, in 1884, Johann Jakob Balmer, working with the emission lines in the visible region of the hydrogen spectrum, succeeded. Later, Johannes Rydberg expressed the following, more complete relationship with which it is possible to calculate the wavelength of any line in the entire emission spectrum of hydrogen

$$\frac{1}{\lambda} = R\left(\frac{1}{n_1^2} - \frac{1}{n_2^2}\right) \tag{4–7}$$

R is known as the *Rydberg constant,* and has a value of 109,678 cm^{-1}, and n_1 and n_2 are integers that depend on which line (wavelength) in the spectrum you want to calculate. The first line in the ultraviolet region can be calculated by letting $n_1 = 1$ and $n_2 = 2$; the second line in this region by letting $n_1 = 1$ and $n_2 = 3$; and so on. The lines of the visible region can be calculated by letting $n_1 = 2$, and $n_2 = 3, 4, \ldots$ for as many lines as you desire to calculate.

The observations that the emission lines have definite positions that are related mathematically, and that light of a particular wavelength represents a definite amount of energy, set the stage for a new model of the atom. It seemed that somehow there must be discrete energy states in the atom. It was this clue that led Niels Bohr to his development of a quantum theory of atomic structure. We shall discuss this topic in Chapter 5.

Example 4–7. Calculate the wavelength of the line in the emission spectrum of hydrogen that is associated with the values of $n_1 = 1$ and $n_2 = 2$.

Solution. The Rydberg equation states, after substitution of values for R, n_1, and n_2, that

$$\frac{1}{\lambda} = 1.097 \times 10^5 \text{ cm}^{-1}\left(\frac{1}{1^2} - \frac{1}{2^2}\right)$$

$$\frac{1}{\lambda} = 1.097 \times 10^5 \text{ cm}^{-1}\left(1 - \frac{1}{4}\right)$$

$$= 1.097 \times 10^5 \text{ cm}^{-1}\,(0.75)$$

$$\frac{1}{\lambda} = 8.228 \times 10^4 \text{ cm}^{-1}$$

$$\lambda = \frac{1}{8.228 \times 10^4 \text{ cm}^{-1}} = 1.215 \times 10^{-5} \text{ cm} = 1215 \text{ Å}$$

4–12 Summary

1 Atoms are composed of even smaller constituent particles.
2 For most chemical purposes, an atom can be considered to consist of a positively charged nucleus and a number of electrons equal to the charge on the nucleus.
3 The atomic number of an element is equal to the number of positive charges on the nucleus, and defines the position of the element in the periodic table.
4 Isotopic atoms have the same atomic numbers, but different atomic masses.
5 Light consists of photons. Each photon has a clearly defined energy that depends on the frequency (or wavelength) of the light.
6 β Particles are electrons.
7 α Particles are helium nuclei, or ions, having a charge of $+2$ and a mass of 4 amu.
8 x Rays and γ rays are high-energy photons.

4–13 Postscript: Common and Uncommon Sense

Science is sometimes accused of making things more complex than, or unreasonably different from, the dictates of common sense. This point of view strikes us as being the real nonsense. Neither science nor common sense *does* anything; they both *are* loosely structured systems for thinking about things, based upon experience. The systems of thought may seem contradictory, but that is because the models used in both systems are models, not absolute reality. Common sense models are most likely to break down when a scientist, or, for that matter, anyone, does a careful, uncommon experiment.

Common sense told Rutherford, Geiger, and Marsden what to expect when they fired α particles at a gold foil. Their expectations were based upon the fact that previous examination of the foil had been done visually, and visual observation depends upon the use of light having wavelengths lying in the visible region. Alpha particles have much shorter wavelengths, so very different things were "seen" when they were used (see postscript to Chapter 7). This uncommon experiment required modification of a model that was perfectly compatible with common experimental evidence.

Are electrons mysterious? Actually, many common phenomena are much more easily explained in terms of electrons than with earlier models, based on common experience. We believe that passage of an electric current through a copper wire involves the flow of electrons. In metals, some of the electrons are bound loosely to the atoms and require only a little energy to be promoted to excited states, in which they are free to flow through the sea of metal atoms. In other substances, called semiconductors, a minimum "threshold" voltage must be applied before electrons flow and the substance conducts electricity. Silicon and germanium, when extremely pure, are semiconductors. In other substances, called insulators, no reasonable applied voltage can make a current flow. This description is an extremely oversimplified version of a much more complex theory of electronic conduction in solids. Is it too complicated? No, because it leads to predictions that can be verified and to useful results. From such theories have come first, vacuum tubes, and then the transistors that now replace them in radios, television sets, and other electronic devices.

We can almost "see" electrons jump through space from a negatively charged terminal to a positively charged one, if the terminals are sufficiently close together and the electrical potential difference is high enough. If the wires from the poles of a storage battery are brought close enough together, a spark will jump from the negative lead to the positive one. A larger version of the electric spark is a lightning bolt. Until recently, we simply said that lightning occurs when a cloud builds up an unusual excess of electrons and is discharged by a giant spark, jumping from cloud to earth, and that thunder results from the inrush of air in the wake of the electrical discharge. High-speed photography now tells a more complicated story, and it permits the formulation of a more detailed and interesting model.

The lightning discharge begins in steps. A trail of electrons from the cloud rushes down about 50 meters at a sixth the speed of light, pauses for about 50 microseconds (50×10^{-6} sec), and takes another 50-meter step, not necessarily in the same direction. This "step leader" is only a small bright spot of light on the photograph, and is not the brilliant flash that is associated with a lightning discharge. The leading electrons cause ionization of molecules of the nearby atmosphere, making them ionic conductors. As soon as the step leader touches the ground, we have a conducting pipe of charged particles from the cloud to the ground. At this time, the excess electrons in the cloud rush to the ground. The flash is light emitted from excited positive ions along the track. Heat produced expands the air suddenly and produces the thunderclap. On the photographs, a so-called "return stroke" starts at the *bottom* and works its way progressively up to the cloud. The dark leader falls from cloud to earth, but the lightning flash rises from earth to cloud. After a few hundredths of a second, a new dark leader can be emitted down the old path, still cluttered with ionic debris, and a new lightning bolt can rise from the earth. This can happen once, twice, or a dozen times, or sometimes many more, all in one rapid "flash of lightning."

Radiation of heat and light from the sun is not only a part of the common experience of all living things on this planet, but it is also the primary source of energy to sustain life itself. Models of the sun are as old as written history, and they change continuously. Our present description of the sun depends heavily on experiments with electrons, ions, and gas molecules here on earth. Most of our knowledge of the sun comes from study of the light that it gives off. Nearly all the light comes from excited hydrogen atoms and He^{2+} ions. Emission from ions of other elements can be seen, but they are only present as minor "impurities." We also know that, at the temperature of the sun (ranging from about 6000° at the surface to several million degrees in the interior), hydrogen and helium would be ionized almost completely. We thus have a picture of the sun as a gas made up of H^+ (83%), He^{2+} (17%), electrons, and small bits of other matter. The solar furnace burns by producing energy from the fusion of hydrogen nuclei to form helium nuclei.

We know the mass of the sun from the law of gravity, and we know its size by direct observation. This allows us to calculate its average density (1.6 g cm^{-3}). If the sun is a gas, the density should vary in the same way as that of the earth's atmosphere. This calculation leads to the conclusion that, at the sun's center, the density is roughly 100 g cm^{-3}. How can we say that a material is a gas when the density is higher than that of any solid material known on the surface of the earth? Once again, Rutherford's experiment helps us. We believe that any solid material is mostly space. If the nuclei and electrons in a gold foil were given enormously higher energies, such as 10^6 degrees, they would move freely, like molecules in a gas, even though they were crowded still closer together by gravitational forces.

Imagine the kinds of models that you would make of the structure of matter on earth if your only experience came from watching the inside of the sun. Solids, liquids, and molecules would all seem ridiculously remote from common experience. Description of a human would surely blow the mind of any conceivable sun-dwelling intelligence.

New Terms

Alpha ray: Emission from a radioactive material. The ray is composed of α particles, which are helium nuclei. Each α particle has a charge of $+2$, being made up of two protons and two neutrons.

Ampere: A measure of the electric current, or the rate at which charge is flowing. One ampere is equal to a flow rate of one coulomb per second.

Anode: The positively charged electrode in a cathode ray tube or an electrolysis cell; the electrode to which the negatively charged ions (anions) migrate. Electrons come to the anode.

Atomic number: The number according to which the elements are arranged in the periodic table. It is equal to the number of protons in the nucleus of an element and the number of electrons in the neutral atom.

Atomic weight: The weight of an element relative to the carbon isotope ^{12}C, which is assigned the value exactly 12.

Beta ray: Emission from a radioactive substance. The ray is composed of β particles (electrons); each particle has a charge of -1.

Calorie: The energy required to raise the temperature of one gram of water from 14.5°C to 15.5°C.

Canal rays: The stream of positively charged particles that moves toward the cathode inside a cathode ray tube.

Cathode: The negatively charged electrode in a cathode ray tube or an electrolysis cell; the electrode to which positively charged ions (cations) migrate. Electrons leave the cathode.

Cathode ray: The stream of negatively charged particles that moves toward the anode inside a cathode ray tube; the stream is composed of electrons.

Coulomb: A basic unit of electrical charge, equivalent to the charge of 6.28×10^{18} electrons.

Current: The flow of electricity or electrons, expressed in terms of the ampere.

Electrolysis: The decomposition of a substance by passing an electric current through it.

Electromagnetic radiation: Light and all other similar forms of radiation including x rays, γ rays, ultraviolet, visible, infrared, radio waves, and microwaves.

Electron: A basic particle of atomic structure bearing a negative electrical charge and weighing 1/1835 as much as a proton; a β particle.

Electron volt: The energy acquired by a charged particle carrying one unit of electronic charge when it moves from one electrode to another while the electrical potential difference between the electrodes is one volt.

Erg: An energy unit equivalent to gram cm^2 sec^{-2}. One erg is the kinetic energy of a mass of one gram moving at the velocity 1 cm sec^{-1}.

Faraday: The amount of charge, 96,500 coulombs, required to electrolyze one gram equivalent weight of substance; the total charge of one mole of electrons.

Frequency: The number of complete waves that pass a point per unit time (usually per second).

Gamma ray: High-energy electromagnetic radiation emitted from radioactive materials.

Gram equivalent weight: The weight of substance that will react with one gram-atom of hydrogen, or that is electrolyzed by one faraday of electrical charge.

Ion: An atom or molecule with an electrical charge.

Isotopes: Forms of the same element (same atomic number) that are essentially identical chemically, but which have different atomic weights.

Neutron: A nuclear particle that has an atomic mass of 1.0087 amu (roughly the same as the proton's mass) and is electrically neutral.

Nucleus: The small, dense part of the atom composed of protons and neutrons.

Photoelectric effect: The emission of electrons from a metal surface produced by beaming light of a critical wavelength on the surface of the metal.

Photon: The packet or unit of electromagnetic radiation containing energy equal to the product of the frequency of the radiation times Planck's constant (or $h\nu$).

Planck's constant: A fundamental physical constant, h, relating the energy of a photon of electromagnetic radiation to its frequency. The value of h is 6.626×10^{-27} erg sec.

Proton: A nuclear particle that has an atomic mass of 1.0073 amu (roughly the same as the neutron's mass) and has unit positive charge.

Quantum theory: Energy is packaged in discrete units, which we refer to as *quanta.*

Radioactivity: The spontaneous decomposition of atomic nuclei accompanied by the emission of α, β, or γ rays.

Rydberg constant: The proportionality constant, R_H, which relates the wavelengths of the several lines in the emission spectrum of atomic hydrogen in the equation $1/\lambda = R_H(1/n_1^2 - 1/n_2^2)$. The value of R_H is 109,678 cm^{-1}.

Spectroscope: An instrument that separates light into its various components, or bands.

Wavelength: The distance between equivalent points on a wave.

Wave number: The reciprocal of the wavelength. The wave number, $\bar{\nu} = \lambda^{-1}$, is proportional to frequency, ν, and is often used to describe wave properties.

x Ray: High-energy electromagnetic radiation emitted by elements under bombardment by high-energy electrons. The x-ray frequency is characteristic of the atomic number of the element.

Questions and Problems

1   What are α, β, and γ rays? Which of them are composed of particles? Which are waves? Why is this an unfair question?

2   How are Thomson's and Rutherford's models of the atom different, and how does the scattering of α particles differentiate between them?

3   How did Faraday's work in electrochemistry assist in determining the structure of atoms?

4   A flow of electrons at the rate of one coulomb per second is a current of one ampere. How many moles of metallic aluminum can be obtained by passing a current of 1.00 amp through molten $AlCl_3$ for 5 h 30 min?

5   How many faradays and coulombs of electricity are required to reduce 0.782 g of Cu^{2+} to metallic copper?

6   Often, the amount of charge passed through a circuit is determined by measuring the mass of solid silver deposited by the electrolysis of Ag^+ solution. If a cathode increases in mass by 0.197 g, how many coulombs have passed through the electrolysis cell?

7   What evidence is there for the assertion that one faraday of charge contains 6.022×10^{23} electronic charges?

8   Suppose that a series of electrochemical cells are set up, each of which produces different elements as current is passed through them. If a current of 1.5 amp is passed through each

cell for one hour, how many grams of each of the following substances could be produced in its cell: oxygen, potassium, aluminum, and bromine? (*Hint:* See Table 4–1).

9   Calculate the charge in coulombs that is necessary to change one hydrogen ion (H^+) to one hydrogen atom. Recall that it takes 96,500 coulombs to convert one gram-atom of hydrogen atoms into H^+ ions.

10 Assume that the charge associated with one H^+ being converted to one H atom in Problem 4–9 is the charge of one electron. Calculate the mass of one electron from the charge/mass ratio determined in Thomson's experiment, $e/m = 1.76 \times 10^8$ coulombs g^{-1}.

11 A certain piece of sodium weighs 10.0 g. How many electrons, protons, and neutrons are there in this sample? What is the total weight of the electrons? If all the electrons were removed, how many grams of substance would be left?

12 Indicate how the following isotopes differ with regard to the number of protons, neutrons, and electrons in each atom: carbon-12 and -13; oxygen-16, -17, and -18; chlorine-35 and -37.

13 Which of the following is proportional to energy in electromagnetic radiation: speed, wave number, or wavelength?

14 If wavelengths are normally the measured quantities in spectroscopy, why are wave numbers preferable to frequencies when a quantity proportional to energy is desired?

15 What color would a solution appear if it absorbs green light?

16 Calculate the total energy (in calories) of one mole of helium ions (α particles) that have a velocity of 8.4×10^8 cm sec^{-1}.

17 When a photon of light strikes an electron in a metal surface, it may cause the electron to be ejected from the surface. What determines whether a photon can cause this effect? Where does all of the energy of the photon go?

18 If the frequency (ν) of a particular radiation is 5.0×10^{14} sec^{-1}, what is its wavelength (λ)? What is the color of the light?

19 The wavelength of blue-green light is approximately 460 nanometers. What is the frequency of this light?

20 x Rays typically have wavelengths of 1 Å to 10 Å. Calculate the energy in ergs of photons with a 2-Å wavelength. Calculate the energy in kcal $mole^{-1}$ of such 2-Å photons, and compare this with the bond energy of a carbon–carbon single bond, 83 kcal $mole^{-1}$. Would you expect x rays to be able to produce chemical reactions?

21 Calculate the energy of photons, in erg $photon^{-1}$ and in kcal $mole^{-1}$, for 1000-kilocycle broadcast band radio waves. (One kilocycle is a frequency of 10^3 sec^{-1}.) What is the wavelength of such photons? How does the energy compare with that for a carbon–carbon single bond? Would you expect radio waves to be able to produce chemical reactions?

22 When a photon strikes a metal surface, a certain minimum energy is required to eject an electron from the metal. This minimum energy is known as the threshold energy of the metal. Any energy in the original photon above this minimum is translated into kinetic energy for the ejected electron. The threshold wavelength for photoelectric emission from Li is 5200 Å. Calculate the velocity of electrons emitted as the result of absorption of light at 3600 Å.

23 Calculate the energy difference between the two lines in the emission spectrum of hydrogen that have wavelengths of 4861 Å and 4340 Å.

24 Calculate the energy of two lines that will appear in the emission spectrum of hydrogen in the visible range of the spectrum. The visible, or Balmer, series of lines is calculated from the Rydberg equation by letting $n_1 = 2$ and $n_2 = 3, 4, \ldots, n$.

ELECTRONIC STRUCTURE
OF ATOMS

5

In Chapter 4, we presented evidence to show that atoms consist of dense, positively charged nuclei and enough electrons to make the atoms electrically neutral. This chapter deals with the way in which electrons arrange themselves about the nucleus, and there is a brief account of early attempts to explain atomic structure, using a model based on analogy to the solar system. This classical mechanical model was replaced by a description of atoms in terms of modern quantum theory, or wave mechanics. We will show that modern quantum theory leads to a picture of atoms that provides good explanations of some of the properties of atoms. The most important result of the theory is that it provides a basis for understanding the existence of families, or groups, of chemically similar elements. The methods of quantum mechanics depend upon rather advanced mathematical skills. Consequently, we will simply state some rules derived from mathematical analysis, because the rules can be used easily to construct a systematic method for describing the behavior of electrons in atoms.

Chapters 4 and 5 may perplex you, because of the extraordinary differences between the behavior of objects large enough to see and feel, and that of the tiny bits of matter that make up an atom. However, the ideas presented should provide an invaluable basis for understanding many important facts about the structure and reactivity of compounds. In a sense, these two chapters lay a foundation for the rest of the book.

5–1 Quantum Theory of Atoms

In 1913, Niels Bohr, a Danish physicist, proposed the quantum model of the hydrogen atom. He reasoned that most of the energy of an excited atom is associated with the motion of the electron around the nucleus, and he visualized electrons as having orbits, similar to the orbits that the planets follow around the sun (Figure 5–1).

The first problem in describing the motion of the electron in Bohr's model for the hydrogen atom is similar to the problem of describing the motion of planets in the solar system. Revolution of a small body in a circular orbit around a larger body requires a balance of forces. Centrifugal force tends to make the smaller moving body fly in a straight line, which would make it fly away from the large body. Opposing the centrifugal force are the forces tending to pull the two bodies together (Figure 5–2).

Everyone is familiar with centrifugal force. If you tie a ball to a string and swing it around you, the tug on the string is due to centrifugal force. If the string breaks, the ball will fly away in a straight line. In a solar system, the principal force of attraction between the sun and its planets is gravity. In a hydrogen atom, electrostatic attraction between the positive nucleus and the negative electron is most important. The gravitational attraction between the nucleus and the electron is negligible, because their masses are very small. The two opposing forces can be formulated exactly using simple laws of physics. The relationship is

centrifugal force $=$ electrostatic force of attraction

$$\frac{m_e v^2}{r} = \frac{e^2}{r^2} \tag{5-1}$$

where m_e is the mass of the electron; v, the velocity of the electron; r, the radius of the orbit; and e, the unit of electrical charge (i.e., $-e$ for the electron and $+e$ for the proton).

The term on the right of Equation 5–1 is the usual expression for the force of interaction of two electrically charged particles, and the term on the left is the mathematical expression for centrifugal force. Such forces have direction as well as magnitude. In this instance, the centrifugal force is acting in a direction opposite to the electrical force of attraction between the two particles. When the two opposing forces are equal, as required by Equation 5–1, a balance exists that allows the electron to stay in orbit.

Bohr proposed that the different energy states of the hydrogen atom, which were suggested by the distinct lines in its atomic emission spectrum (Figure 4–14), involve different orbits with different energies for the electron. The lowest energy state is the one in which the radius of the orbit is smallest, thus the force of attraction between the nucleus and the electron is largest. Energy must be put into the atom to move the electron to an orbit with a larger radius, since work is required to separate positive and negative charges. Emission of light was believed to be associated with loss of energy when the electron "jumps" from one orbit to another of smaller radius.

The real problem was to account for the discrete emission spectrum. Equation 5–1 can be made to balance for any value of r by choosing the right value of v. Then why aren't there an

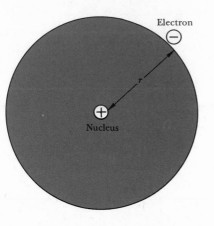

Figure 5–1.   Bohr's model of the hydrogen atom.

infinite number of energy states associated with a continuous variation of *r*? Bohr made the brilliant suggestion that the energy of the electron is *quantized;* that is, the electron can only gain or lose certain definitely prescribed units of energy. With this as a restriction, we realize that there is a limited number of stable orbits available to the electron. Using the quantum hypothesis and ordinary laws of physics, Bohr developed the following formula to describe the quantum states of a hydrogen atom

$$E = -\frac{2\pi^2 e^4 m_e}{h^2 n^2} \qquad\qquad (5\text{–}2)$$

where *n* is an integer, called the *principal quantum number,* with allowed values of 1, 2, 3, . . . , ∞; *h* is Planck's constant. The negative sign indicates a more stable (lower energy) state when the electron is close enough to interact with the nucleus than when the electron is removed

Figure 5–2.   Representation of opposing forces in the Bohr model.

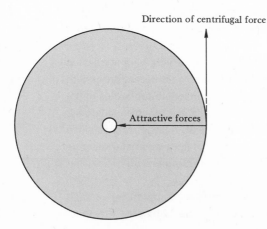

completely from the nucleus. When the electron is removed completely, the energy of the system is zero. Equation 5–2 may be rewritten in a simpler form

$$E = -\frac{R'}{n^2} \tag{5-3}$$

Every term in Equation 5–2, except n, is a constant, and these constants are combined in R'. Equation 5–3 gives the total energy of the electron in the atom, including its kinetic and potential energies. The value of E is determined by the value of n. If the electron in the lowest energy state, E_1 (called the *ground state,* in which $n = 1$), absorbs energy from some outside source, it will go to the next higher state ($n = 2$), and the atom will have a new energy, E_2. The energy difference (ΔE) between these two states will be

$$\Delta E = E_2 - E_1 = -R'\left(\frac{1}{n_2^2}\right) - \left(-R'\frac{1}{n_1^2}\right)$$

$$\Delta E = R'\left(\frac{1}{n_1^2} - \frac{1}{n_2^2}\right) \tag{5-4}$$

If we substitute for ΔE in Equation 5–4, using the Planck equation for light energy,

$$\Delta E = h\nu = h\frac{c}{\lambda} \tag{5-5}$$

then

$$\Delta E = h\frac{c}{\lambda} = R'\left(\frac{1}{n_1^2} - \frac{1}{n_2^2}\right)$$

and

$$\frac{1}{\lambda} = \frac{R'}{hc}\left(\frac{1}{n_1^2} - \frac{1}{n_2^2}\right) \tag{5-6}$$

Compare this final expression with the Rydberg equation given in Section 4–11. The ratio R'/hc can be evaluated, since all of the terms in R' are known, and h and c are known constants. The value of R'/hc is very close to the experimentally determined Rydberg constant. This agreement between experiment and theory was a major success for the Bohr atomic model.

The values of n, the principal quantum number, determine the energy of the atom. If we set n equal to one, we can calculate the radius of the hydrogen atom in its ground state, which we find to be 0.529 Å. As n becomes larger, the energy of the system increases, and the electron is farther from the nucleus.

To explain further the energetics of this simple system, let us return to Equation 5–3. The minus sign is an indication that attractive forces predominate in the hydrogen atom. For example, let n be very large, approaching infinity. This is the same as saying that the electron is so far from the nucleus that there are no effective attractive forces. However, as the electron moves closer to the nucleus (n becomes smaller), the attractive forces start to operate, and the electron–nucleus system becomes more stable, or lower in energy. This condition is represented by a minus sign. Substitution of the numerical values for the terms comprising R' in Equation 5–3 yields a value of −13.6 eV (electron volts) when $n = 1$. This is the amount of energy required to remove the electron from the hydrogen atom.

Example 5–1. Calculate the wavelength of the emission line of hydrogen when an electron moves from the $n = 4$ level to the $n = 2$ level.

Solution.

$$\frac{1}{\lambda} = \frac{R'}{hc}\left(\frac{1}{n_1^2} - \frac{1}{n_2^2}\right)$$

$$= 109{,}678 \text{ cm}^{-1}\left(\frac{1}{2^2} - \frac{1}{4^2}\right)$$

$$= 109{,}678 \text{ cm}^{-1}\left(\frac{1}{4} - \frac{1}{16}\right)$$

$$\cong 109{,}700 \text{ cm}^{-1} \times 0.1875$$

$$\frac{1}{\lambda} \cong 20{,}569 \text{ cm}^{-1}$$

$$\lambda = 4.86 \times 10^{-5} \text{ cm, or } 4860 \text{ Å}$$

This line is in the visible (Balmer) series of emission lines for hydrogen.

Example 5–2. Calculate the energy, in calories, required to ionize one mole of hydrogen atoms.

Solution. The energy required to ionize one hydrogen atom can be calculated directly, in ergs, from the relationship

$$\Delta E = hc\,\frac{1}{\lambda} = hcR\left(\frac{1}{n_1^2} - \frac{1}{n_2^2}\right)$$

We may think of the ionization process as taking an electron from the ground state, $n = 1$, to an infinite distance from the nucleus, which may be described by letting $n = \infty$. Substitution into the equation yields

$$\Delta E = (6.626 \times 10^{-27} \text{ erg sec}) \times (3.00 \times 10^{10} \text{ cm sec}^{-1}) \times (109{,}700 \text{ cm}^{-1}) \times \left(\frac{1}{1^2} - \frac{1}{\infty}\right)$$

$$\Delta E = 2.18 \times 10^{-11} \text{ erg atom}^{-1}$$

$$\Delta E \text{ (in calories)} = \frac{2.18 \times 10^{-11} \text{ erg atom}^{-1}}{4.18 \times 10^7 \text{ erg cal}^{-1}} = 0.521 \times 10^{-18} \text{ cal atom}^{-1}$$

The total energy for a mole of hydrogen atoms is

$$\Delta E_{\text{mole}} = (6.022 \times 10^{23} \text{ atoms mole}^{-1}) \times (0.521 \times 10^{-18} \text{ cal atom}^{-1})$$
$$= 3.14 \times 10^5 \text{ cal mole}^{-1}$$
$$= 314 \text{ kcal mole}^{-1}$$

Figure 5–3 shows how the calculated *energy levels* of hydrogen atoms are arranged for quantum numbers up to $n = \infty$. An excited hydrogen atom with its electron in the orbit with $n = 4$ could emit light if the electron jumped to an orbit with a smaller quantum number ($n = 1, 2,$ or 3). These transitions would give rise to three different emission lines, in different regions of the spectrum. When the Bohr model was presented, it became a rather simple matter to explain the various sets of atomic emission lines observed years earlier by the spectroscopists.

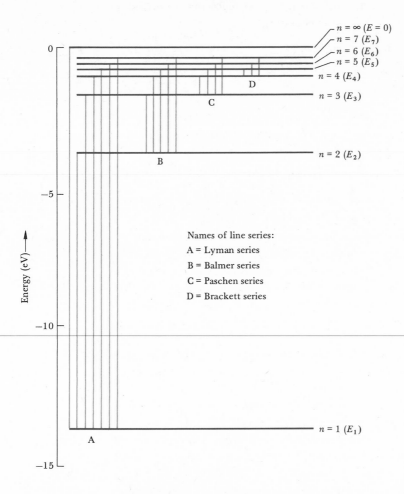

Figure 5–3.   Energy levels for the hydrogen atom.

Figure 5–3 shows a larger energy difference between the ground state, E_1, and the higher energy states than between any other two levels. It follows, then, that an electron making a transition to the E_1 level from E_2, E_3, . . . , E_n would give up more energy than, say, a transition from E_3, E_4, . . . , E_n down to the E_2 level. Since more energy is released in transitions to E_1, the accompanying photon that is emitted must have a higher frequency. In fact, all of the transitions to E_1 involve photons in the ultraviolet region of the spectrum. Transitions to the E_2 and E_3 levels from higher states correspond to the emission of photons in the visible and infrared regions, respectively. The groups of lines shown in Figure 5–3 were named after spectroscopists whose pioneering work provided the basis for testing the Bohr theory.

Example 5–3. Assign the following electronic transitions of the hydrogen atom to the appropriate series, and arrange them in order of increasing energy:

$$n_3 \rightarrow n_2; \quad n_4 \rightarrow n_3; \quad n_4 \rightarrow n_1; \quad n_2 \rightarrow n_1; \quad n_5 \rightarrow n_4$$

Solution. $n_3 \rightarrow n_2$ (Balmer); $n_4 \rightarrow n_3$ (Paschen); $n_4 \rightarrow n_1$ (Lyman); $n_2 \rightarrow n_1$ (Lyman); $n_5 \rightarrow n_4$ (Brackett). The order of the transitions as they increase in energy is

$$n_5 \rightarrow n_4; \quad n_4 \rightarrow n_3; \quad n_3 \rightarrow n_2; \quad n_2 \rightarrow n_1; \quad n_4 \rightarrow n_1$$

5-2 Modern Quantum Theory

The Bohr theory worked beautifully with hydrogen atoms, but it had serious defects for all other atoms, which have more than one electron. Numerous attempts were made to improve the theory by making a variety of assumptions, including the postulate that some electrons move in elliptical orbits, rather than the simple circular orbits postulated by Bohr. The modifications are not worth considering in detail. But it is worth noting that the principal reason for introducing new models was the need for more quantum numbers to accommodate the observation of many unexpected lines in the emission spectra of atoms more complex than hydrogen.

Attempts to modify Bohr's theory ended, not because of complete failure, but because of the appearance of a new theory. In 1923, Louis de Broglie made the very important suggestion that particles, such as electrons or protons, might have wave properties. One way to illustrate the meaning of the wave properties of a particle is to imagine the motion of an electron in flight. Although the flight path would, on the average, be a straight line (see Figure 5–4), it is even conceivable that the actual motion might involve "wobbling" of the electron around the average path.

Now we know that it is not possible to know the exact path of a moving electron well enough to determine whether the wobble is real. We can say only that the way in which the electron interacts with other particles along the path indicates that there could be a wobble. De Broglie proposed that the motion of any body follows a wave equation, rather than classical laws of motion. However, the wobble becomes smaller as the mass of the body increases. The uncertainty in the instantaneous line of flight of an electron is large enough to be important in explaining its behavior. However, the uncertainty is very small for a particle the size of a proton, and it is entirely negligible for bodies large enough to see.

The wave theory of matter leads to the *Heisenberg uncertainty principle.* This principle states that there are theoretical limits to the accuracy with which the motion of any object can be known. The concept is not easy to grasp, because it deals with matter beyond the range of our ability to make direct observations. However, certain "imaginary" experiments can give us a feeling for the meaning of the principle. Imagine trying to "look at" small objects by spraying beams of electrons at them. This would not be greatly different from shining a beam of light on the object, since light may also be thought of as being particlelike. If the objects are much larger than electrons, we can observe ways in which electrons bounce off them. We know that on impact an electron must transfer some momentum to the larger object, causing it to move. We can neglect the motion of the heavier object as long as it is much larger than the electron. However,

Figure 5–4. A representation of an electron in flight with a wavelike motion, or wobbling, superimposed on its average straight line path.

Average line
of e^- flight

if the target is another electron, the impact is likely to be enough to set the target in rapid motion. Consequently, the very act of detecting the target will drive it away, creating large uncertainty in its position. Mathematical analysis using equations for wave motion leads to the following statement of the uncertainty principle as first described by Heisenberg

$$\left(\begin{array}{c}\text{uncertainty of}\\\text{position}\end{array}\right) \times \left(\begin{array}{c}\text{uncertainty of}\\\text{momentum}\end{array}\right) \text{ is greater than } \frac{\text{Planck's constant}}{4\pi}$$

$$(\Delta x) \times (\Delta mv) > \frac{h}{4\pi} \tag{5-7}$$

where x is the coordinate specifying position, m is mass, and v is velocity.

The uncertainty principle leads directly to the conclusion that the Bohr theory is wrong, because the motion of electrons in relationship to their nuclei is specified too accurately. The motion of a particle of light or an electron can be known only vaguely, if we have located its position as accurately as is implied by saying that it is confined to a volume of a few cubic angstroms around the nucleus.

Example 5–4. Let the uncertainty in the location of an electron be 0.01 Å and the mass of the electron be 9×10^{-28} g. Calculate the uncertainty in the velocity of the electron.

Solution.

$$\Delta mv \times \Delta x = \frac{h}{4\pi}$$

$$\Delta mv \ = \frac{6.63 \times 10^{-27} \text{ erg sec}}{4(3.14) \times (1 \times 10^{-10} \text{ cm})}$$

$$\Delta mv \ = 5.28 \times 10^{-18} \text{ g cm}^2 \text{ sec}^{-2} \text{ sec cm}^{-1}$$

$$m(\Delta v) = 5.28 \times 10^{-18} \text{ g cm sec}^{-1}$$

The uncertainty in the velocity is

$$\Delta v = \frac{5.28 \times 10^{-18} \text{ g cm sec}^{-1}}{9 \times 10^{-28} \text{ g}}$$

$$\Delta v = 5.8 \times 10^{9} \text{ cm sec}^{-1}$$

This represents a very large uncertainty in the velocity of the electron.

The formulation of *wave equations* to describe the motion of an electron in a hydrogen atom gets rid of the unwanted feature of overspecification, and, at the same time, leads automatically to the introduction of additional quantum numbers. The solutions that are obtained from the wave equations are called *wave functions,* which are mathematical functions that resemble the functions used to describe the motion of waves over the surface of a body of water. The functions are not really in the form of a trajectory, or path. However, they can be interpreted as a measure of the *probability* of finding the electron at various points in space. The modern model of a hydrogen atom is built using the following steps, but you should keep in mind that these steps involve some complex mathematical relationships:

1   Write a wave equation for a negative electron moving under the influence of the electrical force of the positive nucleus.

2   Solve the equation. The solution is a wave function involving the distance between the electron and the nucleus.

3   Evaluate the function for many points in space close to the nucleus.

4   Square the values of the function at the various points. The square of the wave function is proportional to the probability of finding an electron at a given point.

5   Plot the results by using dots to indicate relative probability densities. A plot of the wave function squared for the hydrogen atom in its lowest energy state is shown in Figure 5–5. The outer boundary encloses a volume that includes approximately 90% of the electron's probability of residence.

Because of the form of the results, scientists discuss the position of electrons in hydrogen atoms (and other atoms), in terms of *probability density* or *cloud density.* In other words, we may as well discuss the electron as though it is a large diffuse cloud of varying density, as to attempt to concoct some fictitious story about a very rapid motion that cannot, in theory, be tested.

As in the Bohr theory, the different solutions to the wave equation for hydrogen atoms give the same fine agreement with the separations between the energy states indicated by the emission spectrum of hydrogen. The additional quantum numbers introduced by requirements of the mathematical methods help solve the problems encountered in comparing hydrogen atoms with other, more complex atoms. The solutions of the wave equations are called *orbitals,* because they bear some resemblance to the equations of the Bohr orbits. An orbital can be described graphically by plotting its probability density. Orbitals have a variety of shapes, and most non-mathematical discussions of the electronic structures of atoms and molecules are given in terms of properties of orbitals that can be shown graphically. In the next section, we will discuss shapes associated with various quantum numbers. Since we have not attempted to present the mathematics of modern quantum mechanics, we must ask you to accept the quantum numbers and their meaning, with our assurance that no trickery is involved in the mathematics. The basic assumption on which the whole scheme rests is the one already stated—that matter has wave properties.

Figure 5–5.   Plot of electron density for the lowest energy state of the hydrogen atom.

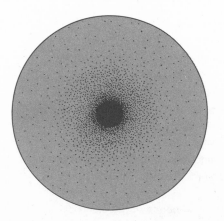

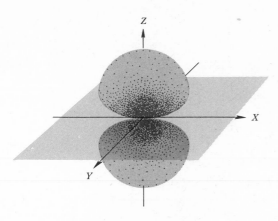

Figure 5-6. Electron density plot for an orbital with n = 2 and l = 1.

5-3 The Quantum Numbers

Three quantum numbers arise automatically in the solution of the wave equation for a hydrogen atom. A fourth is added to account for special properties that are observed when an electron is in a magnetic field.

THE PRINCIPAL QUANTUM NUMBER, *n*

The first quantum number corresponds rather closely to Bohr's orbital quantum number. It can have any integral value, beginning with one. The energies of hydrogen orbitals are determined only by the quantum number, *n*

$$E = -\frac{13.6}{n^2} \text{ eV} \tag{5-8}$$

As the values of *n* become larger, the energy increases, and the orbital becomes more spread out; thus the average separation of the electron and proton increases.

THE ORBITAL SHAPE QUANTUM NUMBER,

The orbital shown in Figure 5-5 is symmetrical; that is, the electron density depends only on the distance from the nucleus. There are other kinds of orbitals that give electron density clouds that are selectively directed in space. The shape of orbitals is determined by the value of the *orbital shape quantum number, l,* which can have values 0, 1, 2, 3, . . . , and so on. The mathematical requirements that give rise to the quantum number require that *l* must be smaller than *n*. Figure 5-6 shows an electron density plot for an orbital with *n* = 2 and *l* = 1.

Example 5-5. What values of *l* are allowed for an electron in the orbitals having principal quantum numbers (*n*) of 1, 2, and 3?

Figure 5–7.   Electron flow creating
a magnetic field.

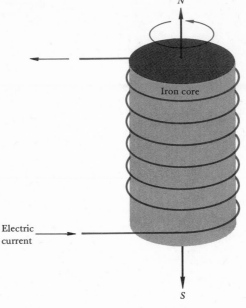

Solution. For $n = 1$, l can have only the value of zero, since l can be only as large as $(n-1)$, which in this case is $1-1$, or zero. For $n = 2$, l can have the values 0 and 1, but not 2, since the limiting value is $n-1$, or $2-1$, which is 1. For $n = 3$, l can have values of 0, 1, and 2, but not 3, since, as we have said, l cannot equal n.

THE MAGNETIC QUANTUM NUMBER, m

A moving electron creates a magnetic field. This is the principle underlying the operation of electromagnets, in which an electric current is passed through a wire wound around an iron core, as shown in Figure 5–7.

The magnetic properties of electrons do not contribute much to their energies under ordinary conditions, but they become important when atoms are placed in a magnetic field. The field always has a certain direction, which tends to create preferred areas for high electron probability.

The magnetic quantum number, m, describes the number of orientations in space a particular type of orbital may have. The magnetic quantum number can have both positive and negative integral values, as well as the value of zero (i.e., $m = \ldots -3, -2, -1, 0, 1, 2, 3, \ldots$).

The range of possible values of m again is restricted by mathematical requirements; m can have any integral value, including zero, between $-l$ and $+l$. When $l = 0$ and the orbital is spherical, the orientation in space has no effect on the orbital, and there is only one value of m, which is 0. However, when $l = 1$ and the electron density is directional (see Figure 5–6), there are three orientations in space for these orbitals, along the X, Y, and Z directions. These correspond to the three values ($+1$, 0, and -1) that m may have when $l = 1$.

Example 5–6. What values may the quantum numbers l and m have if the principal quantum number of an electron in an atom is 3?

Solution. The range of values of l is 0 to 2, and the possible values of m depend on the particular value of l. All possibilities are shown in Table 5–1.

THE SPIN QUANTUM NUMBER, s

Wave mechanics gives three quantum numbers associated with the motion of an electron about a nucleus. However, many kinds of experiments show that these three numbers are not enough to account for the properties of atoms, or even of free electrons. For example, there are more changes in the hydrogen spectrum caused by a magnetic field than can be accounted for by the states having different values of m. In fact, there appear to be just twice as many states as predicted. This kind of observation led to the idea that there must be a tiny magnetic field associated with an electron, even when it is not in flight. Such a magnetic field would be created if the electron were a little charged sphere spinning around an axis, in the same way the earth turns about its axis. Of course, since we cannot really see electrons, we cannot see whether they spin. However, the additional quantum number is called the *spin quantum number, s.* We assume that the spin quantum number of any electron can have either of two values. For convenience in relating the spin properties of an electron to the magnetic properties arising from its motion through space, the values of s are given as $+\frac{1}{2}$ and $-\frac{1}{2}$.

Example 5–7. Describe the ground state hydrogen atom with an acceptable set of values for the four quantum numbers, *n, l, m,* and *s*.

Solution. The term "ground state" means the lowest energy state. Thus, for the hydrogen atom, $n = 1$. When $n = 1$, the only value l can have is zero. Since m is limited to values of $+l \cdots 0 \cdots -l$, its only allowed value is 0 when $l = 0$. The spin quantum number, s, may have a value of $\pm\frac{1}{2}$. An acceptable set of quantum numbers is, therefore, $n = 1$, $l = 0$, $m = 0$, and $s = \pm\frac{1}{2}$.

THE USE OF QUANTUM NUMBERS

In 1924, a young German physicist, Wolfgang Pauli, went to Bohr's laboratory to set the old master straight. Furthermore, he succeeded in doing so! A major problem in the theory of atomic structure was the description of atoms having many electrons. The electrons are all attracted by the positive nucleus, but the electrons repel each other. The mathematical problems presented by models in which there are many particles interacting with each other simultaneously are still impossible to solve. Consequently, theoreticians work hard to develop "partial theories"

Table 5–1.   Allowed Values of l and m for n = 3

l	m	Number of states
0	0	1
1	−1, 0, 1	3
2	−2, −1, 0, 1, 2	5
	Total number of states	9

Table 5–2.   Ionization Energies of Three Elements

Element	Atomic number	Ionization energy (eV)
H	1	13.6
He	2	24.5
Li	3	5.4

to account for the patterns of behavior of complex systems. An especially difficult problem in the 1920's was the pattern established by a study of the energy required to remove an electron from various atoms; that is, their *ionization energies.* Consider the first three elements as shown in Table 5–2.

Such behavior seems crazy. Each increase in the atomic number adds one positive charge to the nucleus and increases the number of electrons by one. We can see that both electrons in helium must be bound much more tightly than the one in a hydrogen atom, since the ionization energy measures the energy required to tear one electron out of the atom. How can we explain this result? Obviously, the +2 helium nucleus will attract each electron with a stronger force than is supplied by the +1 nucleus in hydrogen. Somehow the two electrons in helium avoid each other, so the repulsion between electrons is not as important as the increased attraction of the nucleus. However, for lithium, the change in ionization energies is in the opposite direction. For lithium, we must say that the overwhelming effect is electron repulsion, despite the fact that the nuclear charge has increased.

Pauli made a very simple suggestion that did much to solve the dilemma. He suggested that *only one electron can be assigned the same four quantum numbers.* If we assume that the orbitals used to describe electrons in complex atoms are similar to those of the hydrogen atom, we can then proceed to build a model of electrons in atoms that provides a basis for most current thinking about chemical structure and reactivity. The suggestion is known as the *Pauli exclusion principle.*

5–4 Electronic Structures of Atoms

In the lowest energy state of the hydrogen atom, the electron will be assigned to an orbital with $n = 1$; the values of l and m must be 0, and s can have a value of either $+\frac{1}{2}$ or $-\frac{1}{2}$. Thus, there will be two kinds of ground state (most stable) hydrogen atoms. However, they will appear to be identical in any experiment not involving a magnetic field. In a magnetic field, the hydrogen atoms with $s = +\frac{1}{2}$ and $s = -\frac{1}{2}$ will behave differently. The same situation will arise in any atom or molecule containing an odd number of electrons; such atoms or molecules can be studied by a technique known as *electron spin resonance.*

In the helium atom, both electrons can be assigned to the orbital with $n = 1$, by using both spin quantum numbers. The assigned quantum numbers for the ground state helium atom (He) are

electron 1: $n = 1$ $l = 0$ $m = 0$ $s = +\frac{1}{2}$
electron 2: $n = 1$ $l = 0$ $m = 0$ $s = -\frac{1}{2}$

This description implies that the electrons occupy the same general volume of space. However, the electrons must avoid each other, so they do not crowd simultaneously into any given small volume. It is possible for two electrons in the same orbital to avoid each other as long as their spin quantum numbers are not the same.

In a lithium atom (Li), one of the three electrons must be assigned to an orbital with $n = 2$. For the ground state lithium atom, the quantum numbers are

electron 1: $n = 1$ $l = 0$ $m = 0$ $s = +\frac{1}{2}$
electron 2: $n = 1$ $l = 0$ $m = 0$ $s = -\frac{1}{2}$
electron 3: $n = 2$ $l = 0$ $m = 0$ $s = \pm\frac{1}{2}$
   or
electron 3: $n = 2$ $l = 1$ $m = 0, \pm1$ $s = \pm\frac{1}{2}$

Regardless of the assignments chosen for l, m, and s, the chief difference between lithium and helium is the necessity for assigning a principal quantum number of two to one of the electrons. This electron is, on the average, farther from the nucleus than the two electrons having $n = 1$. The electron having $n = 2$ is the electron that is lost with the low ionization energy shown in Table 5–2. The easy loss of this electron is the reason that lithium is highly reactive. And the fact that there is only one loosely bound electron is why the valence of lithium is one. The loosely bound electrons in atoms have the highest values of n, and they are often called *valence electrons*, since they are involved in chemical reactions. Note also that the loss of an electron from a lithium atom leaves a lithium ion, Li^+. The two remaining electrons should be bound even more strongly in Li^+ than the two electrons in helium, since the nuclear charge is greater in Li^+. We would anticipate that lithium ions should be very inert chemically, and experiments show this to be the case. More examples of the relationship between electronic structure of atoms and chemical reactivity are in the next chapter.

Other problems in assigning the electronic structure of a lithium atom involve the choice of quantum numbers l and m. In the hydrogen atom, the energies of the orbitals with $n = 2$, $l = 0$ and $n = 2$, $l = 1$ are identical. In a polyelectronic atom, the former orbital has a lower energy, and we can generalize by saying that, where the choice is otherwise free, the orbital of lowest energy will be the one with l having the lowest allowed value (i.e., $l = 0$). Orbitals with $l = 0$ are known as *s* orbitals. The *s* has nothing to do with spin quantum number, and the name "*s* orbital" is used for historical reasons, important in spectroscopy, but not of interest to us now.

A compact way of indicating the electronic structure of lithium is $(1s)^2(2s)$. The exponent indicates that two electrons having $+\frac{1}{2}$ and $-\frac{1}{2}$ spin quantum numbers are in a 1s orbital; no exponent is used with the 2s symbol, because only one electron occupies that orbital.

The configuration of a beryllium atom is $(1s)^2(2s)^2$. No new concepts are involved with beryllium. However, it is worth noting that the first ionization energy of beryllium is 9.3 eV, and the second electron can be removed by adding 18.2 eV additional energy. Ionization of two electrons would leave the doubly charged beryllium cation, Be^{2+}. Two electrons can be removed from beryllium atoms with a total input of 27.5 eV, only slightly higher than the 24.6 eV required for the single ionization of helium. Thus, the configuration $(1s)^2(2s)^2$ does not have the great stability found in helium with the $(1s)^2$ configuration.

The next element, boron, must start using the orbitals with $n = 2$ and $l = 1$. There are three such orbitals, with m values -1, 0, and $+1$. In any atom, the energies of all these $l = 1$ orbitals are equivalent. Since there are three such orbitals, a total of six electrons can be assigned to

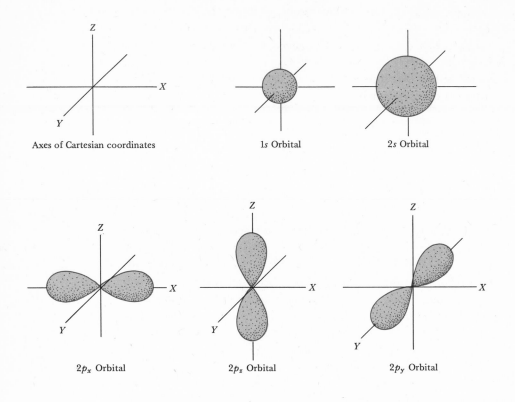

Figure 5–8.　The shapes and spatial orientations of the s and p orbitals.

the $n = 2$, $l = 1$ orbitals. Any orbital with $l = 1$ is called a p orbital, again a term having historical significance of no immediate importance to us. The electron density cloud of a p orbital does not have the same spherical (or ball-like) symmetry as s orbitals. The three p orbitals of any set are concentrated in different directions in space. Each is symmetrical about an axis, and the three axes are perpendicular to each other. The members of a set of three are usually referred to as p_x, p_y, and p_z orbitals. The subscripts refer to the axes of a three-dimensional Cartesian coordinate system.

Figure 5–8 shows the kinds of drawings often used to represent s and p orbitals. The charge density drawings attempt to show the three-dimensional characteristics of the orbitals. The outlines attempt to show directional characteristics, but such drawings give a false impression, because a wave function does not have a sharp boundary. However, most of the value of the function describing the probability of the electron's location lies within a few angstroms of the nucleus, and, for most purposes, an orbital can be thought of as a definite volume, within which an electron can almost certainly be found.

The electronic configurations of atoms with atomic numbers of five to ten follow directly from the ideas presented in discussion of lithium and beryllium, and they are given in Table 5–3.

The ionization energies tend to rise as the atomic numbers increase, although there is a discontinuity in the trend with nitrogen and oxygen. This apparent anomaly can be explained by

Table 5–3. Electronic Configurations and First Ionization Energies for Elements with Atomic Numbers 5–10

Element	Atomic number	Electronic structure	First ionization energy (eV)
Boron	5	$(1s)^2(2s)^2(2p)$	8.3
Carbon	6	$(1s)^2(2s)^2(2p)^2$	11.3
Nitrogen	7	$(1s)^2(2s)^2(2p)^3$	14.5
Oxygen	8	$(1s)^2(2s)^2(2p)^4$	13.6
Fluorine	9	$(1s)^2(2s)^2(2p)^5$	17.4
Neon	10	$(1s)^2(2s)^2(2p)^6$	21.6

considering the order in which the three $2p$ orbitals are used. In the nitrogen atom, each of the three p orbitals accommodates a single electron, and in oxygen, two electrons are forced to use the same p orbital. The general rise in ionization energies is because each additional increase in the nuclear charge leads to tighter binding of all the electrons. Note also that the ionization energy of neon is especially high. All of the noble gases (helium, neon, argon, krypton, xenon, and radon) have ionization energies considerably higher than those of preceding elements. Furthermore, each noble gas is followed by an element having a very low ionization energy. For example, sodium (atomic number 11) has an ionization energy of only 5.1 eV.

Example 5–8. Illustrate the distribution of the electrons in the p orbitals of carbon, nitrogen, and oxygen.

Solution. We can write the following electronic configurations for the elements carbon, nitrogen, and oxygen

$$\text{C:} \quad (1s)^2(2s)^2(2p)^2 \qquad \text{N:} \quad (1s)^2(2s)^2(2p)^3 \qquad \text{O:} \quad (1s)^2(2s)^2(2p)^4$$

Schematically, we can represent these configurations as

$1s \quad 2s \quad 2p_x \; 2p_y \; 2p_z$ $1s \quad 2s \quad 2p_x \; 2p_y \; 2p_z$ $1s \quad 2s \quad 2p_x \; 2p_y \; 2p_z$

     carbon   nitrogen   oxygen

Note that when electrons are added to orbitals of the same energy (such as the three $2p$ orbitals), they will occupy each of the available orbitals with one electron before doubling up in any one orbital. This is *Hund's rule,* which states that in a collection of orbitals of the same energy, electron spins remain unpaired, if possible.

Helium and neon are elements in which there are just enough electrons to make full use of all orbitals having the principal quantum number $n = 1$ for helium and $n = 2$ for neon. We conclude that the remarkable chemical inertness of the family of noble gases is because they have a completely filled set of orbitals. The fact that a very reactive element always follows the noble gases is logical, because the next added electron after the filled set of orbitals must occupy a new orbital, in which the electron is much less firmly bound.

5–5   The Heavier Elements

Examination of the rules for allowed quantum numbers shows a simple relationship between the principal quantum number and the maximum number of electrons associated with it. The maximum number of electrons having quantum number n is $2n^2$. This relationship predicts two electrons for the $n = 1$ shell and eight electrons for the completely filled $n = 2$ shell. It indicates that for $n = 3$ there should be 18 electrons. We might expect that the next inert element after neon would have an atomic number of $10 + 18$, or 28. However, the next member of the noble gas family is argon, with atomic number 18. Furthermore, element 28 is nickel, a metallic element that is found in chemical combination in many compounds. The especially stable configuration of argon is associated with the filling of the $3s$ and $3p$ orbitals, leaving empty the orbitals with $n = 3$, $l = 2$. The latter orbitals are not filled until some electrons enter the orbitals with $n = 4$. The result is a little disturbing, but it arises from the fact that electrons in orbitals having $l = 2$ are fairly strongly repelled by other electrons in the same atom.

Whereas s and p are used to represent orbitals with values of $l = 0$ and 1, respectively, d is used to represent orbitals with $l = 2$. The d orbitals become important in the fourth period of the periodic table. There are five d orbitals of equal energy when $l = 2$, just as there were three p orbitals of equal energy when $l = 1$. These five orbitals correspond to the five values that m may have when $l = 2$; that is, $m = +2, +1, 0, -1, -2$. Since each orbital may accommodate two electrons ($\pm\frac{1}{2}$ spin), the set of five d orbitals may accommodate a maximum of ten electrons. The electronic configurations of the elements are given in Table 5–4.

Example 5–9. Write the electronic configuration for iron, element 26.
Solution.

$$Fe = 1s^2 2s^2 2p^6 3s^2 3p^6 4s^2 3d^6$$

Notice that some of the electrons are not placed in the $n = 3$ level before the $n = 4$ level begins to fill. That is, $3d^6$ follows $4s^2$.

5–6   Summary

1   The motion of electrons cannot be treated by classical laws of motion; the very small size of electrons makes consideration of their wave properties crucial.

2   Mathematical methods of quantum mechanics, which have not been discussed, show that the behavior of electrons in atoms can be described by characteristic wave functions called orbitals.

3   Orbitals are identified by three characteristic quantum numbers, n, l, and m. A fourth quantum number, s, is also used.

4   According to the Pauli exclusion principle, no two electrons can have the same four quantum numbers.

5   Since the spin quantum number, s, can have two values, any orbital can accommodate two electrons.

6   Ionization energies of atoms provide a good guide to the changes in binding of electrons in atoms as atomic numbers are changed.

7   The chemical character of an element is determined largely by the arrangement of the valence electrons (those most weakly bound).

Table 5–4.   The Electronic Configurations and Ionization Energies of the Elements

Z	Atom	Orbital electronic configuration	First ionization energy (eV)
1	H	$1s$	13.60
2	He	$1s^2$	24.48
3	Li	$(He)2s$	5.39
4	Be	$(He)2s^2$	9.32
5	B	$(He)2s^22p$	8.30
6	C	$(He)2s^22p^2$	11.26
7	N	$(He)2s^22p^3$	14.54
8	O	$(He)2s^22p^4$	13.61
9	F	$(He)2s^22p^5$	17.42
10	Ne	$(He)2s^22p^6$	21.56
11	Na	$(Ne)3s$	5.14
12	Mg	$(Ne)3s^2$	7.64
13	Al	$(Ne)3s^23p$	5.98
14	Si	$(Ne)3s^23p^2$	8.15
15	P	$(Ne)3s^23p^3$	10.48
16	S	$(Ne)3s^23p^4$	10.36
17	Cl	$(Ne)3s^23p^5$	13.01
18	Ar	$(Ne)3s^23p^6$	15.76
19	K	$(Ar)4s$	4.34
20	Ca	$(Ar)4s^2$	6.11
21	Sc	$(Ar)4s^23d$	6.54
22	Ti	$(Ar)4s^23d^2$	6.82
23	V	$(Ar)4s^23d^3$	6.74
24	Cr	$(Ar)4s3d^5$	6.76
25	Mn	$(Ar)4s^23d^5$	7.43
26	Fe	$(Ar)4s^23d^6$	7.87
27	Co	$(Ar)4s^23d^7$	7.86
28	Ni	$(Ar)4s^23d^8$	7.63
29	Cu	$(Ar)4s3d^{10}$	7.72
30	Zn	$(Ar)4s^23d^{10}$	9.39
31	Ga	$(Ar)4s^23d^{10}4p$	6.00
32	Ge	$(Ar)4s^23d^{10}4p^2$	7.88
33	As	$(Ar)4s^23d^{10}4p^3$	9.81
34	Se	$(Ar)4s^23d^{10}4p^4$	9.75
35	Br	$(Ar)4s^23d^{10}4p^5$	11.84
36	Kr	$(Ar)4s^23d^{10}4p^6$	14.00
37	Rb	$(Kr)5s$	4.18
38	Sr	$(Kr)5s^2$	5.69
39	Y	$(Kr)5s^24d$	6.38
40	Zr	$(Kr)5s^24d^2$	6.84

Table 5–4 (continued)

Z	Atom	Orbital electronic configuration	First ionization energy (eV)
41	Nb	$(Kr)5s4d^4$	6.88
42	Mo	$(Kr)5s4d^5$	7.10
43	Tc	$(Kr)5s^14d^6$	7.28
44	Ru	$(Kr)5s4d^7$	7.36
45	Rh	$(Kr)5s4d^8$	7.46
46	Pd	$(Kr)4d^{10}$	8.33
47	Ag	$(Kr)5s4d^{10}$	7.57
48	Cd	$(Kr)5s^24d^{10}$	8.99
49	In	$(Kr)5s^24d^{10}5p$	5.79
50	Sn	$(Kr)5s^24d^{10}5p^2$	7.34
51	Sb	$(Kr)5s^24d^{10}5p^3$	8.64
52	Te	$(Kr)5s^24d^{10}5p^4$	9.01
53	I	$(Kr)5s^24d^{10}5p^5$	10.45
54	Xe	$(Kr)5s^24d^{10}5p^6$	12.14
55	Cs	$(Xe)6s$	3.89
56	Ba	$(Xe)6s^2$	5.21
57	La	$(Xe)6s^25d$	5.61
58	Ce	$(Xe)6s^24f^2$	5.6
59	Pr	$(Xe)6s^24f^3$	5.46
60	Nd	$(Xe)6s^24f^4$	5.51
61	Pm	$(Xe)6s^24f^5$	—
62	Sm	$(Xe)6s^24f^6$	5.6
63	Eu	$(Xe)6s^24f^7$	5.67
64	Gd	$(Xe)6s^24f^75d$	6.16
65	Tb	$(Xe)6s^24f^9$	5.98
66	Dy	$(Xe)6s^24f^{10}$	6.82
67	Ho	$(Xe)6s^24f^{11}$	—
68	Er	$(Xe)6s^24f^{12}$	6.08
69	Tm	$(Xe)6s^24f^{13}$	5.81
70	Yb	$(Xe)6s^24f^{14}$	6.2
71	Lu	$(Xe)6s^24f^{14}5d$	5.0
72	Hf	$(Xe)6s^24f^{14}5d^2$	5.5?
73	Ta	$(Xe)6s^24f^{14}5d^3$	7.88
74	W	$(Xe)6s^24f^{14}5d^4$	7.98
75	Re	$(Xe)6s^24f^{14}5d^5$	7.87
76	Os	$(Xe)6s^24f^{14}5d^6$	8.5
77	Ir	$(Xe)4f^{14}5d^9$	9
78	Pt	$(Xe)6s4f^{14}5d^9$	9.0
79	Au	$(Xe)6s4f^{14}5d^{10}$	9.22
80	Hg	$(Xe)6s^24f^{14}5d^{10}$	10.43

Table 5–4 (continued)

Z	Atom	Orbital electronic configuration	First ionization energy (eV)
81	Tl	$(Xe)6s^2 4f^{14} 5d^{10} 6p$	6.11
82	Pb	$(Xe)6s^2 4f^{14} 5d^{10} 6p^2$	7.42
83	Bi	$(Xe)6s^2 4f^{14} 5d^{10} 6p^3$	7.29
84	Po	$(Xe)6s^2 4f^{14} 5d^{10} 6p^4$	8.43
85	At	$(Xe)6s^2 4f^{14} 5d^{10} 6p^5$	9.5
86	Rn	$(Xe)6s^2 4f^{14} 5d^{10} 6p^6$	10.75
87	Fr	$(Rn)7s$	4
88	Ra	$(Rn)7s^2$	5.28
89	Ac	$(Rn)7s^2 6d$	6.9
90	Th	$(Rn)7s^2 6d^2$	6.95
91	Pa	$(Rn)7s^2 5f^2 6d$	—
92	U	$(Rn)7s^2 5f^3 6d$	6.1
93	Np	$(Rn)7s^2 5f^4 6d$	—
94	Pu	$(Rn)7s^2 5f^6$	5.1
95	Am	$(Rn)7s^2 5f^7$	6.0
96	Cm	$(Rn)7s^2 5f^7 6d$	—
97	Bk	$(Rn)7s^2 5f^9$	—
98	Cf	$(Rn)7s^2 5f^{10}$	—
99	Es	$(Rn)7s^2 5f^{11}$	—
100	Fm	$(Rn)7s^2 5f^{12}$	—
101	Md	$(Rn)7s^2 5f^{13}$	—
102	No	$(Rn)7s^2 5f^{14}$	—
103	Lr	$(Rn)7s^2 5f^{14} 6d$	—

8 Special stability is characteristic of completely filled *s* and *p* orbitals that are associated with the largest value of *n* that is used in the atom.

5–7 Postscript: Analytical and Synthetic Science

Who really cares about atoms? The only ones found free in nature in significant amounts, at least on this planet and in its atmosphere, are the noble gases. These elements exist in the atomic state because they are almost entirely unreactive. Consequently, they are almost chemical nonentities. Almost all chemistry of real significance deals with molecules—their structure and dynamic behavior in chemical reactions. We can properly ask why we do not go directly to the study of molecules, instead of first devoting three chapters to playing with ideas about elementary particles and atoms. To be honest, we must admit that there would be some virtue in first presenting some information about molecules, and then asking whether concepts of the properties of simpler forms of matter are of any use in understanding the behavior of molecules. As a matter of fact, this approach is really the one that has been used in the development of chemistry during the past 150 years. A great deal was known about molecules before elementary particles were discovered.

To some extent, we can say that the relationship between particle physics and molecular properties is a rationalization of basic facts of chemistry, after those facts were already known. However, this seemingly cynical view is only a part of the story. Formulation of models for particle behavior and atomic structure from 1896 to 1930 led to new models of molecular structure that were basically laid out in the 1930's. This insight was followed by an explosion of chemical knowledge that has been accelerating during the 30-year era between 1940 and 1970. Clearly, the quantum mechanical model of atoms has had a profound influence on the ways in which chemists study molecules.

The relationships among models for elementary particles, atoms, and molecules can teach us a great deal about the way in which science works, at least at this time. There are really two forms of science, analytical and synthetic. In analytical science, emphasis is placed upon dividing nature into small pieces and trying to learn everything possible about the pieces. Synthetic science consists of studying and modeling complex systems containing many components. During the past 50 to 70 years, principal emphasis has been placed upon analytical science. The study of smaller and smaller things has been so successful that some chemists seem to feel that the analytical mode is really the only form of science. We disagree, and believe that there is a great challenge to develop systematic methods for modeling complicated systems. After all, the world we live in is complicated. We have not developed methods of synthetic science nearly as well as we would like. Although this fact seems unfortunate to us, the authors, it should provide assurance to students that there is plenty of science left to do.

In the study of matter, the smallest bits that we can find are called elementary particles. There are many elementary particles now known, but those of direct interest to chemists were discussed in Chapter 4. In Chapter 5, we showed how ideas about the behavior of elementary particles have been used to build new models for the structure of atoms. In Chapter 7, we will turn to models for molecular structure. The entire cycle of scientific modeling is involved. Matter was torn apart with the discovery of negative electrons and various kinds of positively charged atomic nuclei. The properties of these particles were studied and thought about in the analytical stage. The use of ideas about particles to formulate new models for atoms illustrates the synthetic phase of science.

We easily could give the impression that the model of atomic structure is derived in a straightforward way from theory of the behavior of the particles in the atoms. Reflection on the subjects treated in Chapter 5 will show that this is far from the truth. Approximations, shrewd guesses, and new hypotheses were needed to make the new atomic model. The same features will be apparent in Chapter 7, when we take the next synthetic step and move on to consider molecular structure. We will end with a highly detailed and useful model of molecules. The model will contain many references to electrons and nuclei. However, the molecular model, not electrons and nuclei, will be used in the discussions of chemical structure and reactions. When we think of water, we will think of H_2O, not a seething mass of electrons and nuclei.

Many people have the naive idea that if we pursue the analytical phase of science long enough, the synthetic phase will take care of itself. At first glance, it may seem that, if we knew enough about the parts of a system, we ought to be able to predict automatically the behavior of the whole system. This ambition will be thwarted for two reasons. First, in any system of really interesting size there will be so many elementary parts that even the largest electronic computers would have no possibility of recording all the motions of all the parts. The second problem arises from the fact that the parts of the system will usually interact strongly with one another and influence each other's behavior. To deal with large numbers of interacting elements will, as far as we can see, always require ingenious synthetic work to create appropriate models for real systems.

We can see an analogous problem in social science. We can imagine that some very wise psychologist might develop a detailed and nearly perfect description of an individual person—how he thinks, feels, and responds to stimuli. Even with infinitely detailed knowledge of an individual, the psychologist would be unable to give an accurate prediction of the behavior of a large group of people, because of the complex pattern of interactions among the individuals. He might know exactly how a man would react to every kind of smile and frown on the faces of other people, but he could not predict the man's behavior during a day, simply because his behavior would depend on the number of smiles and frowns encountered, the order in which they were experienced, and so on.

New Terms

De Broglie wave theory: Particles have wave properties associated with them. The wave properties influence significantly the behavior of atomic particles, but become insignificant in the case of macroscopic particles.

Heisenberg uncertainty principle: It is not possible to know with great accuracy both the momentum and the location of an atomic particle. The limits of accuracy are determined by the relationship $(\Delta mv)(\Delta x) > h/4\pi$

Hund's rule: In a collection of orbitals of the same energy, electron spins remain unpaired, if possible.

Ionization energy: The energy required to remove an electron from a gaseous atom; that is, the energy associated with the process $M \rightarrow M^+ + e^-$.

Magnetic quantum number (m): This quantum number describes the different spatial orientations an orbital can have. It is limited to the values $+ l \cdots 0 \cdots -l$.

Orbital shape quantum number (l): This quantum number is related to the shape of the orbital and is limited to the values $0 \cdots (n - 1)$.

Orbitals: Wave functions.

Pauli exclusion principle: No two electrons in the same atom can have the same four quantum numbers.

Principal quantum number (n): This quantum number corresponds to the one described by Bohr. It describes the major energy levels in the atom, and can have integer values from one to infinity.

Probability density: The electron may be thought of as spread out over a volume in space, and the probability of finding the electron in a given volume is known as the probability density. It is proportional to the square of the wave function, which describes the wave properties of the electron.

Quantized energy: Energy that can assume only discrete values.

Spin quantum number (s): This quantum number derives from the fact that the electron behaves as if it were a spinning, charged particle. It is limited to values of $\pm\frac{1}{2}$.

Valence electrons: Those electrons in an atom that have the highest value of *n* and are in the outermost orbitals.

Wave function: The mathematical function that describes the wavelike behavior of a small particle, such as an electron.

Questions and Problems

1 Describe the Bohr model of the hydrogen atom. What new concept did Bohr introduce in this model? What does the quantum number *n* mean in the Bohr model?

2 Calculate the energy required to excite an electron in a hydrogen atom from the ground state ($n = 1$) to the $n = 3$ state. What wavelength of light would be required to accomplish this? Into what part of the spectrum would this light fall?

3 A small communication satellite weighs 1470 g and is traveling at a rate of 7.5×10^3 cm sec^{-1}. If the uncertainty in its velocity (Δv) is 10 cm sec^{-1}, what uncertainty in its location (Δx) will be associated with the satellite, according to the Heisenberg uncertainty principle? Is the uncertainty significant in this case?

4 How are the Bohr model and the Heisenberg uncertainty principle mutually contradictory?

5 Describe the quantum numbers that arise from the wave mechanical model of the atom, giving the possible numerical values each can have. State briefly what each quantum number describes in the atom.

6 How can the same atom of hydrogen, in quick succession, emit a photon in the Brackett, Paschen, Balmer, and Lyman series? Can it emit them in the reverse order? Why, or why not?

7 Why are the energies of electrons in atoms always negative?

8 Write the electronic configuration for the element argon. How many *s* and how many *p* electrons are there in this atom? Write a set of acceptable values for the four quantum numbers *n*, *l*, *m*, and *s* for the eighteenth electron in argon.

9 How does the Pauli exclusion principle explain the recurrence of an ns^1 electronic configuration and the beginning of new rows in the periodic table, starting with elements 3 and 11?

10 Write the electronic configurations ($1s^2 2s^2 \ldots$) for the elements oxygen, sulfur, selenium, and tellurium. What do these elements have in common regarding their configurations, and what can you predict about their chemical properties?

11 Explain why ionization energies increase across the periodic table from lithium (5.4 eV) to neon (21.6 eV). Then explain the decrease to a value of 5.1 eV for sodium, the element following neon in the periodic table.

12 How many d levels are there in a quantum level?

13 How many electrons will there be in the completely filled $n = 4$ level? When $n = 4$, what is the maximum value that l can assume? What letter is used to represent electrons in the $n = 4$ shell that have the maximum value of l? (You may consult Table 5–4.)

14 Write the electronic configuration ($1s^2 2s^2$ . . .) for the following: F^-, Na^+, Ne, O^{2-}, and N^{3-}. What would you predict about the relative sizes of these species?

15 Describe the electron distribution in the p orbitals of the elements Al, Si, P, and S.

16 Write electronic configurations (without consulting a table) for Cr ($Z = 24$) and Cu ($Z = 29$). After writing your version, compare it with the configuration given in Table 5–4. Can you conclude anything regarding the stability of half-filled and completely filled d orbitals?

17 The general electronic configuration ns^1 is characteristic of the outer electron in the alkali metal group of elements. What general configuration is characteristic of the halogens?

18 In Problem 17, the Group IA and Group VIIA elements were described in terms of their outer electronic configurations. Do this for Groups IIA through VIA. What relationship exists between the group number of a family of elements and the number of outer valence electrons?

19 There is a tendency for atoms to form (by gaining or losing electrons) charged species (ions) that have the electronic configuration of a nearby noble gas. Using this as a rule of thumb, what ions would you expect the following elements to form, and what would their electronic configurations be, both as atoms and as ions: oxygen, chlorine, sodium, and calcium?

20 What is incorrect about the following electronic configurations as ground states (lowest energy)? Write the correct configurations in each case.

$$Mg = 1s^2 2s^2 2p^5 3s^2 3p^1$$
$$K = 1s^2 2s^2 2p^6 3s^2 3p^7$$
$$Al^{3+} = 1s^2 2s^2 2p^6 3s^2 3p^4$$

21 A more detailed way of specifying the distribution of electrons in orbitals of equal energy, such as the three p orbitals, is to label them $p_x p_y p_z$. Using this more detailed notation, write the various possible equivalent ways of representing the ground state electronic configuration of atomic N and O.

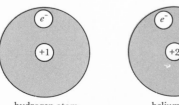

BONDS IN MOLECULES

Chemists have long known the common rules of valence, the periodic nature of chemical behavior, and many general rules about the structures and reactions of chemical compounds. All this knowledge developed from examining and thinking about a large volume of experimental evidence. In Chapter 3, we constructed a periodic table from a limited knowledge of chemical properties. Then, in Chapter 5, we saw that this organization of the elements is a simple consequence of the theory of atomic structure. Now we seek a theory that can describe the way atoms are bonded together to form molecules. As we develop the theory, we will keep in mind that all the things we already know to be true about composition, structure, and periodic behavior must be explained by, or at least be consistent with, our description of the chemical bond. Since the theory of atomic structure explained the periodic table so beautifully, we use it as the basis for our discussion of the chemical bond.

6–1 Hydrogen and Helium

Because chemists have known the properties of hydrogen and helium for many years, little fuss is made about the extraordinary difference between these elements. However, the contrast between the chemical properties of the first two elements can, by itself, provide a basis for the development of much of the theory of structural chemistry.

A hydrogen atom consists of a positively charged nucleus and one electron. A helium atom has two electrons and a nucleus bearing two positive charges

hydrogen atom   helium atom

Although their electronic structures differ by a single electron, helium forms no compounds, whereas hydrogen is found in millions of compounds. Elemental hydrogen itself consists of diatomic molecules (H_2), and the majority of the other elements form hydrides (Chapter 3).

Why do H_2 molecules form at all? Like the helium atom, hydrogen molecules contain two electrons, and the two nuclei provide a total charge of $+2$

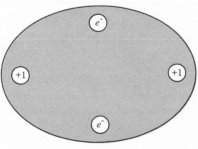

hydrogen molecule

If the two nuclei were jammed together, they would have the same effect on the electrons as a single helium nucleus. Apparently, even when the two positive charges are separated, as in the H_2 molecule, they have an effect on the two electrons similar to that of a He nucleus.

The hydrogen atom, the helium atom, and the hydrogen molecule can be represented in the following way:

H·	He:	H:H
hydrogen atom	helium atom	hydrogen molecule

These are the *electronic formulas* of the atoms and molecules. Formulas of this type are often called *Lewis structures,* after the great American chemist G. N. Lewis, who suggested, in 1916, that a shared electron pair could bind two atoms together. In drawing Lewis structures for more complicated molecules, we find it convenient to indicate a bonding pair of electrons by a line between the attached atoms. The electronic formula $H:H$ thus becomes H—H.

Another way to look at the hydrogen molecule is suggested by the way we have represented the electronic structure. If we count the electrons *shared* by the two nuclei as if they belonged to the individual nuclei, then *each* hydrogen atom shares two electrons. Such an arrangement simulates the stable electronic structure of helium. This point of view is another way of saying that both electrons interact strongly with each nucleus, thus each nucleus effectively has two electrons. The realization that electrons may interact strongly with two nuclei, thereby bonding the nuclei together, has been part of the basic foundation of theoretical chemistry. Applied to hydrogen, the theory states that hydrogen atoms will form bonds to many other atoms that can supply a single electron. The fact that the addition of one more electron to the hydrogen atom produces a stable electronic structure, like that of helium, accounts for the fact that hydrogen almost invariably has a valence of one.

Bonds formed between atoms by sharing pairs of electrons are called *covalent* bonds. The strength of a covalent bond is measured in terms of the energy required to break the bond to give isolated atoms: H—$H + bond\ energy \rightarrow 2H$. The bond energy of H_2 is 103 kcal mole^{-1}, which means that it requires 103 kcal of energy to change one mole of H_2 to two gram-atoms of H.

The distance between bonded atomic nuclei in a molecule, usually referred to as the *bond length,* also indicates, in a relative way, something about the strength of a bond. For a given pair of atoms, the bond energy increases as the bond length decreases. In its stable configuration, the H—H bond length is 0.74 Å, and its bond energy is 103 kcal mole^{-1}. An electron can be knocked out of H_2 by bombardment with energetic particles in the gas phase, giving the much less stable species $H_2{}^+$. The two hydrogen nuclei are now held together by a single electron $(H \cdot H)^+$, and the 61 kcal mole^{-1} bond energy $[(H \cdot H)^+ \rightarrow H \cdot + H^+]$ is much less than that for H_2. The bond length for $H_2{}^+$ illustrates the point made previously. It has increased to 1.06 Å, from the very short length of 0.74 Å in the more tightly bonded H_2 molecule.

The primary rules of valence for elements, especially those having low atomic numbers, can be formulated easily by assuming that pairs of electrons can be shared by two or more nuclei. We can expect a compound to be stable if the sharing process gives each of the individual atoms the electronic configuration of one of the noble gases. This concept is of great importance and will be amplified by the examples in the following sections.

6–2   Fluorine and Neon

The fluorine atom is in some ways like the hydrogen atom. We have said already that it, like hydrogen, invariably has a valence of one. Just as hydrogen can attain the inert helium structure by adding one electron, so fluorine can attain the inert neon structure by adding one electron.

Let us consider the electronic structure of the fluorine atom

$$:\!\overset{\cdot\cdot}{\underset{\cdot\cdot}{F}}\!\cdot$$
fluorine atom

Although the fluorine atom has nine electrons, only seven are shown. This is because the two electrons in the $1s$ orbital form a stable (helium) closed shell of electrons that has little effect on the chemical properties of the atom. The other seven electrons are located in the $n = 2$ shell and are called valence electrons, because they are the electrons that determine the valence of the atom.

The fluorine molecule, F_2, can be represented by a Lewis structure with a single covalent bond

$$:\overset{..}{F}—\overset{..}{F}:$$
fluorine molecule

Each fluorine atom in the molecule has a share in eight valence electrons, which gives it the stable neon electronic structure. The valence electrons not involved in the bond between the two fluorine atoms are called *unshared pairs;* sometimes they are called nonbonding, or lone pairs. However, the total number of valence electrons is the most important factor in determining the number and kinds of bonds that the element forms. The tendency of atoms to achieve a completed, or closed, valence shell of eight electrons as they form chemical bonds is known as the *octet rule.*

Here are some simple rules to follow when drawing a Lewis structure:

1 Draw the atomic skeleton of the molecule. This may require some special knowledge concerning how the atoms of a molecule are arranged.

2 Determine the number of electrons each atom contributes to the structure. The number of electrons contributed is equal to the periodic table group number to which that element belongs. For example, carbon, in Group IV, contributes four electrons; oxygen, in Group VI, contributes six electrons.

3 Arrange the electrons in the atomic skeleton in such a way that each atom has a share in an inert gas configuration (the octet rule). (The octet rule does not apply to hydrogen. Why?)

We shall apply these steps in constructing the Lewis formulas in the following sections.

6–3   The Hydrides and Fluorides

The molecule HF can be formulated in a similar way

H·	·$\overset{..}{F}$:	H—$\overset{..}{F}$:
hydrogen	fluorine	hydrogen fluoride
atom	atom	molecule

The shared pair of electrons gives the hydrogen atom the stable helium electronic structure, and gives the fluorine atom the stable neon electronic structure.

This procedure can be extended to the hydrides and fluorides of oxygen, nitrogen, and carbon with no difficulty. The Lewis electronic structure of the water molecule is

$$H—\overset{..}{\underset{..}{O}}—H$$

The shared pairs of electrons give each hydrogen atom the helium electronic structure, and the oxygen atom, the neon electronic structure.

Example 6–1. Write the electronic formula for the oxygen difluoride molecule, OF_2.

Solution. Arrange the three atoms in some suitable array. It is not necessary to know the exact shape of the molecule, just the sequence in which the atoms are attached. In this case, the choices are FFO and FOF. Since the valence of fluorine is always one, the second choice seems more suitable. Fluorine is a Group VII element, and oxygen is in Group VI. This indicates a total of 14 electrons from the two fluorine atoms and six from oxygen, making a total of 20 electrons to be used for the electronic formula. When we distribute the 20 electrons throughout the molecule, being sure to place at least one pair of electrons between adjacent atoms, we obtain the following structure

$$:\ddot{F}—\ddot{O}—\ddot{F}:$$

Similarly, nitrogen with five valence electrons

$$\cdot\ddot{N}\cdot$$

forms ammonia, NH_3

$$
\begin{array}{c}
H \\
| \\
:N—H \\
| \\
H
\end{array}
$$

with a stable closed-shell electronic configuration.

Example 6–2. Write the electronic formula for the nitrogen trifluoride molecule, NF_3.
Solution.

    1. Draw an atomic skeleton

$$
\begin{array}{ccc}
F & N & F \\
  & F &
\end{array}
$$

    2. Determine the number of electrons for the structure

$$
\begin{array}{rl}
N \text{ (Group V)} = & 5 \text{ electrons} \\
3 F \text{ (Group VII)} = & \underline{21 \text{ electrons}} \\
\text{Total} = & 26 \text{ electrons}
\end{array}
$$

    3. Distribute the electrons to obey the octet rule

$$
\begin{array}{c}
:\ddot{F}—\ddot{N}—\ddot{F}: \\
| \\
:\ddot{F}:
\end{array}
$$

The next element in the second row of the periodic table is carbon, which has four valence electrons. Carbon forms the stable hydride, methane (CH_4)

$$
\begin{array}{c}
H \\
| \\
H—C—H \\
| \\
H
\end{array}
$$

We will return to carbon later to discuss the more complex hydrides of this element.

Example 6–3. Write the electronic formula for the carbon tetrafluoride molecule, CF_4.
Solution.

    1. Skeleton

$$
\begin{array}{ccc}
 & F & \\
F & C & F \\
 & F &
\end{array}
$$

2. Number of valence electrons

$$C = 4 \ e^-$$
$$4 \ F = \underline{28 \ e^-}$$
$$Total = 32 \ e^-$$

3. Distribution of electrons

6–4   Boron Trihydride

We cannot write a simple electronic formula for boron hydride that corresponds to those given for methane (CH_4), ammonia (NH_3), and water (H_2O). Boron atoms have three valence electrons, so the simple sharing formulation allows us to write a formula for the trihydride, BH_3

·B·

boron atom

H—B—H
|
H

boron trihydride

All evidence indicates that the BH_3 molecule is extremely reactive. Attempts to prepare the molecule lead to formation of a material called *diborane,* which has the molecular formula B_2H_6. The tendency of BH_3 to form B_2H_6 may be explained by our simple theory of electronic structure. As shown in the electronic formula, the boron atom in BH_3 has not completed its valence shell; it needs two electrons to achieve the neon structure.

The existence of B_2H_6 illustrates two accomplishments of the electronic theory of molecular structure. First, the theory suggests that each boron atom should be able to bind three hydrogen atoms, much as a nitrogen atom binds three hydrogen atoms. This is in accordance with the fact that both boron and nitrogen have principal valences of three. Second, the theory predicts that BH_3 should be very reactive, since the boron atoms in no way can be considered to have the electronic configuration of a noble gas. However, the simple theory does not predict that two BH_3 molecules can be bound together to form the stable molecule B_2H_6.

The most direct method of determining the structure of B_2H_6 involves electron diffraction. A beam of electrons is passed through a sample of gaseous diborane, and some of the electrons are scattered, or diffracted, in much the same way that light from the sun is scattered in passing through a foggy atmosphere. Analysis of the electron scattering pattern shows the arrangement of the boron and hydrogen atoms in diborane molecules.

A model of the actual structure of diborane, as well as a three-dimensional sketch, are shown in Figure 6–1. Surprisingly, the hydrogen atoms are not all equivalent. Four are bound to single boron atoms, but the other two are bound to both boron atoms. This structure has a "three-center" bond. If we try to translate this structure into an electronic formula, we arrive at the following:

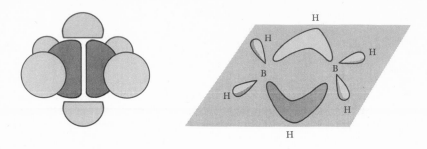

Figure 6–1.   A model of the molecule B_2H_6 *and a three-dimensional representation show-ing the "three-center" bond.*

If we count the *four electrons* associated with the two bridging hydrogen atoms (those placed between the borons) as belonging to the valence shells of both boron atoms, we see that we have provided the desired number of eight valence electrons.

The simplest models for electronic structure of molecules give satisfactory results, if we simply discuss the sharing of pairs of electrons by pairs of atoms, as in H_2, F_2, H_2O, and so on. Now we find an example, B_2H_6, in which a pair of electrons apparently binds *three atoms together.* Diborane can be thought of as containing two bonds that extend over *three* atoms, with each bond formed by simultaneous interaction of two electrons with two borons and one bridging hydrogen. This type of three-center, electron-pair bond is commonly denoted with a curved line connecting the three atoms

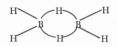

6–5   Boron Trifluoride

From our discussion of the hydrides and fluorides of other elements, we might have anticipated that boron trifluoride would form "double" or "dimeric" molecules, B_2F_6, analogous to diborane. However, it does not. The boron atoms in the single molecules must find some other way of satisfying the octet of valence electrons in the single molecule. An obvious difference between BH_3 and BF_3 is the presence of nonbonding valence electrons attached to fluorine in boron trifluoride

We believe that some interaction of the unshared-pair fluorine electrons helps to satisfy the electronic needs of the boron. The nature of this interaction will be discussed in Section 6–15.

Many other boron compounds, in which the boron atoms are bound to atoms having unshared electron pairs, also exist as single molecules. Furthermore, boron compounds react with molecules that possess unshared valence electrons to form stable *addition compounds.* For example, boron trifluoride reacts with ammonia to form a very stable compound

$$:NH_3 + BF_3 \rightarrow H_3NBF_3$$

The unshared valence electrons shown in the formulas of many molecules often play an important role in chemical reactions.

Example 6–4. Write the electronic formula for H_3NBF_3.
Solution.

1. Since the compound is made by combining NH_3 and BF_3, a reasonable structure might be

```
    H   F
H   N   B   F
    H   F
```

2. The total number of valence electrons is

$$
\begin{aligned}
N \text{ (Group V)} &= 5\ e^- \\
B \text{ (Group III)} &= 3\ e^- \\
3\ H \text{ (Group I)} &= 3\ e^- \\
3\ F \text{ (Group VII)} &= \underline{21\ e^-} \\
\text{Total} &= 32\ e^-
\end{aligned}
$$

3. Distribute the electrons

```
     H  :F:
     |   |     ..
 H—N—B—F:
     |   |     ..
     H  :F:
          ..
```

6–6 Ionic Bonds

So far, we have discussed the electronic structures of compounds of hydrogen and fluorine, which are nonmetals, with themselves and with the other nonmetals oxygen, nitrogen, carbon, and boron. Now we will discuss the compounds of fluorine and oxygen with a metallic element, beryllium.

In contrast to the situation for a nonmetal, the valence electrons are bound loosely to a metal atom. This factor, combined with the strong affinity of fluorine for electrons, suggests that, in the compound BeF_2, electron pairs are not shared equally in the Be—F bonds. In fact, the electron pairs are attracted so strongly by the fluorine atoms that it is better to say that in BeF_2 the beryllium atom has lost its two valence electrons to form the stable (helium electronic structure) *beryllium ion*, Be^{2+}. Notice that the beryllium ion has a charge +2, and each of the fluorine atoms has gained one valence electron to form the stable (neon electronic structure) *fluoride ion*, F^-

Be ·	·F̈:	$\left[\,:\ddot{F}:\,\right]^- Be^{2+} \left[\,:\ddot{F}:\,\right]^-$
beryllium atom	fluorine atom	beryllium difluoride

The resulting molecule, BeF_2, is held together not by mutually shared electron pairs, but by the electrostatic attraction of the *negative fluoride ions* for the *positive beryllium ion*. The strong electrostatic attraction of oppositely charged ions is responsible for *ionic bonds.* As we might expect, beryllium also forms an oxide, BeO

$$Be^{2+}\left[:\overset{..}{\underset{..}{O}}:\right]^{2-}$$

Example 6–5. Using Lewis diagrams, write an equation showing the formation of an ionic compound between calcium and chlorine.

Solution. Calcium is a Group II metal, and chlorine is a Group VII nonmetal. Thus, we can write

$$\overset{.}{Ca}+2\left[\overset{..}{\underset{..}{Cl}}:\right]\rightarrow\left[:\overset{..}{\underset{..}{Cl}}:\right]Ca^{2+}\left[:\overset{..}{\underset{..}{Cl}}:\right]^{-}$$

Brackets are used for the sake of clarity when more than one ion is involved and when an ionic charge must be indicated.

Lithium has but one loosely bound valence electron and is more metallic than beryllium. The helium inert electronic structure is easily attained by forming the lithium ion, Li$^+$

$$\overset{.}{Li}\rightarrow Li^{+}+e^{-}$$

Lithium forms an ionic fluoride LiF

$$\overset{.}{Li}+\overset{..}{\underset{..}{F}}:\rightarrow Li^{+}\left[:\overset{..}{\underset{..}{F}}:\right]^{-}$$

and an ionic oxide Li$_2$O

$$Li^{+}\left[:\overset{..}{\underset{..}{O}}:\right]^{2-}Li^{+}$$

Compounds formed between elements from the left of the periodic table (alkali and alkaline earth metals) and the halogens and oxygen are essentially ionic in character. In general, these compounds, of which BeF$_2$, BeO, LiF, and Li$_2$O are representative, are high-melting solids. Individual molecules are not found in ionic solids, because the compound is built by packing positive and negative ions close together in an essentially infinite network. These structures will be discussed in more detail in Chapter 8.

6–7   Electronegativity

Our models of covalent and ionic bonding have been highly idealized. Equal sharing of an electron pair by two atoms can be strictly true only when the atoms are exactly the same; and the ionic model is always an exaggeration, because even with very dissimilar atoms, transfer of an electron pair cannot be 100% complete. We must face the fact that the great majority of molecules have bonds that are somewhere between the two limits that we have discussed.

To discuss this more general case, it is helpful to introduce the concept of *electronegativity.* Electronegativity is a term relating the relative ability of an atom to attract electrons to itself in a bond. For example, in the beryllium difluoride molecule, the fluorine atom holds electrons tightly, and has a very high electronegativity. The beryllium atom does not pull electrons in so well, therefore it has a low electronegativity. The large difference in the two electronegativities results in nearly a complete transfer of the two valence electrons from beryllium to the two fluorines in the molecule. The American chemical physicist Robert S. Mulliken (1896–   ) proposed that electronegativity (*EN*) be defined as proportional to the sum of the first ionization

Table 6–1. Some First Ionization Energies and Electron Affinities

First ionization energies (kcal mole^{-1})		Electron affinities (kcal mole^{-1})	
H $\rightarrow$ H$^+$ + e^-	313.5	H$^-$ $\rightarrow$ H + e^-	17
C $\rightarrow$ C$^+$ + e^-	259.6	C$^-$ $\rightarrow$ C + e^-	29
O $\rightarrow$ O$^+$ + e^-	314.0	O$^-$ $\rightarrow$ O + e^-	34
F $\rightarrow$ F$^+$ + e^-	401.8	F$^-$ $\rightarrow$ F + e^-	79.5
Na $\rightarrow$ Na$^+$ + e^-	118.5	Na$^-$ $\rightarrow$ Na + e^-	19
Cl $\rightarrow$ Cl$^+$ + e^-	300.0	Cl$^-$ $\rightarrow$ Cl + e^-	83.3
K $\rightarrow$ K$^+$ + e^-	100.1	K$^-$ $\rightarrow$ K + e^-	16
Br $\rightarrow$ Br$^+$ + e^-	273.0	Br$^-$ $\rightarrow$ Br + e^-	77.5
I $\rightarrow$ I$^+$ + e^-	241.1	I$^-$ $\rightarrow$ I + e^-	70.6

energy (*IE*) and a quantity known as the electron affinity (*EA*) of an atom (*c* is a proportionality constant)

$$EN = c(IE + EA) \tag{6–1}$$

The atomic ionization energy (Section 5–3) is a measure of the energy required to remove an electron from a neutral gaseous atom (first *IE*) or a positively charged ion (second *IE*, etc.)

$$Na(g) \rightarrow Na^+(g) + 1e^- \quad \text{118.5 kcal mole}^{-1}, \text{ first } IE$$
$$Na^+(g) \rightarrow Na^{2+}(g) + 1e^- \quad \text{1090 kcal mole}^{-1}, \text{ second } IE$$

The electron affinity (*of an atom*) is equal to the energy needed to detach an electron from a gaseous *negative ion* to give a neutral atom. For example,

$$H^-(g) \rightarrow H(g) + 1e^- \quad EA = \text{17 kcal mole}^{-1}$$

Atoms with the largest *EA* values have the greatest ability to hold an extra electron.

Some ionization energies and electron affinities are given in Table 6–1. Notice that the elements at the right of the periodic table have the largest *IE* and *EA* values. It follows that these elements have the highest electronegativities. A quantitative Mulliken scale for the atoms can be obtained by assigning one atom a specific *EN* value, thus fixing the proportionality constant *c* (Equation 6–1). Unfortunately, only a few *EN* values can be calculated in this way, because not many atomic electron affinities are known accurately.

Example 6–6. Use *IE* and *EA* values for chlorine and fluorine given in Table 6–1 to calculate which of these two elements is the more electronegative, according to the Mulliken definition.

Solution. Mulliken defined electronegativity as

$$EN = c(IE + EA)$$

so that

$$\frac{EN_F}{EN_{Cl}} = \frac{c(402 + 80)}{c(300 + 83)} = 1.3$$

Therefore, $EN_F = 1.3\ EN_{Cl}$. That is, fluorine is 1.3 times as electronegative as chlorine.

A more widely applied quantitative treatment of electronegativity was introduced by the American chemist Linus Pauling (1901–), in the early 1930's. The Pauling electronegativity value for any given atom is obtained by comparing the bond energies of certain molecules containing that atom. If the bonding electrons were equally shared in a molecule AB, it would be reasonable to assume that the bond energy of AB would be the mean of the bond energies of the molecules A_2 and B_2. However, it is a very general result that *the bond energy of a molecule AB is almost always greater than the geometric mean of the bond energies of A_2 and B_2.*

A striking example that illustrates this point is the HF molecule. The bond energy of HF is 134 kcal mole^{-1}, whereas the bond energies of H_2 and F_2 are 103 kcal mole^{-1} and 36 kcal mole^{-1}, respectively. The geometric mean of the latter two values is $(36 \times 103)^{1/2} = 61$ kcal mole^{-1}, which is far less than the observed bond energy of HF. This "extra" bond energy, Δ, in an AB molecule is assumed to be a contribution due to the *partial ionic character* of the bond. In this model, the electronegativity difference between two atoms A and B is defined as

$$EN_A - EN_B = 0.208\Delta^{1/2} \qquad\qquad\qquad (6\text{--}2)$$

where EN_A and EN_B are the electronegativities of atoms A and B, Δ is the extra bond energy in kilocalories per mole, and the factor 0.208 converts from kilocalories per mole to electron volts. The square root of Δ is used because it gives a more nearly consistent set of electronegativity values for the atoms. Since only differences are obtained from the application of Equation 6–2, one atomic electronegativity is assigned a specific value, then the other values are calculated from the equation. On the Pauling scale, the most electronegative atom, fluorine, is assigned an electronegativity of approximately 4. A recent compilation of *EN* values, calculated using the Pauling idea as expressed in Equation 6–2, is given in Table 6–2. The alkali metal atoms lithium, sodium, potassium, and so on, which have very small ionization energies and very small electron affinities, have the least ability to attract electrons in bonds. The halogen family shows the largest electronegativities.

Example 6–7. Estimate the bond energy (*BE*) of the molecule HCl, using the electronegativity table and the bond energies given for H_2 and Cl_2, respectively, as 103 kcal mole^{-1} and 57.2 kcal mole^{-1}.

Solution. The actual bond energy of HCl can be expected to be greater than the geometric mean of the H_2 and Cl_2 bond energies by an amount Δ. Stated in a mathematical form

$$BE_{(HCl)} = [BE_{(H_2)} \times BE_{(Cl_2)}]^{1/2} + \Delta$$

We can obtain Δ from Equation 6–2

$$EN_{Cl} - EN_H = 0.208\Delta^{1/2}$$

or

$$\Delta = \left(\frac{EN_{Cl} - EN_H}{0.208}\right)^2$$

Table 6–2. Pauling Electronegativity Values[a]

I	II	III	II	II	II	II	II	II	II	I	II	III	IV	III	II	I
H 2.20																
Li 0.98	Be 1.57											B 2.04	C 2.55	N 3.04	O 3.44	F 3.98
Na 0.93	Mg 1.31											Al 1.61	Si 1.90	P 2.19	S 2.58	Cl 3.16
K 0.82	Ca 1.00	Sc 1.36	Ti 1.54	V 1.63	Cr 1.66	Mn 1.55	Fe 1.83	Co 1.88	Ni 1.91	Cu 1.90	Zn 1.65	Ga 1.81	Ge 2.01	As 2.18	Se 2.55	Br 2.96
Rb 0.82	Sr 0.95	Y 1.22	Zr 1.33		Mo 2.16			Rh 2.28	Pd 2.20	Ag 1.93	Cd 1.69	In 1.78	Sn 1.96	Sb 2.05		I 2.66
Cs 0.79	Ba 0.89	La 1.10			W 2.36			Ir 2.20	Pt 2.28	Au 2.54	Hg 2.00	Tl 2.04	Pb 2.33	Bi 2.02		
		Ce 1.12	Pr 1.13 (III)	Nd 1.14 (III)		Sm 1.17 (III)		Gd 1.20 (III)		Dy 1.22 (III)	Ho 1.23 (III)	Er 1.24 (III)	Tm 1.25 (III)		Lu 1.27 (III)	
					U 1.38 (III)	Np 1.36 (III)	Pu 1.28 (III)									

[a] *Roman numerals refer to the common valences of the atoms in the molecules used in the calculation of atomic electronegativity values.*

Substituting the electronegativities of H and Cl, we get for Δ

$$\Delta = \left(\frac{3.16 - 2.20}{0.208}\right)^2$$

$$\Delta = \left(\frac{0.96}{0.208}\right)^2 = 21.3 \text{ kcal mole}^{-1}$$

$$BE_{(HCl)} = (103 \times 57.2)^{1/2} + 21.3$$

$$= 76.7 + 21.3 = 98 \text{ kcal mole}^{-1}$$

The experimentally observed value is 102 kcal mole^{-1}.

Electronegativity is a useful concept in describing qualitatively the sharing of the electrons in a bond between two different atoms. For either beryllium or lithium and fluorine, the difference in electronegativities is so large that there is effectively complete transfer of the bonding electron pair. We recognize, however, that there are a very large number of molecules in which the bonds between dissimilar atoms are better described as covalent, with some ionic character.

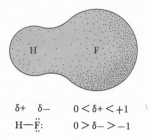

$$\delta+ \quad \delta- \qquad 0 < \delta+ < +1$$
$$H—\ddot{\underset{..}{F}}: \qquad 0 > \delta- > -1$$

Figure 6–2.   Model of the HF molecule showing the large electron density near fluorine.

6–8   The HF Molecule; A Covalent Bond with Ionic Character

The next molecule we will study is such an intermediate case. The bond in a molecule composed of a hydrogen atom and a fluorine atom, HF, is neither purely covalent nor, even in an approximate sense, ionic. It is helpful to draw an acceptable Lewis structure of HF. We make a Lewis structure representation of this molecule by associating two electrons with hydrogen and eight with fluorine. This is done by forming an electron-pair bond between H and F to give H—$\ddot{\underset{..}{F}}$: . This shared-pair Lewis structure is better than an ionic structure, because the electronegativity of a hydrogen atom is sufficiently high that only partial transfer of the bonding pair of electrons to the fluorine atoms takes place. The electronic density in the bond is more concentrated in the region of the fluorine than in the region of the hydrogen atom. This inequality in electron sharing is illustrated in Figure 6–2.

6–9   Dipole Moments

Molecules having unequal distribution of electron density are *electric dipoles,* since there is a negative end (pole) and positive end (pole) associated with the molecule. The magnitude of the *polarization* of a molecule can be demonstrated in the following manner. When two small metal plates are located at some fixed distance from each other, and a voltage is applied across the plates, there is a tendency for charge to build up on the two plates, as shown in Figure 6–3(a). If a polar substance such as HF is placed between the two plates, the molecules will tend to orient themselves as shown in Figure 6–3(b). The capacity of the two plates to "store" charge is enhanced when the space separating the plates is filled with a polar substance. When H_2 is placed between the plates, the capacity of the plates is essentially the same as if there were a vacuum between the plates. Hydrogen is a nonpolar molecule; thus it does not enhance the capacity of the two plates to store charge.

The electrical polarity of the hydrogen fluoride molecule may be shown by two kinds of notation

In the first formula, the arrow and the + and − signs show the direction of the electron drift. The same drift is shown in a different way in the second formula. The Greek letter delta (δ) indicates a small, or fractional charge. In this case, $\delta+$ written by the hydrogen shows that a small positive charge has developed in the region of the hydrogen nucleus, because the electrons in the molecule have been attracted toward the fluorine.

We express the polarity of a molecule in terms of its dipole moment, μ, which is given by the product of the magnitude of the separated charge, qe, and the distance of separation, R:

$$\mu = (qe)R \tag{6-3}$$

We usually consider the separated charges as centered at the positions of the atomic nuclei. Thus, for a diatomic molecule, R is the bond length. For example, if in a diatomic molecule, one electronic charge of 4.8×10^{-10} esu were separated by a 1-Å (10^{-8} cm) bond length, its dipole moment would be

$$\mu = (4.8 \times 10^{-10} \text{ esu})(10^{-8} \text{ cm}) = 4.8 \times 10^{-18} \text{ esu cm}$$

For convenience, scientists report dipole moments in debye units (D) in honor of a Dutch-American chemist, Peter Debye, who made both experimental and theoretical contributions to a wide range of chemistry. The conversion from esu cm to D units is

$$1 \text{ D} = 10^{-18} \text{ esu cm}$$

The dipole moment just calculated may be expressed as 4.8 D.

The dipole moment of HF is 1.82 D. Since the length of the bond is 0.92 Å, the dipole moment would be 4.42 D, if a full electronic charge were transferred from hydrogen to fluorine. Since 1.82 is 41% of 4.42, the result suggests that the covalent bond in hydrogen fluoride has 41% ionic character.

The dipole moment often can give a clue to the geometry of the molecule, when the molecule is larger than diatomic. Previously, we gave a Lewis structure for water

Figure 6–3. Charge stored between two plates with (a) a vacuum between the plates and (b) polar molecules of HF *between the plates. Polar molecules tend to align in an electrical field to maximize the electrostatic attraction between the plates.*

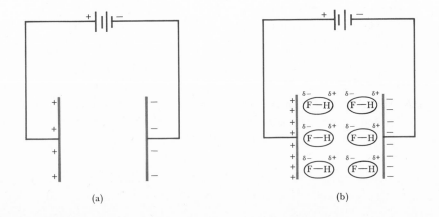

(a)   (b)

H—Ö—H

but simple valence rules are also satisfied if the molecule is drawn in the bent configuration

H—Ö:
|
H

Each H—O bond is polar, because of the unequal distribution of electrons, $H^{\delta+}$—$O^{\delta-}$. If the molecule were bent, there would be a resultant drift to electrons of one "end" of the molecule, and the entire molecule would act as a dipole. If the molecule were linear, the two H—O bond dipoles would cancel each other, and no net dipole would result

bent structure adds the two
dipole contributions and gives
a net dipole moment

linear structure cancels the two
dipole contributions and results
in no net dipole moment

Experiments show that H_2O has a dipole moment of 1.85 D. Thus, the molecule must have a bent structure.

We shall see in Chapter 8 that knowledge of molecular dipole moments is of considerable value in understanding the properties of liquids and solids.

6–10 Self-Bonding Elements

We have seen that atoms of the elements hydrogen and fluorine can "self-bond" to give H_2 and F_2 molecules, respectively. Logically, we would expect the same phenomenon to occur with atoms of other elements, and so it does. Complex hydrides of carbon, nitrogen, and oxygen were mentioned briefly in Chapters 1 and 3. All of these compounds contain bonds between equivalent atoms. Electronic structures of such compounds can usually be formulated straight-forwardly. Consider hydrogen peroxide, H_2O_2, as an example. We can imagine formation of the molecule by first bonding single hydrogen atoms to oxygen atoms

H—Ö·

The HO units would be unstable, because the valence shell of oxygen is incomplete. Addition of a second hydrogen atom would give water. However, if pairs of HO groups are joined together, the valence shells of both oxygen atoms can be completed by forming an oxygen–oxygen bond

H—Ö—Ö—H
hydrogen peroxide

137

BONDS IN MOLECULES

Complex hydrides of nitrogen and carbon can be formulated in the same way

$$4H\cdot\ +\ 2\cdot\ddot{N}\cdot\ \rightarrow\ 2\left[H{-}\overset{\displaystyle ..}{\underset{\displaystyle H}{N}}\cdot\right]\ \rightarrow\ H{-}\overset{\displaystyle ..}{\underset{\displaystyle H}{N}}{-}\overset{..}{\underset{\displaystyle H}{N}}{-}H$$

hydrazine

$$6H\cdot\ +\ 2\cdot\dot{C}\cdot\ \rightarrow\ 2\left[H{-}\overset{\displaystyle H}{\underset{\displaystyle H}{C}}\cdot\right]\qquad H{-}\overset{\displaystyle H}{\underset{\displaystyle H}{C}}{-}\overset{\displaystyle H}{\underset{\displaystyle H}{C}}{-}H$$

ethane

In principle, there could be an infinite number of carbon hydrides formed first by binding carbon atoms together in chains (or rings), then adding hydrogen atoms in sufficient numbers to give each carbon atom a stable electronic structure. As far as we know, the principle is correct, although testing it by preparation of an *infinite* number of compounds is not feasible. Several million compounds of carbon are known. Although the majority contain elements other than carbon and hydrogen, they can all be thought of as being derived from a vast family of carbon hydrides, known as hydrocarbons.

Example 6–8. Write an electronic formula for the hydrocarbon C_3H_8 and for the compounds C_2H_5F and $C_2H_4F_2$ (two different compounds have this formula).

Solution. One skeleton can be drawn that satisfies the valences of carbon (4) and hydrogen (1).

```
    H  H  H
 H  C  C  C  H
    H  H  H
```

There are 20 electrons to distribute, three carbons with four valence electrons each, and eight hydrogens with one electron each. Distributed according to the octet rule, the electrons give the following pattern:

$$H{-}\overset{\displaystyle H}{\underset{\displaystyle H}{C}}{-}\overset{\displaystyle H}{\underset{\displaystyle H}{C}}{-}\overset{\displaystyle H}{\underset{\displaystyle H}{C}}{-}H$$

Similarly, for C_2H_5F

$$H{-}\overset{\displaystyle H}{\underset{\displaystyle H}{C}}{-}\overset{\displaystyle H}{\underset{\displaystyle H}{C}}{-}\ddot{\underset{..}{F}}:$$

For $C_2H_4F_2$, a slightly different situation arises, because there are two possible arrangements of atoms

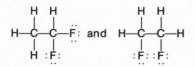

In principle, we would expect a similarly complex system of nitrogen compounds such as

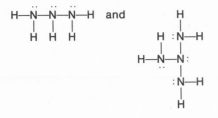

and a series of compounds containing three or more oxygen atoms such as

$$H—\overset{..}{\underset{..}{O}}—\overset{..}{\underset{..}{O}}—\overset{..}{\underset{..}{O}}—H \quad \text{and} \quad H—\overset{..}{\underset{..}{O}}—\overset{..}{\underset{..}{O}}—\overset{..}{\underset{..}{O}}—\overset{..}{\underset{..}{O}}—H$$

However, N–N bonds and O–O bonds are much *weaker* than C–C bonds, so the higher hydrides of nitrogen and oxygen are too unstable to exist, under ordinary conditions. The difference between carbon and its neighboring elements is not predicted by *simple* electronic theory, although slightly more advanced considerations, which invoke the repulsive interactions of unshared pairs on neighboring atoms, provide fairly easy rationalization of the facts.

A considerable number of complex boron hydrides are known. However, their structures are not easily related to those of hydrocarbons. The electron-deficient character of the compounds once again is shown, for the complex hydrides all contain a number of "three-center" bonds like those in diborane (Section 6–4). Figure 6–4 shows structural formulas of two interesting and representative boron hydrides.

6–11   More Practice in Writing Lewis Structures

Most compounds contain many atoms and even more electrons. Consequently, the step-by-step building up of their electronic structural formulas, as was done for hydrogen peroxide,

Figure 6–4.   *Models of (a)* B_4H_{10} *and (b)* $B_{10}H_{14}$. *The shaded area shows the planes formed by the boron atoms. The hydrogens that are not shaded form bridges between boron atoms.*

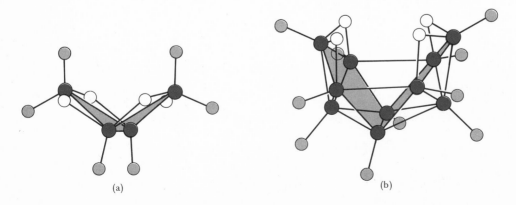

(a)   (b)

hydrazine, and ethane is tedious. Given the molecular formula of a compound, a chemist usually writes trial structural formulas, using lines for bonds, guided by the common valences of the atoms involved. This procedure is equivalent to building a molecular ball-and-stick model. For example, there is a compound that has the molecular formula NOH_3. To construct a formula, we use the common valence rules

Element	Valence	Part formula
N	3	$-\overset{\textstyle\vert}{N}-$
O	2	$-O-$
H	1	$H-$

Inspection shows that there is only one way to fit these elements together to produce a formula in which each atom has its usual number of bonds

$$H-\underset{\underset{\textstyle H}{\textstyle\vert}}{N}-O-H$$

Suppose that we try other combinations

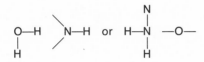

In each case, we have written formulas for smaller molecules and have an *unsatisfied fragment* left over.

After a trial structure has been written using common valence rules, a full Lewis electronic structure is written showing both the bonds and the unshared valence pairs. The electronic formula may be useful as a check on the fact that the electronic requirements for each atom have been met. The correct Lewis structure for hydroxylamine is

$$H-\underset{\underset{\textstyle H}{\textstyle\vert}}{\overset{\textstyle..}{N}}-\overset{\textstyle..}{\underset{\textstyle..}{O}}-H$$

hydroxylamine

Example 6–9. Write structural formulas for H_4CO (methanol) and H_6C_2O (two compounds have this formula, ethanol and dimethyl ether).

Solution. Only one arrangement of bonds will satisfy the valences of C, H, and O for the molecule with the formula H_4CO

$$H-\underset{\underset{\textstyle H}{\textstyle\vert}}{\overset{\textstyle\overset{\textstyle H}{\textstyle\vert}}{C}}-O-H$$

There are 14 valence electrons in all, so two unshared pairs should be added to the oxygen

H—C—O—H (with H above and below the C)

methyl alcohol (or methanol)

The formula H_6C_2O may assume two satisfactory atomic arrangements. Each has 20 valence electrons, giving these two structures

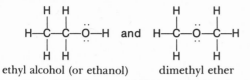

ethyl alcohol (or ethanol)   dimethyl ether

Electronic structures may also be useful in predicting chemical behavior of the substance. As an example of the relationship of electronic structure and chemical behavior, recall the molecules BF_3 and NH_3 (see Example 6–4). At first glance, we might not expect BF_3 and NH_3 to form the addition compound H_3NBF_3. However, when the electronic structures of BF_3 and NH_3 are written showing the unshared electron pair on nitrogen and the unsatisfied valence shell of boron, it is clear that the unshared pair on nitrogen can be shared between boron and nitrogen to form a B–N bond between BF_3 and NH_3.

Occasionally, electronic formulas show interesting structural possibilities not predicted by common valence rules. Consider H_3NO again. If we look closely at the electronic formulas of NH_3 and O, we can see a way of putting them together to form a new structure that seems to fulfill electronic rules of valence

$$\text{H—N: } + \text{ O: } \rightarrow \text{ H—N—O:}$$

If the unshared pair of electrons that originally belonged to the nitrogen of ammonia is shared between oxygen and nitrogen, each of the atoms will have eight valence electrons. The N–O bond would be much like the N–B bond in the addition compound formed from ammonia and boron trifluoride. Although the formula does not fit common valence rules, which state that oxygen should form *two* bonds, we cannot dismiss lightly the possible existence of a compound having such a structure. In this particular case, the compound is not known, and there are good reasons for thinking that, if it were ever produced, the atoms would rapidly rearrange themselves to form hydroxylamine. However, related compounds, such as F_3NO, are known

:F:
:F—N—O:
:F:

6–12   Formal Charges

Unusual formulas, such as those discussed in the preceding section, present a problem in book-keeping. Ammonia is a neutral molecule and so is the oxygen atom. If these two neutral entities are put together, the resulting molecule would also be neutral. However, since the shared pair of electrons was donated by nitrogen to oxygen, some reallocation of charges must occur within the molecule. We frequently assign electrical charges to atoms in molecular formulas to help keep track of the electrons. Such *formal charges* are determined by assigning a positive number equivalent to the number of valence electrons in the neutral atom (+5 for nitrogen) and adding −2 for every unshared electron pair around the atom and −1 for every shared electron pair. Let us do this for the formulas of hydroxylamine and the nonexistent $H_3N—O$

Consider the hydroxylamine molecule first. Each hydrogen has a charge of +1 to which we add −1 for one shared electron pair, giving a formal charge of zero. Nitrogen has a charge of +5, to which we add −2 for the unshared electron pair and −3 for the three shared electron pairs. Thus, nitrogen has a formal charge of zero. Similarly, oxygen has a formal charge of zero. In the same way, we can determine that in $H_3N—O$ nitrogen has a formal charge of +1 and oxygen, a formal charge of −1. (Test yourself to see if you obtain this answer.)

Example 6–10. What is the formal charge assignment in the addition compound of ammonia and boron trifluoride?

Solution. The Lewis structure for this compound is

$$
\begin{array}{ccc}
\text{H} & :\!\ddot{\text{F}}\!: & \\
| & | & \\
\text{H—N—B—F}: & & \\
| & | & \\
\text{H} & :\!\ddot{\text{F}}\!: &
\end{array}
$$

The formal charge on each H is calculated by adding −1, for the shared pair of electrons, to the number of valence electrons contributed by hydrogen, which is 1. Therefore, (1 − 1) is zero for the formal charge on each hydrogen. For nitrogen (Group V), we have +5 minus 4 (for the four pairs of shared electrons), or a formal charge of +1. Boron has the same number of shared electron pairs as nitrogen, but has only three valence electrons. Its formal charge is 3 − 4 = −1. Each fluorine has seven valence electrons and three unshared pairs, for which you must add −2 each, or a total of −6. Each fluorine also has one shared pair, requiring the addition of another −1. This gives (7 − 6 − 1) = 0. The formal charge distribution is then

$$
\begin{array}{ccc}
\text{H} & :\!\ddot{\text{F}}\!: & \\
| & | & \\
\text{H—N}^{\oplus}\text{—B}^{\ominus}\text{—F}: & & \\
| & | & \\
\text{H} & :\!\ddot{\text{F}}\!: &
\end{array}
$$

6–13 Molecules with Double and Triple Bonds

A large number of compounds have compositions that seem, at first glance, to defy representation by structural formulas, using simple valence rules. Among the complex hydrides of carbon, the problem is indicated by the fact that some substances contain too little hydrogen. For example, there are three hydrocarbons that contain two atoms of carbon per molecule. One, ethane,

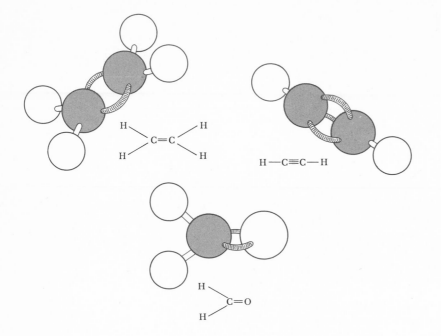

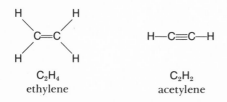

Figure 6–5. Ball-and-stick models showing multiple bonding in some common molecules.

has the formula C_2H_6. As was shown before, ethane can be assigned a reasonable structure simply by assuming that the carbon atoms are bonded to each other

$$H-\overset{\overset{\displaystyle H}{|}}{\underset{\underset{\displaystyle H}{|}}{C}}-\overset{\overset{\displaystyle H}{|}}{\underset{\underset{\displaystyle H}{|}}{C}}-H$$

ethane

The other two hydrocarbons have the formulas C_2H_4 and C_2H_2. We have been able to find only one device for formulating these compounds so that common valence rules appear to be preserved. The system is maintained if we say that the carbon atoms are bound by a *double bond* in C_2H_4 and a *triple* bond in C_2H_2

$$\overset{H}{\underset{H}{\diagdown}}C=C\overset{\diagup H}{\underset{\diagdown H}{}} \qquad\qquad H-C\equiv C-H$$

C_2H_4 C_2H_2

ethylene   acetylene

Counting electrons indicates that both formulas assign eight valence electrons to the carbon valence shells.

There are many compounds that require the postulate of double or triple bonds. The following are important examples of the occurrence of multiple bonds:

$\begin{array}{c}\diagdown\diagup\\ \text{C}=\text{C}\\ \diagup\diagdown\end{array}$ carbon–carbon double bond

—C≡C—   carbon–carbon triple bond

—N̈=N̈—   nitrogen–nitrogen double bond

:Ö=Ö: oxygen–oxygen double bond

:N≡N: nitrogen–nitrogen triple bond

$\begin{array}{c}\diagdown\\ \text{C}=\ddot{\text{N}}—\\ \diagup\end{array}$ carbon–nitrogen double bond

—C≡N: carbon–nitrogen triple bond

$\begin{array}{c}\diagdown\\ \text{C}=\ddot{\text{O}}:\\ \diagup\end{array}$ carbon–oxygen double bond

These formulas include that of the nitrogen molecule, N_2. The fact that elementary nitrogen is quite unreactive is well known. Nitrogen makes up about 80% of the earth's atmosphere, and it is regarded generally as being almost entirely unreactive. The inert character of nitrogen results because atoms of the element are very strongly bound together to form diatomic molecules.

Figure 6–5 shows a way in which ball-and-stick models can be modified to build molecules with multiple bonds. Sticks are replaced by springs so that two or three connections between a pair of atoms can be made.

6–14 Bonding to Heavier Elements

The octet rule has been extremely valuable as a guide in writing electronic formulas. For second-row nonmetallic elements (B, C, N, O, F), exceptions to the rule are very rare. It is easy to rationalize why this is so. The second-row elements have stable 2s and 2p orbitals, and the "magic number" of eight corresponds to the closed valence ($n = 2$) configuration $2s^2 2p^6$. Adding a ninth, tenth, or larger number of electrons is unacceptable, because the next orbital available to a second-row element is the highly energetic 3s.

Beyond the second row in the periodic table, however, the octet rule is not obeyed with such satisfying regularity. It remains a useful rule, however, because heavier elements form compounds exhibiting the principal valences of the group. Examples are PH_3, PF_3, H_2S, and SF_2

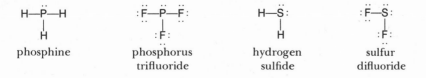

The heavier elements do more than obey the octet rule. Some of them show a surprising ability to bind more atoms (or associate with more electron pairs) than would be predicted from

the noble gas configuration rule. For example, phosphorus and sulfur form the compounds PF_5 and SF_6, respectively. Lewis structures for these compounds use all the valence electrons of the heavy element in bonding

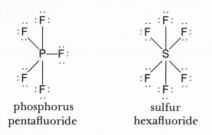

phosphorus
pentafluoride

sulfur
hexafluoride

That phosphorus shares ten electrons and sulfur shares twelve obviously violates the octet rule. The theory of atomic structure helps us see why the violation has occurred. The noble gas in the third row with phosphorus and sulfur is argon. The argon electronic structure fills the $3s$ and $3p$ orbitals, but leaves the five $3d$ orbitals vacant. If some of these $3d$ orbitals can be used in sharing electron pairs, extra bonds are possible. The atomic theory thus provides an explanation of the enhanced bonding versatility of elements in the third row and beyond.

Perhaps the most important consequence of the use of d orbitals is the existence of a very important series of so-called oxyacids. The most well-known examples are phosphoric acid (H_3PO_4), sulfuric acid (H_2SO_4), and perchloric acid ($HClO_4$). It is possible to write a Lewis structure for sulfuric acid that obeys the octet rule

Examination of this structure, however, reveals that a formal charge of +2 is placed on the sulfur atom. Development of a large positive formal charge on an electronegative nonmetal atom is not very attractive. The situation can be corrected by allowing the sulfur to share 12 electrons by making two double bonds

sulfuric acid

Similar Lewis formulas can be written for other oxyacids

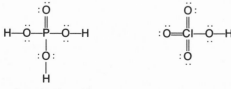

phosphoric acid   perchloric acid

6–15 Resonance

There are molecules for which more than one satisfactory Lewis formula can be drawn. For example, the nitrite ion, NO_2^-, can be formulated as either

In either case, the octet rule is satisfied. If either of these structures were the "correct" one, the molecule would have two distinguishable nitrogen–oxygen bonds, one single and one double. Double bonds are shorter than single bonds, but structural studies of NO_2^- show that the two bonds are indistinguishable.

Consideration of NO_2^- and many other molecules and ions shows that our simple scheme for counting electrons and assigning them to the valence shells of atoms as bonds or unshared pairs is not entirely satisfactory. Fortunately, the simple model is fairly easily altered to fit many of the awkward cases. The problem with NO_2^- is that the ion is actually more symmetrical than either one of the Lewis electronic structures that we wrote. However, if we took photographs of the two formulas shown previously and superimposed the pictures, we would obtain a new formula, having the same symmetry as the molecule. The photographic double exposure method is the same as writing a formula such as

This formula would imply, "NO_2^- is a symmetrical ion, having partial double bond character in each of the N–O bonds." For some purposes, the formula is adequately informative. However, keeping track of the electrons in such a formula requires the addition of some rather special notation. What we actually do most of the time in such situations is to write two or more Lewis formulas and connect them with a symbol that means, "Superimpose these formulas to get a reasonable representation of the molecule." Applied to NO_2^-, the formulas are

The double-headed arrow is a symbol reserved for this purpose. It should not be confused with the symbol consisting of two arrows pointing in opposite directions, $\leftrightarrows$, which indicates that a reversible chemical reaction occurs. The double-headed arrow conveys no implication of dynamic action.

The method of combining two or more structural formulas to represent a single chemical species is called the *resonance method*. The method is used not only for construction of electronic formulas, but also as the basis of one method for doing approximate quantum mechanical analysis of molecular structure.

When we consider the compound benzene, C_6H_6, which has its six carbon atoms arranged in a ring, we can draw two formulas that are equally satisfactory

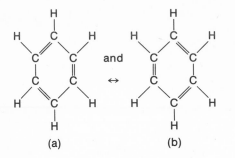

(a)   (b)

Both formulas show the ring to be composed of alternating single and double bonds. However, structural studies reveal that all of the carbon–carbon bond distances are equal. The full symmetry of the molecule is indicated by connecting the two formulas with a double-headed arrow.

Resonance notation is required in many cases other than those in which it is demanded by symmetry. For example, compare two well-known anions, nitrate (NO_3^-) and the nitroamide ion ($O_2NNH^{\ominus}$). Nitrate has threefold symmetry, so we can write a set of three equivalent resonance formulas

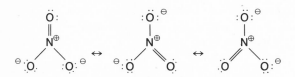

For the nitroamide ion we can write two equivalent formulas, plus a third, which is not equivalent to the other two

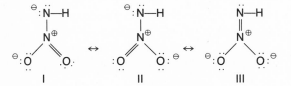

I   II   III

Common sense tells us that all three formulas should contribute to our description of the ion. Since the formulas are not equivalent, the resonance symbol no longer means, "Mix these formulas equally in your thinking." It merely means, "Mix them." When we become semiquantitative in our description, we will end up saying that Formula III "contributes" more to the structure of the nitroamide ion than either of the equivalent Formulas I and II. However, no quantitative implications are intended in the double-headed arrow.

We have used the term "resonance formula" in our discussion. Most authors use "resonance structure." We prefer the former term, because the word "structure" carries with it an implication of physical reality. Regardless of what terms are used, remember that the model we have in mind is a single molecular structure with electrons spread around in it, not two structures that interconvert by allowing electrons to jump between alternate positions. In the next chapter, we will have more to say about the spreading around or "delocalization" of electrons in molecules and ions.

Finally, we return to the molecule BF_3. In Section 6–5 we suggested that some interaction of the unshared pairs on fluorine atoms with boron might possibly account for the stability of the molecule and the nonexistence of a dimeric B_2F_6 molecule. The resonance method allows us to rationalize this and still be consistent with the fact that all three B–F bonds are equivalent. Resonance Formulas II, III, and IV must mix equally. When II, III, and IV are mixed with I, we can say that the boron has a partial share in eight electrons

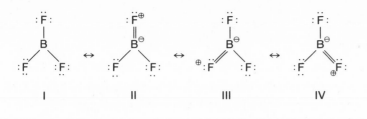

6–16 Summary

1 Atoms usually bond to each other by sharing one, two, or three pairs of electrons.
2 Atoms in molecules tend to surround themselves with enough electrons to achieve a noble gas configuration.
3 The ability of an atom of an element to attract electrons to itself is called electronegativity.
4 Elements that have very different electronegativities commonly form ionic bonds.
5 There is a separation of charge in a diatomic molecule composed of unlike atoms. The extent of this electrical charge separation is measured by determining dipole moments, and is described in terms of ionic character in covalent bonds.
6 Some molecules and ions are not well represented by a single Lewis formula. The description can usually be improved by using two or more formulas connected by the resonance symbol, ↔.

6–17 Postscript: Models and Reality

One of the dangers of models to which we have already alluded is the temptation to believe that they are real. Our explanations of many things about the composition and shape of molecules by using ball-and-stick models do not mean that molecules are like ball-and-stick models. The simple octet rule used to predict molecular composition does not mean that molecules are like the convenient electron-dot representation. An important fact that both these models ignore is molecular motion.

We asserted in Section 2–4 that gas molecules are in motion, and that this motion increases as the absolute temperature increases. We can generalize this statement by saying that *all* matter is in motion in all possible ways at every temperature above absolute zero. That is, every atom is moving in ways determined by the temperature and by its environment. A helium atom in a dilute gas simply undergoes translation; that is, it moves in a straight line until it hits something and ricochets. The hotter it is, the faster it moves.

A diatomic molecule such as hydrogen has motions that are far more complex. In addition to its translational motion, it can tumble in space and it can vibrate

vibrating

tumbling

More complex molecules, such as sulfur dioxide (SO_2), have more complicated vibrations

asymmetric stretch   symmetric stretch   breathing

The more atoms in the molecule, the more different kinds of vibration and rotation are possible, and the more the molecule may appear to be practically shaking itself apart. Many molecules undergo decomposition or dissociation in the gas phase. Although this is a complicated process, a simple view is that the molecule accumulates so much energy in its vibrations that it flies apart.

Another thing to remember when thinking about molecules is that atoms are mostly empty space. One way to visualize a molecule is as a sea of electrons in which atomic nuclei are embedded. The next chapter gives a refined view of this picture. But the image of an electron sea conveys a feeling for the amorphous, blobby, always changing form of a molecule that is missed completely, if the molecular reality is replaced by balls and sticks and by dots and dashes in one's mind.

Of course, we must admit that the picture we have given of molecular vibrations is mechanical and stylized, and that the image of a molecule as a sea of electrons is just an image, nothing more. The reality of molecules persists quite independently of our attempts to describe them, and our perception of that reality is, at best, science, not reality itself.

New Terms

Bond energy: The energy required to separate two atoms held together by a chemical bond. It may also be thought of as the energy released when two atoms come together to form a chemical bond.

Bond length: The distance between two bonded nuclei in the stable configuration of a molecule.

Covalent bond: The bond that results when two atoms are bound together through the sharing of a pair of electrons.

Debye unit: The unit used to express dipole moments. One debye (D) is equal to 1×10^{-10} Å esu or 1×10^{-18} cm esu.

Double bond: The covalent bond that results when two pairs of electrons are shared between two atoms.

Electric dipole: A pair of electric charges of the same magnitude but opposite sign separated by a small distance.

Electron-deficient molecule: A molecule that does not contain enough electrons to form electron-pair bonds between its constituent atoms and still obey the octet rule.

Electronegativity: A measure of the relative ability of an atom to attract electrons to itself in a bond.

Formal charge: The charge assigned to an atom in a molecule, as determined by subtracting one for each shared pair of electrons and one for each unshared electron from the number of valence electrons of that atom.

Ionic bond: The type of bond formed when two oppositely charged particles (ions) are bound together through their mutual electrostatic attraction.

Lewis formula: A simple diagram showing the placement of shared and unshared valence electrons in a molecule.

Octet rule: For atoms other than hydrogen in molecular combination, the total number of shared-pair and unshared-pair electrons should be equal to eight, which represents the stable, closed valence shell of a noble gas element.

Questions and Problems

1   Write electronic formulas (Lewis structures) for the atoms K, C, N, O, F, and Ne.

2   Write Lewis structures for the atoms or ions Ca^{2+}, K^+, Ar, Cl^-, and S^{2-}.

3   Write Lewis structures for the compounds C_2H_4O, HNO_3, H_2SO_4, $NH_4^+Cl^-$, $Na^+ClO_4^-$, and C_4H_{10}.

4   What unusual bonding situation do you encounter in the molecule B_2H_6?

5   What is incorrect about the following Lewis structures?

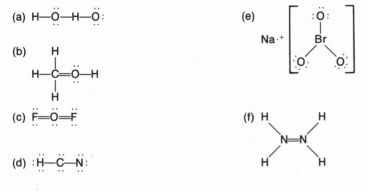

6   What problem do you encounter in writing Lewis structures for the molecules NO_2 and NO?

7   Describe as either ionic, polar, or nonpolar the bonds between the atoms of the compounds NaI, IF, MgO, Cs_3N, S_2, and NO.

8   Using arrows ($\leftrightarrow$) or $\delta+$, $\delta-$, indicate the direction of polarity in the molecules HF, ClBr, NaI, HCl, and CsI.

9   Arrange the molecules in Problem 8 in a probable order of increasing polarity.

10   Describe the trend of electronegativity values from left to right across the periodic table in a given row, and then from top to bottom in a given group. Which element is the least, and which the most, electronegative?

11   Suppose that the two atoms in the molecule A—B are separated by a distance of 1.27 Å. If B is more electronegative than A, there will be a separation of charge and a dipole moment will result. Calculate the ionic character in the A—B bond for a dipole moment of 2 D. Which atom has the partial negative charge?

12   (a) Write three acceptable Lewis structures for the compound HNNN, using only this arrangement of atoms. (b) Assign formal charges to each atom in each of the three structures. (c) Using the formal charges, would you say that any one of these electronic configurations is less stable than any other one? (Are two atoms of like charge adjacent to each other more, or less, stable than two atoms of opposite charge adjacent to each other?)

13 In diazomethane (H_2CNN), one nitrogen is attached directly to the carbon, and the second nitrogen is attached to the first. Draw Lewis structures for this molecule in which (a) the two N atoms are joined by a triple bond and (b) the middle nitrogen forms two double bonds to C and to N. When correctly drawn, each C or N atom should have eight electrons in its valence shell. What is the formal charge on each atom in structures (a) and (b)?

14 Draw Lewis structures for the following compounds:

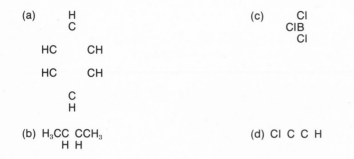

(a)   H
        C
    HC   CH

    HC   CH

        C
        H

(b) H_3CC CCH_3
      H H

(c)   Cl
      ClB
        Cl

(d) Cl C C H

15 Carbon dioxide, CO_2, has *no* dipole moment. What does this indicate about its molecular structure? Water, H_2O, *has* a dipole moment. What does this indicate about its structure?

16 Draw the resonance formulas for: NO_3^-, CO, NO_2, SO_4^{2-}, SO_3, and O_3.

17 The electronegativity of iodine is 2.66. Calculate the electronegativity of bromine. (The

ionization energies of Br and I are 273 kcal mole^{-1} and 241 kcal mole^{-1}, and the electron affinities of Br and I are 77.5 kcal mole^{-1} and 70.6 kcal mole^{-1}, respectively.)

18 Using their electronic configurations as the basis for your answer, predict the most likely charge for the following atoms to assume as ions: P, Li, Sr, S, I, Al, Cu, Sc, and Fr.

19 What geometry do you expect for each of the following molecules, given their dipole moments?

(a) CCl_4, 0.0 D

(b) H_2S, 0.92 D

(c) SO_2, 1.60 D

(d) BF_3, 0.0 D

(e) NH_3, 1.49 D

(f) PH_3, 0.55 D

20 Write electronic formulas for the three sulfur fluorides SF_2, SF_4, and SF_6. Which formulas violate the octet rule? Can you offer a rationalization for these violations? How many fluorides of oxygen would you expect?

21 Draw Lewis structures, and then suggest a probable order of carbon–carbon bond lengths, for the organic compounds acetylene (C_2H_2), ethylene (C_2H_4), and ethane (C_2H_6). Which molecule will have the largest carbon–carbon bond energy? Which will have the smallest? Why?

22 Try to develop a simple theory to explain the fact that self-bonding is strong in the case of C–C single bonds and relatively weak for N–N and O–O single bonds. Extend your theory to account for the small single bond energy of F_2.

23 Do you expect a covalent or an ionic model to be a more appropriate description of the electronic structure of LiH? Does the observed dipole moment of 5.88 D support your model?

MOLECULAR GEOMETRY
AND MOLECULAR ORBITALS

In Chapter 6, we showed, using concepts of atomic structure and electron counting, how a simple model can provide a rational basis for discussion of the composition of many molecules. We did not say much about molecular geometry, except to state that a water molecule is bent. Although electron counting gives no immediate answers, a rather simple set of rules, based upon the repulsion between electron pairs, can be added to give useful predictions of molecular geometry. Another method for analyzing molecular geometry, which depends upon the use of *hybrid atomic orbitals,* will also be discussed briefly.

The fact that a single Lewis electronic formula for a molecule such as benzene is less symmetrical than the molecule itself was resolved in Chapter 6, using the concept of resonance, in which two or more Lewis formulas are used together to represent a single molecule. Another useful model for treating this problem, known as *molecular orbital theory,* will be introduced in this chapter.

7–1 Electron Repulsion and Molecular Geometry

From various kinds of experimental evidence, we know that water molecules are bent, those of boron trifluoride are planar, and ammonia molecules have the shape of a triangular pyramid

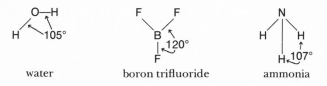

water   boron trifluoride   ammonia

These and many other facts about molecular geometry can be rationalized using a single concept. Electrons are negatively charged, therefore they tend to repel one another. If other factors are equal, the electrons surrounding any atomic nucleus will tend to spread as far apart as possible in three-dimensional space. The geometry of a molecule is determined largely by the repulsions of the valence electrons, both unshared pairs and those shared in covalent bonds. In the water molecule, there are four electron pairs, of which two pairs are involved in bonding. If the electron pairs were as far apart as possible in space, they would outline a regular tetrahedron, a geometric figure having four equivalent triangular faces. This arrangement is shown in Figure 7–1. We can imagine that the valence electrons in a neon atom distribute themselves in this way, although we have no way of verifying the model. In a water molecule, two of the electron pairs become involved in bonds and are attached to two positively charged hydrogen nuclei (protons). Since the protons will attract the electrons, the repulsion between the bonded pairs of electrons may decrease, allowing them to move closer together. If so, the angle between the O—H bonds would shrink to less than the tetrahedral angle of 109°28′. The bent structure of a water molecule and the known H—O—H bond angle of about 105° are in agreement with the model.

In an ammonia molecule, three of the four pairs of electrons are involved in bonding. Thus, we would expect the molecule to be nonplanar, and the H—N—H bond angle to be less than 109°. What we have just predicted agrees with experimental evidence. In methane (CH_4), all four pairs of electrons are involved in bonding the carbon atom to protons. Reasoning from our electron-repulsion model, we would expect all four to be equivalent. In fact, methane has tetrahedral geometry, with all H—C—H angles being 109°28′.

In boron trifluoride, the situation is different, because only six electrons in three bonds are assigned to the valence shell of boron. In this case, maximum separation of the electrons is accomplished by placing them in one plane, but as far apart as possible. The bond angles must be 120° since the sum of the three angles must equal 360°.

Exercise. Either by making a three-dimensional model or drawing pictures, try to find a better way of placing the electrons around the boron atom in boron trifluoride.

Exercise. What would be the most symmetrical arrangement of fielders on a baseball field? Are the players usually arranged exactly symmetrically? Why, or why not?

Carbon dioxide has the following electronic structure

$$:\!O\!\!=\!\!C\!\!=\!\!O\!:$$

The two pairs of electrons in either of the carbon—oxygen double bonds must be counted as a unit, since they are directed generally toward the same oxygen atom. This creates the situation

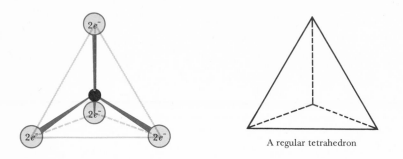

A regular tetrahedron

Figure 7–1. Arrangement of four pairs of electrons in tetrahedral geometry around a central atom.

shown in Figure 7–2. The linear arrangement places the two blobs of electronic charge as far apart as possible.

Example 7–1. Predict the geometry of formaldehyde, $H_2C{=}O$.

Solution. The electronic formula of the molecule is

There are three valence shell electron groups associated with the carbon atom. One group contains four electrons that provide the carbon–oxygen double bond. The other groups are the two pairs of electrons in the C—H bonds. The molecule should be planar, and we might expect the two-electron groups to repel each other less than they are repelled by the four-electron group. If so, the H—C—H bond angle would be less than 120°. However, the experimental value is $121.6° \pm 2.0°$.

Example 7–2. Predict the geometry of the carbonate ion, CO_3^{2-}.

Solution. An electronic formula for the ion is

Applying the same reasoning as we did in the preceding example, we predict that the ion is planar and that the bond angles are unequal. But *experiments reveal that all bond angles are 120°, and that all of the C—O bonds are the same length.* This is clearly a case that requires the resonance method of representation, if we are to use Lewis formulas

Figure 7–2. Linear arrangement of the valence electronic units in carbon dioxide.

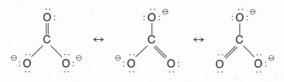

7–2   Orbitals in Molecules

In Chapter 5, we discussed orbitals that are used to describe the behavior of electrons in atoms. In Chapter 6, we showed how a rational picture for bonding in molecules can be drawn by using the idea that electrons in molecules bind nuclei together because the electrons may interact simultaneously with two nuclei. Thus, we presented the following picture to represent a hydrogen molecule

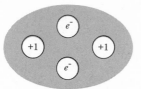

Although the model looks fine in the sense that the idea of mutual attraction is shown, it is unreasonable from another point of view. We know that electrons must be in very rapid motion. When talking about atoms, we made a major point of thinking of electrons as diffuse clouds of negative charge. Now we will think of electrons in molecules in the same way.

The electrons in a hydrogen molecule must be spread through the molecule forming some kind of charge cloud such as

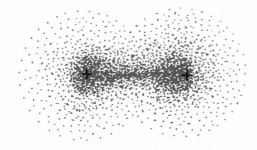

Staying with quantum mechanical notions, we say that there is a wave function that describes the bonding electrons. Moreover, we can see immediately why electron pairs are associated with bonds. If there are many molecular wave functions, describing different kinds of charge clouds in molecules, these wave functions will have different energies associated with them. If there is a wave function of lowest energy, then, applying the Pauli principle (Section 5–3) to the molecule, we would say that two electrons with opposite spins can be assigned to the low-energy wave function.

Although the concepts of molecular wave functions are rather easy to grasp, it is much more difficult to calculate the wave functions by straightforward mathematics. Remember that we said that exact solution of the mathematical problem becomes impossible when we have one nucleus and two or more electrons (Section 5–3). Obviously, the situation will be even worse when we deal with molecules containing not only several electrons, but also several nuclei.

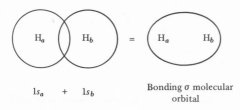

$1s_a$ + $1s_b$ Bonding σ molecular
                                            orbital

Figure 7–3.   Additive combination of 1s orbitals centered on two hydrogen atoms, H_a *and* H_b, *to form a bonding molecular orbital.*

However, mathematical approximation of the solution leads to convenient pictorial models. The usual approximation is based on the idea that *reasonable molecular wave functions can be constructed, using atomic wave functions as a starting point.* The new wave functions are called *molecular orbitals.* This method, which is generated as a means of getting around a mathematical hurdle, has led to all of the common models for the behavior of electrons in molecules. We will present the models and ignore the mathematics, except at the starting point. The starting point is the idea that useful wave functions for molecules can be formulated by adding and subtracting atomic wave functions. Thus, we would write wave functions for the hydrogen molecule

$$\Psi_{H_2} = C(\Phi_{H_a} + \Phi_{H_b}) \tag{7–1}$$

$$\Psi_{H_2}^* = C^*(\Phi_{H_a} - \Phi_{H_b}) \tag{7–2}$$

The labels H_a and H_b refer to the two hydrogen atoms in the molecule, and the Greek letters Ψ (psi) and Φ (phi) represent wave functions. The coefficients C and C^* are called normalization constants and are used for mathematical reasons that need not concern us. The combinations available by adding and subtracting wave functions are such that we generate a number of molecular orbitals equal to the number of atomic orbitals that we use in the problem. Other new functions could be formulated, but they will always turn out to be some combination of the first two. The restriction that two atomic orbitals give only two molecular wave functions is really mathematical, so we will simply assert it as a rule in this nonmathematical development.

The method of molecular orbitals produces much language and many concepts that are commonly used. We talk about orbitals from bonded atoms *combining* or *overlapping* with each other, and we talk about the energies of the orbitals increasing or decreasing in the process of combination. These concepts are useful, but also misleading. Combination, overlap, and orbital energetic changes are terms that refer to mathematical operations and results calculated from them. We tend to treat orbitals as though they were real, physical entities in our models, and this is, of course, incorrect. The physical entity for which we are trying to make a model is the molecule. We might use some different procedure, yet arrive at essentially the same model for a molecule in terms of things that really count, such as molecular geometry and molecular energy.

7–3   Orbitals of the Hydrogen Molecule

Using Equations 7–1 and 7–2, we can formulate the molecular orbitals of hydrogen. The first question is, "What atomic orbitals should we use for the Φ's?" We are guided by realizing that an

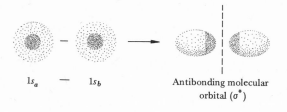

$1s_a$ — $1s_b$ Antibonding molecular
                                                orbital (σ^*)

Figure 7–4.   Subtractive combination of 1s orbitals centered on two hydrogen atoms to form an antibonding molecular orbital.

electron in the molecular orbital will be very close to one or the other of the hydrogen nuclei part of the time. At those times, the electron will "feel" very much like an electron in a hydrogen atom. This local behavior should be described well if hydrogen 1s orbitals are used to build the molecular orbital.

Figure 7–3 shows schematically what happens when we add hydrogen 1s orbitals centered on two neighboring hydrogen atoms. The resulting molecular orbital has some promising properties. It tends to be concentrated in the region between the nuclei, so an electron assigned to the orbital would be attracted by both nuclei. Furthermore, the shape of the orbital in three dimensions is that of a football (an ellipsoid), which is a reasonable shape to imagine for a hydrogen molecule, including its electrons. Since we expect an electron in the orbital to bind the two nuclei together, we call it a *bonding molecular orbital.* It is further called a σ orbital to indicate that it is symmetrical around the bond axis. This is in analogy with an atomic s orbital, which is symmetrical around the atomic nucleus.

The second method of combining the atomic 1s orbitals of hydrogen gives quite a different result. This is shown in Figure 7–4, using shading rather than simple outlines. Subtracting one function from the other gives the molecular orbital a low, or zero, value in the region of overlap between the two nuclei. Intuitively, we might expect that placing an electron in this orbital would make the molecule unstable, because repulsion between the positively charged nuclei would not be compensated by nucleus–electron attractions. This orbital also is symmetrical about the bond axis, so it is called the *antibonding σ orbital.* The antibonding character is often indicated by a star; thus, σ^* means "antibonding σ orbital."

7–4   Energies of the Hydrogen Molecular Orbitals

If we have a wave function for an electron in a molecule, then we can use it, together with known nuclear geometry and the charges on the positive nuclei, to calculate the energy of the electron. We cannot possibly do justice to the details of the calculation in this book, but the calculation can be made. The results can be described by referring to the calculated energies of electrons in isolated atoms. Figure 7–5 shows quantitatively the results for the hydrogen molecular orbitals. The energy calculated for the σ orbital is lower than that for isolated atomic orbitals, and the energy for the σ^* orbital is higher. The energy splitting between the two molecular orbitals is a measure of the interaction of the two atomic orbitals in the molecule.

Now it is easy to formulate the electronic structure of the hydrogen molecule. We simply assign two electrons with opposite spins to the bonding σ orbital, and breathe a sigh of relief to

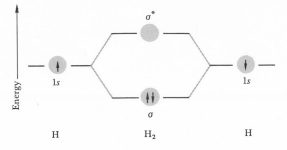

Figure 7–5. Calculated splitting of the energies of the σ and σ* molecular orbitals of a hydrogen molecule.

find that theory allows H_2 to be a stable molecule—a fact known from experimental data since the time of Avogadro. We can represent the electronic structure of H_2 using the shorthand notation $(\sigma)^2$, which is similar to that used for atoms in Chapter 5. Incidentally, we have ignored the fact that the simple calculations dealt with one electron and two nuclei. Clearly, the second electron will introduce a new energetic factor into the problem, because the electrons will repel each other. This factor is much more difficult to deal with mathematically. However, the calculation has been done, and the known stability of the hydrogen molecule remains allowed theoretically.

Example 7–3. It should be obvious from Figure 7–5 that the hydrogen molecule is stable. Explain in words why this is so. What kind of energy level diagram might we expect for an unstable diatomic molecule?

Solution. The diagram shows that the total energy of the electrons in the molecule is less than that in the separated atoms. Therefore, the molecule must be stable. Any energy level diagram and number of electrons that produces an equal or higher total energy for the molecule than for the separated atoms would indicate that the molecule is unstable. (See the next section.)

Treatment of the hydrogen molecule by the molecular orbital method would be a totally sterile exercise, if the results told us nothing except known facts about hydrogen. The great value of the method lies in the fact that we can treat bonds in larger molecules by analogy to bonds in H_2. Just as the theory of the hydrogen atom was extended to give a workable model of polyelectronic atoms, theory of the hydrogen molecule can be extended to produce useful models for molecules containing more electrons and more nuclei.

7–5 Why Not He₂?

In Chapter 6, we used the inertness of helium as a basis for postulating our simple theory of molecular structure. Now we can ask whether molecular orbital theory reveals anything about the inert character of helium. Even our simple theory gives a fairly satisfying answer. Figure 7–6 shows the description of He_2 molecules that we would give, if they existed. The atomic $1s$ orbitals would interact and form σ and σ* molecular orbitals. Following the Pauli principle, we would assign the four electrons in pairs to the two orbitals, giving the electronic structure $(\sigma)^2(\sigma^*)^2$. To

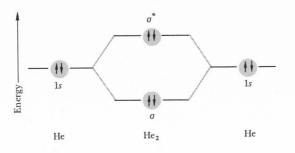

Figure 7–6.   Molecular orbital scheme for He_2.

a first approximation, the effects of the bonding pair and the antibonding pair would just cancel, so the molecule would not be stable.

7–6   Fluorine and Hydrogen Fluoride

We can extend the molecular orbital method to other molecules containing single bonds. Consider the fluorine molecule, F_2 (Section 6–2). The molecule is made from two fluorine atoms, each having the electronic structure

$$\frac{\uparrow\downarrow}{1s} \quad \frac{\uparrow\downarrow}{2s} \quad \frac{\uparrow\downarrow}{2p_x} \quad \frac{\uparrow\downarrow}{2p_y} \quad \frac{\downarrow}{2p_z}$$

As we did in Chapter 6, we will ignore electrons in the $1s$ orbitals and will consider only the valence electrons. Since the $2s$ and two of the $2p$ orbitals are filled in the atom, the simplest approach to making molecular orbitals for the fluorine molecule is to combine the half-filled $2p$ orbitals of the two fluorine atoms. The labeling of the p orbitals is arbitrary, but we usually assign the Z axis along the bond in a diatomic (or linear) molecule. Figure 7–7 shows how p_z orbitals on a pair of atoms can be added to produce a bonding molecular orbital. Since the orbital is symmetrical around the bond axis, it is called a σ orbital. However, the orbital is different from the σ orbitals formed by adding two s orbitals. Consequently, the name of the molecular orbital sometimes includes an indication of the atomic orbitals used to formulate it. The bonding orbital in the fluorine molecule will sometimes be called simply a σ orbital and sometimes a $2p_z\sigma$ orbital.

Figure 7–7.   Formation of a σ *molecular orbital by adding two atomic* $2p_z$ *orbitals.*

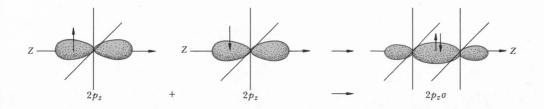

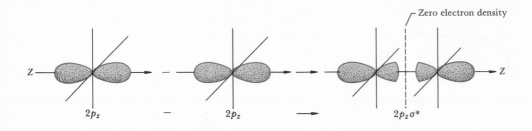

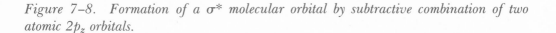

Figure 7–8. Formation of a σ* molecular orbital by subtractive combination of two atomic 2p$_z$ orbitals.

The choice of name will depend upon the amount of information that we want to put into the description.

Figure 7–8 shows how the subtractive combination of two 2p$_z$ orbitals can be used to formulate a σ* orbital. In ordinary fluorine molecules, we would assign a pair of bonding electrons to the σ orbital, and we would not use the antibonding orbital. Thus, we would predict correctly that F_2 has a single bond and is stable. We would assign electrons to the σ* orbital only in electronically excited states of fluorine. We can summarize the electronic configuration of the bonding electrons in the fluorine molecule in a compact manner as $(2p_z\sigma)^2$. Electrons in the 2s, 2p$_x$, and 2p$_y$ orbitals are considered to be unshared pairs in this model

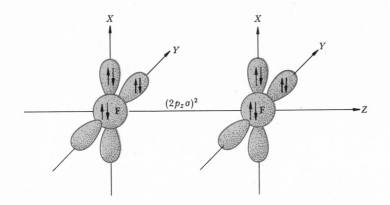

The description of hydrogen fluoride (HF) requires only a simple extension of the ideas developed for H_2 and F_2. Figure 7–9 shows how bonding and antibonding σ-type orbitals can be constructed from hydrogen 1s and fluorine 2p$_z$ orbitals. One new feature is added to the formulation of the molecular orbital that is not shown by the figure. Since the atomic orbitals used in the combination operation are not equivalent, they need not be used in an equivalent way in the two combinations. We can express this in a formal mathematical statement

$$\sigma_{\text{HF}} = C_1(1s)_{\text{H}} + C_2(2p_z)_{\text{F}} \tag{7–3}$$

$$\sigma_{\text{HF}}^* = C_1^*(1s)_{\text{H}} - C_2^*(2p_z)_{\text{F}} \tag{7–4}$$

Because we know that fluorine has a very high electronegativity, the $(2p_z)_F$ orbital will have a greater share of the bonding pair of electrons. This means that C_2 is larger than C_1. This is a way of introducing ionic character (Section 6–8) into our present model of the bond.

Example 7–4. Lithium is a metal with a boiling point of about 1336°C at atmospheric pressure. The vapor contains a large percentage of diatomic molecules. Formulate the electronic structure of Li_2.

Solution. If we ignore the filled 1s shells of the atoms, the problem is exactly analogous to that of H_2, except that the σ molecular orbitals are formed from 2s atomic orbitals. The single bond in Li_2 is then $(2s\sigma)^2$.

7–7 Single Bonds in Polyatomic Molecules

New problems are encountered when molecular orbital theory is extended to most polyatomic molecules. The problems are of two kinds: (1) treatment of single bonds and (2) treatment of double and triple bonds. We will deal with the problem of single bonds in this and the following section. Treatment of multiple bonds will come in Sections 7–12 to 7–14.

Let us consider methane (CH_4). The electronic configuration of a carbon atom is

$$\frac{\uparrow\downarrow}{1s} \quad \frac{\uparrow\downarrow}{2s} \quad \frac{\downarrow}{2p_x} \quad \frac{\downarrow}{2p_y} \quad \frac{}{2p_z}$$

There is very little in its atomic structure to suggest that carbon will form strong bonds to four hydrogen atoms; yet it does. The simple electron-pair repulsion model presented previously in this chapter suggests that methane would have a tetrahedral configuration (Section 7–1). Thus, if we are to keep the molecular orbital approach alive, we must find a way to make four bonding molecular orbitals for a molecule with tetrahedral symmetry.

7–8 Hybrid Atomic Orbitals

All four of the carbon valence orbitals must be used in a compound like methane. Let us try a first step toward building four bonds by moving an electron from the 2s orbital to the vacant

Figure 7–9. Combination of atomic 1s and $2p_z$ orbitals to form σ and σ^* molecular orbitals.

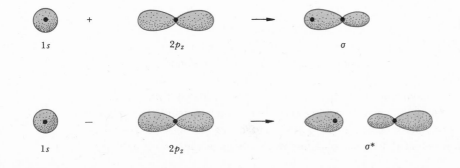

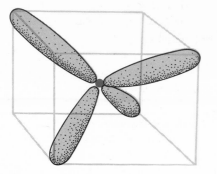

Figure 7–10.   Four sp³ orbitals generated from a single origin.

$2p_z$ orbital. This would, of course, be an excited state (higher energy) of the carbon atom itself

$$\underset{1s}{\underset{\uparrow\downarrow}{\rule{0pt}{0pt}}} \quad \underset{2s}{\underset{\downarrow}{\rule{0pt}{0pt}}} \quad \underset{2p_x}{\underset{\downarrow}{\rule{0pt}{0pt}}} \quad \underset{2p_y}{\underset{\downarrow}{\rule{0pt}{0pt}}} \quad \underset{2p_z}{\underset{\downarrow}{\rule{0pt}{0pt}}}$$

an excited carbon atom

If we try to form four molecular orbitals by combining each of the carbon atomic orbitals in turn with a 1s atomic orbital of one of the hydrogen atoms, we still get into trouble. The three $2p$ orbitals are perpendicular to each other (Section 5–3), so we could make three $2p\sigma$ orbitals with 90° angles between their axes. Because the $2s$ orbital has spherical symmetry, we would not know how to orient any $2s\sigma$ orbitals that we might formulate.

At this point, we make headway if we abandon the idea of using pure $2s$ and $2p$ orbitals to make bonds with the individual hydrogen 1s orbitals. Since we want to make four equivalent bonds, it seems logical to try to make combinations of the 2s and 2p orbitals that will produce four equivalent carbon orbitals. When this mixing is done in proper mathematical fashion, the resulting *hybrid* orbitals are just what we hoped for. The four orbitals are directed to the corners of a tetrahedron, thus they are situated ideally to make four bonds in a tetrahedrally structured molecule such as methane. The individual orbitals look like distorted p orbitals, and they are called sp^3 hybrid orbitals. Each sp^3 hybrid orbital has 25% s-orbital character and 75% p-orbital character

an sp^3 hybrid orbital

Figure 7–10 shows a set of four sp^3 hybrid orbitals originating from a common center. The small extensions of each orbital into the region behind the center are omitted to preserve the clarity of the drawing.

Now we can formulate the bonding orbitals of methane by combining the sp^3 hybrid orbitals with hydrogen 1s orbitals. The operation is shown in Figure 7–11, and is repeated four times, once for each bond. The molecular orbitals produced are called *localized* σ orbitals. They are localized because we have formulated them using atomic wave functions centered at only two

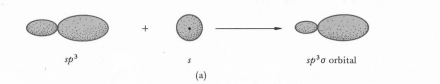

$$sp^3 \qquad\qquad s \qquad\qquad sp^3\sigma \text{ orbital}$$

(a)

Figure 7–11. (a) Combination of sp^3 and s orbitals from different centers to give a bonding molecular orbital. (b) Pictorial representation of the four $sp^3\sigma$ bonds in CH_4.

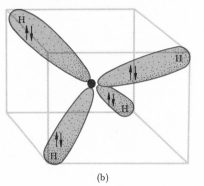

(b)

atoms to make each bond orbital. This procedure corresponds very closely to the scheme developed in Chapter 6 for writing Lewis electronic formulas, using lines to represent pairs of bonding electrons. The method of formulating localized bond orbitals is not the only procedure that could have been used. We could have spread out the molecular orbitals by using all eight of the atomic orbitals together (four from carbon and four from the hydrogens). The resulting molecular orbital picture is more difficult to render pictorially, and it offers no advantage over the localized orbital model for many purposes.

Example 7–5. What are the molecular orbital descriptions of water and of H_3O^+, the protonated water molecule?

Solution. The valence electronic configuration of oxygen is

$$\underset{2s}{\underline{\uparrow\downarrow}} \quad \underset{2p_x}{\underline{\uparrow\downarrow}} \quad \underset{2p_y}{\underline{\uparrow}} \quad \underset{2p_z}{\underline{\uparrow}}$$

The hydrogen atom configuration is $(1s)^1$. Each hydrogen $1s$ orbital can combine with an oxygen $2p$ orbital to form σ and σ^* orbitals, resulting in the bonding electronic configuration $(2s)^2(2p_x)^2(2p_y\sigma)^2(2p_z\sigma)^2$. This would make the H—O—H bond angle 90°. In view of the fact that the angle is actually 105°, and that hybridization works for molecules like CH_4, it is more reasonable to use four sp^3 hybrid orbitals to accommodate the two bonding and two unshared pairs for H_2O. Calling the four sp^3 orbitals sp_a^3, sp_b^3, sp_c^3, and sp_d^3, we would then have the configuration $(sp_a^3)^2(sp_b^3)^2(sp_c^3\sigma)^2(sp_d^3\sigma)^2$. This model predicts a H—O—H bond angle of 109°, which is much closer to the experimental value than is 90°.

Since H_3O^+ is isoelectronic with H_2O, the molecular orbital picture is formulated easily by making a third $sp^3\sigma$ bond from an sp^3 orbital and the third hydrogen $1s$ orbital

$$H_3O^+: \quad (sp_a^3)^2(sp_b^3\sigma)^2(sp_c^3\sigma)^2(sp_d^3\sigma)^2$$

Table 7–1. Some Common Hybrid Orbitals

Hybrids for σ bonds	Number of atomic orbitals used	Geometry	Example
sp	2	Linear	BeH_2, CO_2
sp^2	3	Planar triangle	BF_3, BCl_3
sp^3	4	Tetrahedral	CH_4, CF_4
dsp^2	4	Square planar	$PtCl_4^{2-}$, $Ni(CN)_4^{2-}$
dsp^3 (or sp^3d)	5	Trigonal bipyramidal	$MoCl_5$, PF_5, PCl_5
d^2sp^3 (or sp^3d^2)	6	Octahedral	$Fe(CN)_6^{4-}$, SF_6

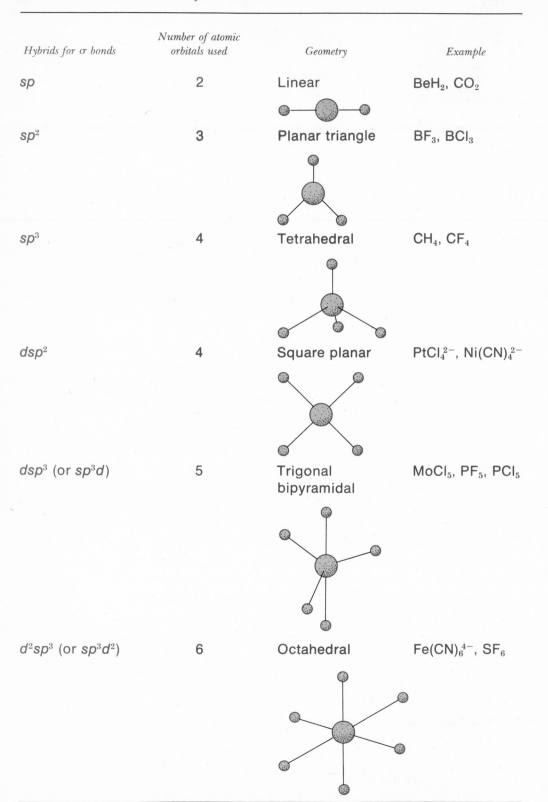

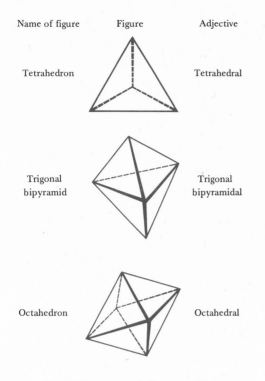

Name of figure	Figure	Adjective
Tetrahedron		Tetrahedral
Trigonal bipyramid		Trigonal bipyramidal
Octahedron		Octahedral

Figure 7–12.   Three-dimensional figures of importance in molecular geometry.

7–9   Other Hybrid Orbitals

A variety of hybrid orbitals can be made by mixing the kinds of atomic orbitals discussed in Chapter 6. Those formed from *s* and *p* wave functions are especially important for discussion of the σ bonds in compounds made from the light elements in the second row of the periodic table (lithium through fluorine). In the heavier elements, *d* orbitals (Section 6–14) become increasingly important. One manifestation of the availability of *d* orbitals for bonding is the appearance of compounds in which more than four groups are bound to a central atom. Table 7–1 lists frequently used hybrid orbitals, and the geometric figures defined by the symmetry axes of the orbitals. There are many known compounds having geometry corresponding to each of the figures.

   Except for the five-orbital case, dsp^3, the hybrid orbitals within a set are all equivalent. Figure 7–12 shows the three-dimensional figures corresponding to the geometric shapes referred to in Table 7–1.

7–10   Boron Trifluoride (BF₃) and Phosphorus Pentachloride (PCl₅)

Consider the problem of predicting the geometry and localized bond orbitals for BF_3. As outlined before (Section 7–1), the concept of electron-pair repulsion indicates that the three pairs of bonding electrons should be in a plane. The logical choice for three localized σ bonds is a set of equivalent sp^2 hybrids. Three such orbitals placed around a common center appear as follows:

There will be another valence orbital of the boron atom that is unused in the hybridized atom. This is a *p* orbital, with its axis perpendicular to the plane of the *sp²* orbitals

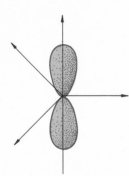

Figure 7–13 shows the way that we can combine the three *sp²* orbitals of boron with *p* orbitals of fluorine to make three localized σ orbitals. The six bonding electrons are assigned to these three orbitals.

   Two other features of the suggested structure of BF$_3$ are of interest. First, the unused *p* orbital of the boron atom does not yet have any assigned role in bonding. This results simply from the way we developed the structure. Remember that, in polyelectronic atoms, the energies of electrons in 2*p* orbitals are higher than the energies of 2*s* electrons. Consequently, we might expect that if any orbital is to be unoccupied, it should be a 2*p* orbital. The second point of interest is that there will be occupied fluorine 2*p* orbitals perpendicular to the one used in forming the σ orbital and *parallel to the vacant 2p orbital of boron.* Looking at boron and any one of the attached fluorine atoms, we see the following situation

   B   F

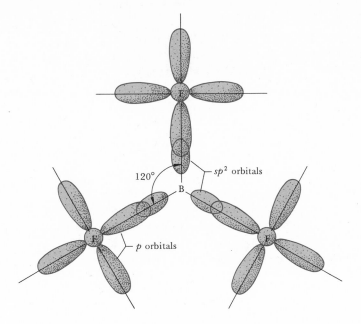

Figure 7–13. Combination of three boron sp^2 orbitals with three p orbitals of fluorine to form three σ bonds. (Note the presence of other p orbitals of the fluorine atoms.)

We can see the possibility that these orbitals will overlap and form some kind of molecular orbital. The assignment of electrons to such an orbital would transfer some electronic charge from fluorine to boron and keep the neglected boron $2p$ orbital from being unused entirely. This kind of interaction has already been suggested in Chapter 6. Without going into detail, we can describe this kind of interaction by writing the following set of orbital resonance formulas for BF_3

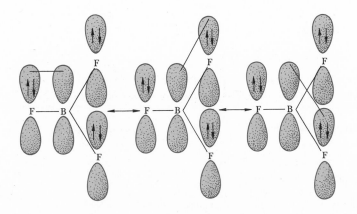

Bonds involving parallel overlap of p orbitals, as shown for BF_3, are called π bonds and will be treated in more detail later.

Phosphorus pentachloride, PCl_5, presents a case in which we obviously need to use more than s and p orbitals, if we are to formulate enough bond orbitals to be able to assign one to each of the five bonds in the molecule. The electronic configuration of a phosphorus atom, ignoring all except the five valence electrons is

The fact that the unused d orbitals are considerably higher in energy[1] than the $3s$ and $3p$ orbitals might lead us to expect that they will be unused in bonding, and that compounds of phosphorus would have the general formula PX_3. There are many such compounds, but there are also many compounds having the general formula PX_5. Faced with the facts, we must conclude that the phosphorus atom finds another "handle" to hold an attached group by using d orbitals. Just as was the case with carbon, we can imagine the phosphorus atom "preparing" itself for bonding by promoting an electron to a d orbital

Once we have taken this step in reformulation, we can write a completely equivalent expression using hybrid orbitals

Note that forming the sp^3d hybrid orbitals is really something that went on in our minds, rather than a physical change in a phosphorus atom. In the final step, we formulate five σ bonding orbitals by overlapping the sp^3d orbitals with a $3p$ orbital on each of the five chlorine atoms.

Looking at Table 7–1, we see that sp^3d hybridization fits a trigonal bipyramidal structure, which is in agreement with the experimentally observed structure of PCl_5.

7–11 Fluorides of the Heavier Elements

Usually, we do not analyze a structural problem in as much detail as we did in the case of PCl_5. Once we have decided on the number of electron pairs to be accommodated by the central atom as σ bonds and unshared pairs, the appropriate set of hybrid orbitals and the molecular geometry can be found in Table 7–1.

Consider the case of SF_6. A sulfur atom has six valence electrons ($3s^2 3p^4$), and each fluorine atom contributes one electron in making the σ-bond framework. There are 12 electrons in all, or six pairs, to go into hybrid bond orbitals centered on sulfur. For six pairs, the choice is clearly a set of six sp^3d^2 hybrids, giving an octahedral structure[2]

---

[1] The lines that we have just used to represent the $3s$, $3p$, and $3d$ orbitals do *not* indicate the relative energies of these orbitals.

[2] Shaded lines are used to indicate that the four F atoms are in a plane; the F atoms top and bottom are in a line perpendicular to the plane.

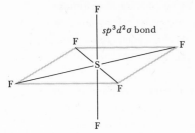

We would have arrived at the same structure using the electron-pair repulsion model. The most spread out arrangement of six electron pairs is octahedral

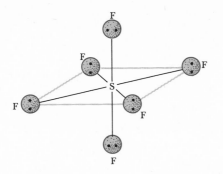

It is slightly more difficult to work out models for BrF_5 and IF_4^-. There are 12 electrons involved to make five σ bonds and one unshared pair in BrF_5. Seven of these electrons are the valence electrons of Br, and one from each F is contributed for σ bonding. For IF_4^-, there are four σ bonds and two unshared pairs. Therefore, the basic problem in both cases is the same as for SF_6; six electron pairs are to be placed in a set of six sp^3d^2 hybrid orbitals. Completion of this step for BrF_5 leads to the prediction that the molecule will have a different structure than PCl_5, because the unshared pair effectively occupies a position around the Br and "pushes" the bonded groups closer together. Of course, it is only our model that places the unshared pair in this hybrid orbital. Experimentally, we cannot locate the unshared pair, but we can determine the positions of the six atoms in the molecule. This placement

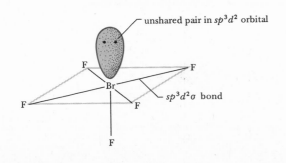

is called a square pyramidal structure. The fact that this model works so well is another indication of the importance of the repulsion of valence electron pairs, both σ bonding and unshared, in determining molecular geometry. The model as we have applied it in this chapter has been given a special set of code letters, VSEPR theory, which stands for valence shell electron pair repulsion theory.

Let us apply VSEPR theory to IF_4^-. There are six electron pairs to distribute octahedrally in sp^3d^2 hybrid orbitals. Four of these electron pairs make the σ bonds, and two are unshared pairs. Now we must make a choice: Do we place the two unshared pairs adjacent (cis) or across (trans) from each other in the octahedral model?

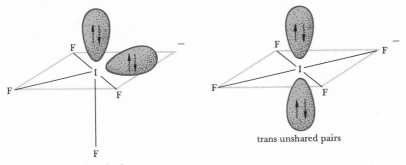

cis unshared pairs

trans unshared pairs

Following the suggestion in Section 7–1 that unshared pairs repel each other more strongly than bonded pairs, we choose the trans arrangement which places them across from each other. The model is consistent with the structure of IF_4^-, which is known to be planar.

7–12 Molecules Containing Double and Triple Bonds

In Figure 6–5, we showed one way of dealing with molecules containing multiple bonds by using springs, rather than sticks, in physical models. We also must use some special device to fit these structures into a molecular orbital scheme. We shall begin with nitrogen (N_2), a molecule with a triple bond. The configuration of the five valence electrons in a nitrogen atom is

$$\frac{\Updownarrow}{2s} \quad \frac{\downarrow}{2p_x} \quad \frac{\downarrow}{2p_y} \quad \frac{\downarrow}{2p_z}$$

Figure 7–14. Combination of two $2p_z$ orbitals to form a $2p\sigma$ molecular orbital. The axes of the $2p_x$ and $2p_y$ orbitals are also shown.

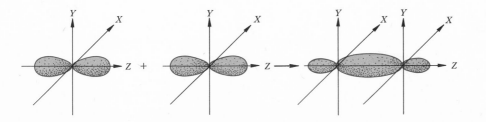

We can use *sp* hybrid orbitals to form σ bonding orbitals, if we like. However, there seems to be no point in starting that way, since the three half-filled *p* orbitals are enough to produce the number of bonding orbitals to make three bonds. The formation of a σ orbital by end-on overlap of the $2p_z$ orbitals of two nitrogen atoms is shown in Figure 7–14. Note that the respective axes of the unused $2p_x$ and $2p_y$ orbitals can be made parallel; thus, overlapping can occur in a way different than in the $2p_z$ overlap. Consideration of the combination of parallel $2p$ orbitals requires consideration of one new feature, raised by the fact that a $2p$ wave function has opposite signs across the axis

The $+$ and $-$ signs have nothing to do with electrical charges. Neither do they have any significance in terms of electron distribution, since electron distribution depends on the *square* of the value of the wave function. The $+$ and $-$ signs indicate the symmetry properties of the orbitals, and they are of use to us in working out the results of additive and subtractive combination. In this book, we will not use all the symmetry properties of molecular orbitals, but it is well to know that they exist. The following schematic drawings indicate what happens when parallel $2p$ orbitals are combined

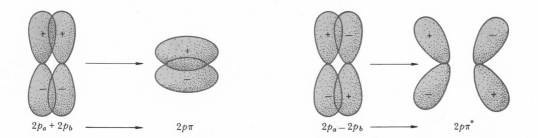

The resulting molecular orbitals are not symmetrical around the internuclear axis. They are called π orbitals, because they retain the same kind of change in sign as in atomic *p* orbitals. Intuition tells us that the π orbital should be bonding, since it is concentrated in the volume between the nuclei, and that the π^* orbital should be antibonding, because it is squeezed out of the internuclear region.

In the nitrogen molecule, we have two sets of parallel $2p$ orbitals that can be combined to give equivalent $2p\pi_x$ and $2p\pi_y$ bonding orbitals, as shown in Figure 7–15. Each $2p\pi$ orbital will have the shape shown in the figure. The overlap is shown by shaded lines, rather than by drawing the $2p\pi$ orbitals. This is done to avoid confusion in the figure.

Now we are ready to formulate the N_2 molecule. The six bonding electrons are assigned in pairs to the $2p\sigma$, $2p\pi_x$, and $2p\pi_y$ orbitals to give the bonding configuration $(2p\sigma)^2(2p\pi_x)^2(2p\pi_y)^2$.

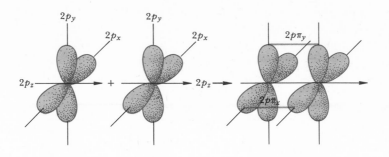

Figure 7–15.   Parallel combination of $2p_x$ and $2p_y$ orbitals to form two $2p\pi$ orbitals.

One more significant problem remains. The overlapping of $2p$ orbitals is different in the σ and π arrangements, and we have not yet given any reasoning that would allow us to decide which lies lower in energy, the $2p\sigma$ electrons or the $2p\pi$ electrons. Many kinds of experiments and calculations both show that the order of the levels in nitrogen is as shown in Figure 7–16. Close inspection may suggest that the diagram is wrong, because you might expect the $2p\sigma$ and $2p\sigma^*$ orbitals to be split evenly around the atomic $2p$ orbitals. This is not the case, because of an important contribution from the $2s$ orbitals, which, in our simple model, we have ignored. To get a more realistic view of the splitting pattern, we also should include the atomic $2s$ orbitals. To do so is not difficult, but too tedious to be worthwhile at this time. The ordering of levels shown in Figure 7–16 is a common one for diatomic molecules formed from second-row elements, but it is not absolutely fixed. The splittings between both $2p\sigma$ and $2p\sigma^*$ levels, and $2p\pi$ and $2p\pi^*$ levels, vary as the internuclear distance changes.

7–13   The Oxygen Molecule, O_2

The treatment of the oxygen molecule is nearly the same as for the N_2 molecules; O_2 has two more valence electrons and one more positive charge in each of the nuclei. Thus, we would expect the molecular orbital energy levels to be similar to those shown in Figure 7–16 for nitrogen. This presumption leads to an easy explanation of one of the properties of molecular oxygen that puzzled chemists for a long time. The lowest lying vacant orbitals in nitrogen are the equivalent pair, $2p\pi_x^*$ and $2p\pi_y^*$. By analogy to the rules developed for the buildup of atoms (Chapter 5), we would expect to assign the two "extra" electrons of O_2 to these two orbitals with parallel electron spins. This would give oxygen molecules the following valence electronic configuration

$$\frac{\Updownarrow}{2s_a} \quad \frac{\Updownarrow}{2s_b} \quad \frac{\Updownarrow}{2p\pi_x} \quad \frac{\Updownarrow}{2p\pi_y} \quad \frac{\Updownarrow}{2p\sigma} \quad \frac{\downarrow}{2p\pi_x^*} \quad \frac{\downarrow}{2p\pi_y^*} \quad \frac{}{2p\sigma^*}$$

Since the electrons in the $2p\pi^*$ orbitals have parallel spins, their individual magnetic moments will add, rather than cancel. Consequently, the molecule will act like a tiny, permanent magnet. A material having a permanent magnetic moment will be drawn into a magnetic field, a property known as *paramagnetism*. Most substances are repelled from a magnetic field and are said to be *diamagnetic*. Figure 7–17 shows schematically a way of measuring magnetic properties of

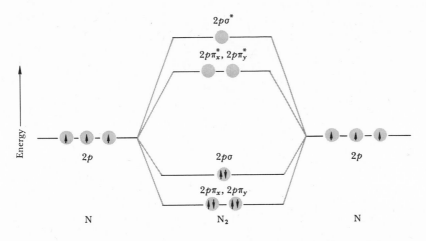

Figure 7–16.   Ordering of energies of molecular orbitals derived from atomic 2p orbitals in N_2.

materials. Very few molecules containing even numbers of electrons, such as oxygen, are paramagnetic. Thus, the paramagnetism of oxygen was an irritation to chemists for years, especially in view of the fact that a perfectly reasonable Lewis formula can be written for the molecule that shows all the electrons paired

$$: \underset{\cdot\cdot}{O} = \underset{\cdot\cdot}{O} :$$

Figure 7–17.   Measurement of diamagnetic or paramagnetic character can be done using a Gouy balance. The figure shows a highly idealized representation of the method.

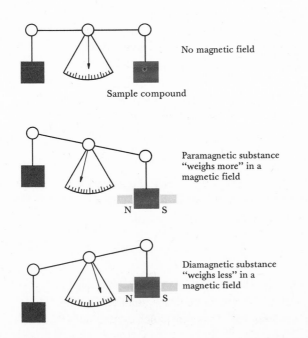

No magnetic field

Sample compound

Paramagnetic substance "weighs more" in a magnetic field

Diamagnetic substance "weighs less" in a magnetic field

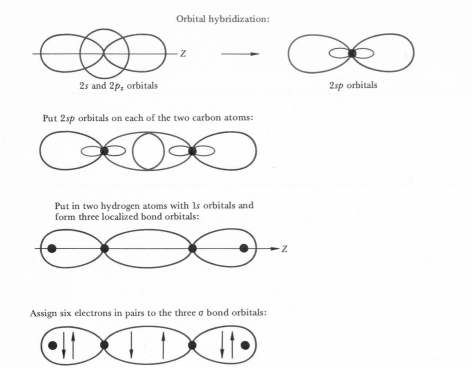

Orbital hybridization:

2s and 2p_z orbitals   2sp orbitals

Put 2sp orbitals on each of the two carbon atoms:

Put in two hydrogen atoms with 1s orbitals and
form three localized bond orbitals:

Assign six electrons in pairs to the three σ bond orbitals:

H —————— C ———————————— C —————— H

Figure 7–18.   *Development of the σ-bond framework for acetylene.*

Other properties of oxygen agree well with the molecular orbital formulation. For example, the energy required to dissociate N_2 into nitrogen atoms is 225 kcal mole^{-1}, whereas the dissociation energy of O_2 is only 118 kcal mole^{-1}. The lower dissociation energy for O_2 is explained nicely by assigning two electrons to antibonding orbitals in molecular oxygen. The two antibonding electrons effectively cancel two of the six bonding electrons, leaving a *net* of four bonding electrons. Thus, it should take less energy to break the double bond in the O_2 molecule than the triple bond in N_2.

Example 7–6. Is NO$^-$ paramagnetic?

Solution. The NO$^-$ ion is isoelectronic with O_2. Nitrogen and oxygen have the same atomic orbitals available for molecular orbital formation. Hence, the electronic configuration of NO$^-$ is the same as that for O_2, and NO$^-$ is paramagnetic.

7–14   Unsaturated Carbon Compounds

All compounds containing double and triple bonds are said to be *unsaturated,* because the atoms involved in multiple bonding could, at least in principle, become attached to other groups. Unsaturated compounds of carbon are very important. The molecular orbital method probably has reached its greatest utility in providing a satisfying electronic structural description of these compounds. We illustrate the method by discussing three important examples: acetylene (C_2H_2), ethylene (C_2H_4), and benzene (C_6H_6).

The usual procedure for developing molecular orbital representations of unsaturated compounds involves two steps: (1) Lay out a σ-bond framework, using localized bond orbitals; (2) Superimpose the π-bonding network with those atomic p orbitals not used in the σ bond orbitals.

Application of the procedure to acetylene, H—C≡C—H, is easy. The molecule is linear, so we use sp hybrid orbitals to describe the σ bonds to carbon, as shown in Figure 7–18. This framework takes care of six of the ten valence electrons in the molecule. Then the unused $2p_x$ and $2p_y$ orbitals on the two carbon atoms can be used, just as they were with nitrogen (Section 7–12), to form $2p\pi_x$ and $2p\pi_y$ orbitals. The other four electrons are assigned to these orbitals. Hence, using lines for the σ bonds, we can draw the electronic structure of acetylene as

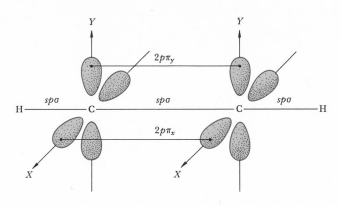

The only conventional structural formula that can be written for the planar ethylene molecule is

The information in Table 7–1 suggests that we use sp^2 hybrid orbitals of carbon to build the σ framework

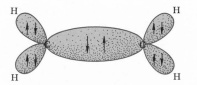

The hybrid bond orbitals take care of ten of the 12 valence electrons. The unused $2p$ orbitals of the two carbon atoms can be combined to produce a $2p\pi$ orbital, to which the remaining electron pair can be assigned. The parallel overlap required for a $2p\pi$ orbital is obtained only if the entire molecule is in a planar configuration, as experiment shows it is

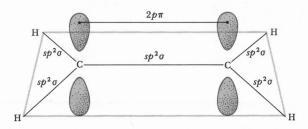

The molecular orbital formulation of ethylene provides an interesting account of some of the known properties of the substance. The order of the energy levels of the various orbitals is given in Figure 7–19. The separation between the $2p\pi$ and $2p\pi^*$ levels is significantly smaller than the spacing between σ and σ^* levels. Furthermore, the σ to σ^* splittings in ethylene are not vastly different from those in saturated hydrocarbons. This helps to explain important differences in the behavior of saturated and unsaturated hydrocarbons (compounds of carbon and hydrogen) in their absorption of light.

When light is absorbed by a molecule, electrons are excited from occupied orbitals to vacant orbitals (Chapter 4). That the $2p\pi \rightarrow 2p\pi^*$ transition in ethylene requires less energy than any transition in saturated hydrocarbons accounts for the fact that ethylene absorbs light of longer wavelengths (less energy) than do saturated hydrocarbons. Ethylene absorbs light of 1850 Å, whereas ethane (C_2H_6) begins to absorb strongly at about 1600 Å. Absorption by ethylene comes at wavelengths too short to be of interest for most purposes. However, if the π-electron system of unsaturated hydrocarbons is made more extensive by stringing unsaturated units together, the energy separation between π and π^* levels becomes smaller, and energy absorption moves

Figure 7–19.   First approximation of molecular orbital energies in ethylene.

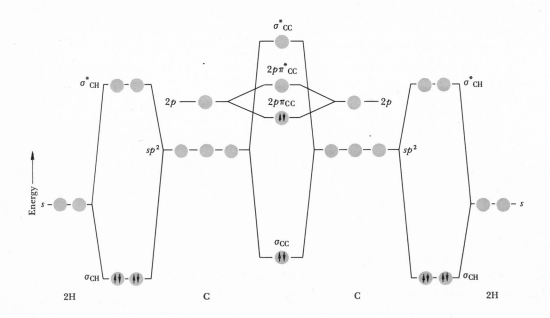

to longer wavelengths. This effect is shown in *conjugated* polyenes (poly = many; ene = C=C bond), compounds in which conventional structural formulas show alternating single and double bonds

$$\ldots \; -C{=}C{-}C{=}C{-}C{=}C{-} \; \ldots$$
conjugated polyene

Polyenes having ten or more conjugated double bonds absorb visible light, hence they are colored. The pigments responsible for light perception in the eye contain long, conjugated polyene chains as do some vegetable pigments, such as carotene, the colored substance in carrots.

Another difference between saturated and unsaturated carbon compounds is the ease with which the molecules can rotate about the carbon–carbon bonds. Rotation about single bonds is so easy that, at room temperatures, the two methyl groups in ethane whirl past each other like windmills in a gale

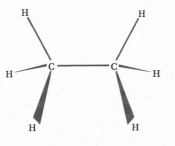

H_3C = methyl group

In contrast, rotation about the double bond in ethylene does not occur at an appreciable rate at temperatures below about 400°C. A rationalization of this restriction is found in the fact that rotation of the CH_2 groups with respect to each other would twist the atomic $2p$ orbitals out of alignment, essentially breaking the $2p\pi$ bond. We must point out that other ways can be found to account for the restriction of rotation, so the phenomenon does not prove the physical reality of the $2p\pi$ bond; it simply illustrates that the model is useful for a discussion of these particular properties.

Benzene shows interesting new features. The molecule is entirely planar, and the σ-bond framework consists of six σ_{CH} and six σ_{CC} bonds

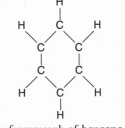

σ framework of benzene

In this framework, three sp^2 orbitals of each carbon atom are used, leaving one $2p$ orbital from each atom for formation of the π orbitals. This arrangement is shown in Figure 7–20. Note that

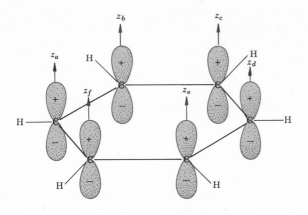

Figure 7–20.   The 2p orbitals of benzene left unused after construction of the σ frame-work.

each atomic $2p$ orbital has two nearest neighbors that would be equivalent if all of the C—C bonds are the same length, which is known to be the case. Clearly, there should be no preference for the formation of localized π orbitals. Instead, we formulate π orbitals for benzene by combining all six of the atomic $2p$ orbitals in a grand mix. This procedure, which is straightforward but requires some mathematics we won't go into, generates six molecular orbitals that have the energy splitting pattern shown in Figure 7–21. Electrons are assigned to the three orbitals of lowest energy. These electrons are not localized between pairs of atoms, but are spread over the entire molecule. The delocalization of the lowest energy orbital is shown in Figure 7–22. Obviously, an electron assigned to such an orbital will contribute equally to binding between all pairs of adjacent carbon atoms.

Benzene is an unusually stable molecule, and the molecular orbital method gives a ready explanation of this fact. Some of the electrons in the molecule are not confined to small volumes, but occupy the whole molecule. One of the principles of quantum mechanics is that the energy

Figure 7–21.   Benzene π orbital energy levels.

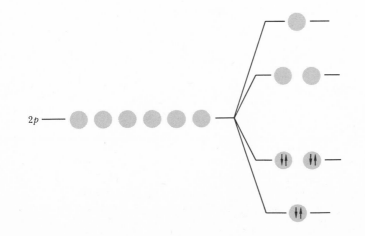

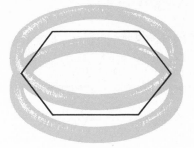

Figure 7–22.   *Lowest energy π molecular orbital of benzene.*

of a particle increases if it is confined in smaller and smaller boxes. Electrons apparently are "more comfortable" in the relatively large box provided by the entire benzene framework.

Similar situations exist in many, but not all, cyclic hydrocarbons, in which classical bond structures indicate alternating single and double bonds. The following are two more examples

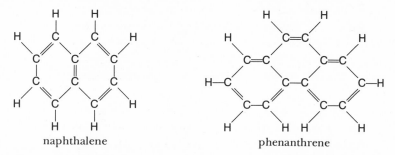

naphthalene   phenanthrene

This class of compounds is called "aromatic," a name given originally because the compounds in this class have distinctive, pungent odors. Now the term aromatic has come to be used for compounds having special stability associated with π-electron delocalization.

The resonance method of notation shows essentially the same thing about benzene as does the molecular orbital method; namely, that six of the electrons are not localized in individual bonds

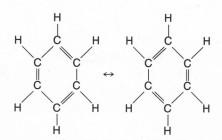

resonance notation for electron delocalization in benzene

7–15   Summary

1   A good approximation of the geometry of a molecule can be made by assuming that the electron pairs in the valence shell of any atom will avoid each other as much as possible.

2   Since we cannot solve exactly the wave equations for molecules, we treat the electronic structures of molecules by approximation methods.

3   A useful approximation is to build molecular orbitals as additive and subtractive combinations of atomic orbitals.

4   Hybrid atomic orbitals can be made as combinations of atomic s, p, and d orbitals.

5   Hybrid atomic orbitals are useful for constructing localized σ bond orbitals in polyatomic molecules.

6   Molecular geometry is the principal criterion for choice of hybrid orbitals for use in Item 5.

7   σ Orbitals are symmetrical about a bond axis.

8   π Orbitals have opposite signs on different sides of the bond axis.

9   π Orbitals can be constructed covering more than two nuclei in some molecules.

10 Delocalization of electrons by spreading them out in a large π orbital lowers their energy.

7–16   Postscript: How Little Can We See on a Clear Day?

In Chapters 4 through 7, we have developed an intricate model for molecules and the particles in them. We have asked you to go through the whole process almost as an act of faith by continuous stress on the idea that we are creating models, rather than discussing absolute physical reality. We think that there must be electrons in molecules, because we can find ways of knocking them out and detecting them. However, it would be enormously satisfying if we could catch a molecule, look into it, and see what is there. It is worthwhile to think briefly about what "seeing" means, and to consider what kinds of experimental methods are used to study directly the fine structure of matter.

We see because light falling on the retina of the eye causes chemical changes that send information along the optic nerve to register a message in the brain. Only light in the visible region of the spectrum, approximately 4000–7500 Å, is registered by the human retina. We see objects, their shape and color, because they reflect light from the sun, or some other source. Reflection can occur only if the object has dimensions larger than the wavelength of the light. Consequently, we can see with our eyes only objects that have surface areas considerably greater than 10^8 Å^2 (area of a square having sides 10,000 Å long). Of course, we cannot see structural details of objects even approaching the theoretical limiting range without the help of optical microscopes having powerful magnification. Our models lead us to believe that molecules of the size of those we have been discussing may have surface areas of about 100 Å^2, and that the largest molecules known are probably no more than a few thousand times this size. If our models are correct at all, a molecule will never be seen by a human eye.

We can look at smaller objects, if we redefine our concepts of seeing. Radiation having much shorter wavelength than visible light is available for use in looking at small structures, if we are willing to accept the use of some detector other than the human eye. For example, we know that x rays blacken a photographic plate; this is the basis of ordinary medical x radiology. Since x rays having wavelengths as short as 0.1 Å can be generated in a laboratory, it should be possible to learn about the details of small structures by observing the ways in which they deflect x rays. The results would have to be recorded on a photographic plate, or some other kind of detector, so the actual visual interpretation would be read from a secondary source. Electrons are also scattered by matter, and can be produced with a wide range of energies (or wavelengths).

High-energy radiation has been used in several ways to study the microstructure of matter. The next stage below the optical microscope is achieved with electron microscopes. An electron gun, which is a hot filament from which electrons are boiled off and accelerated to the desired energy by an electrical field, is used to fire electrons at the target. The scattered electrons are then collected on a photographic plate, where they develop a pattern showing directly the shape of the surfaces from which they are scattered. Electron microscopy has been used to show the structure of many kinds of material: biological specimens, plastics and fibers, metal alloys, and so on. This technique is invaluable to scientists in biology and materials science. Resolution to a few angstrom units is achievable. With the electron microscope, we can "see" individual molecules of very large polymer species, such as the largest protein molecules. No structure can be seen, but when we look at the photographic plates, we feel that we can really see molecules lying like pebbles on a beach. Very small electron guns, called electron microprobes, have been developed recently. The resolution of the image is increased as the size of the electron source is decreased. Since the target area covered by a fine beam is small, the probe is moved around to cover a wide area. The microprobe has given rise to an advanced technique, known as scanning electron microscopy. Recently, the use of scanning electron microscopy with high contrast and 5-Å resolving power has made it possible to see single atoms of uranium and thorium [A. V. Crewe, J. Wall, and J. Langmore, *Science,* **168,** 1338 (1970)].

Radiation of wavelength long enough to be scattered by a whole molecule has a wavelength too long to use in looking inside the molecule. To see parts, we must pass radiation through the molecule that has wavelengths about the same as the distances between the parts to be seen. The radiation emerging will show a scattering pattern that reflects the spacing between the scattering objects. This highly structured pattern is known as a *diffraction pattern.* Interpretation of the diffraction pattern to give information about the distances between atomic nuclei is no easy task, requiring fairly complex mathematical methods and an enormous amount of computation. Fortunately, much of the analysis has been put in a form that can be handled directly by high-speed electronic computers. Now there is an increasing trend toward elimination of photographic recording of diffraction patterns. Electronic detectors are used, and the pattern of intensities is converted by a small computer for structural analysis. The seeing process is therefore relayed from the detector to the human eye on printed sheets. If we believe this is seeing, now we can see distances between atoms in a molecule with precision of better than 0.05 Å.

The two most commonly used diffraction techniques involve x-ray and electron beams. Crystalline solids, and occasionally liquids, are studied using x rays. Electrons are of no use because they nearly all are absorbed by dense materials. However, the structure of gaseous molecules can be studied by electron diffraction. At the the present time, great interest is focused on the determination by x rays of the structures of biologically important molecules such as enzymes, the substances that catalyze reactions in living things.

New Terms

Antibonding orbital: Molecular orbital of higher energy than the atomic orbitals that are used to construct it. The electron density between the bonded atoms in an antibonding orbital is small, and is zero at one point along the bond axis. Electrons placed in antibonding orbitals decrease molecular stability.

Aromatic compound: Compounds having special stability, which can be explained in terms of delocalized π electrons.

Bond axis: The line joining two atoms in a molecule.

Conjugated molecule: Molecules having alternating single and double bonds.

Delocalized molecular orbital: Molecular orbital extending over more than two atoms in a molecule.

Diamagnetic: Slightly repelled by a magnetic field. Diamagnetism is characteristic of compounds with no unpaired electrons.

Electron microscope: An instrument for examining very small objects, using electron scattering.

Excited state: A state of higher energy than the ground state. An electronic excited state occurs when an electron is promoted from a lower to a higher energy level in an atom or molecule.

Hybrid atomic orbital: An orbital that is strongly directed in space and therefore useful for making a localized σ bond to an orbital on an adjacent atom. Hybrid orbitals are constructed by appropriate mixing of atomic *s, p,* and *d* orbitals. Important hybrids are *sp, sp², sp³, dsp², dsp³* (or *sp³d*), and *d²sp³* (*sp³d²*).

Localized molecular orbital: Molecular orbital extending over two atoms in a molecule.

Methyl group: The group —CH_3.

Molecular orbital: A wave function for an electron in a molecule, usually an additive or subtractive combination of atomic orbitals.

Molecular orbital theory: Model of molecular electronic structure based on assigning electrons to molecular orbitals in a way analogous to the buildup of atomic structure by feeding electrons into atomic orbitals.

Orbital overlap: Usually refers to two or more atomic orbitals on different atoms sharing a common region of space. In this region, the atomic orbitals are said to overlap.

Paramagnetic: Drawn into a magnetic field. Paramagnetism is characteristic of compounds with unpaired electrons.

π Orbital: A molecular orbital that has principal lobes on either side of the bond axis, and is unsymmetrical around the bond axis.

Polyene: A molecule with many conjugated double bonds. A molecule of the type . . .

$$(-\overset{\displaystyle |}{C}=\overset{\displaystyle |}{C}-\overset{\displaystyle |}{C}=\overset{\displaystyle |}{C}-\overset{\displaystyle |}{C}=\overset{\displaystyle |}{C}-) \ldots$$

σ Orbital: A molecular orbital that is symmetrical around a bond axis.

Unsaturated molecule: A molecule that has one or more multiple bonds.

Valence shell electron pair repulsion theory (*VSEPR theory*):   A model for predicting molecular geometry based on the principle that electron pairs in the valence shell of a central atom spread out in space as much as possible. Both unshared pairs and bonding pairs are considered in the model; unshared pairs are assumed to have larger mutual repulsions than have bonding pairs.

Questions and Problems

1   Using the concept of electron pair repulsion, predict the geometry of each of the following molecules or ions: NF_3, $CH_3^{\oplus}$, $CH_3^{\ominus}$, SO_2, PF_6^-.

2 There are many similarities between carbon and silicon, but there are also significant differences. A well-known ion has the structure SiF_6^{2-}. Formulate the electronic configuration
   of the ion. Do you think that CF_6^{2-} would be stable? Why?

3 Formulate molecular orbital descriptions of the following species: $^{\oplus}CH_2$—CH=CH_2, NO_2,
   NO_3^-, H_2CO.

4 Write a set of resonance formulas for each of the species in Problem 3. Do you see any
   connection between the implications of the molecular orbital and resonance formulations?

5 Sometimes we speak, or write, about molecular orbitals as though they were pots in which
   to store electrons. Comment on this concept.

6 Can you use the molecular orbital method to "prove" that XeF_4, a known compound, should
   not exist? Can you use the method to show that nature may be correct in letting the compound exist? How would you use VSEPR theory to formulate its molecular geometry?

7 A pentagonal bipyramid is a figure with ten faces and seven corners. It can be thought of
   as being constructed by gluing together the bases of two equal pentagonal pyramids. A
   pentagonal pyramid is a pyramid having a five-sided base. Draw a pentagonal bipyramid.

8 Which do you expect to have the greater ionization energy, ethane (C_2H_6) or ethylene (C_2H_4)?
   If you do not remember what ionization energy means, refer to Chapter 5.

9 Why is the energy required to dissociate N_2 to atoms greater than that required to dissociate
   O_2?

10 Describe the molecular orbitals for F_2, following the scheme we have used to describe N_2
   and O_2. Compare the result with the description of F_2 given in Section 7–6.

11 Draw all of the sets of hybrid atomic orbitals that you can formulate by mixing s and p wave
   functions.

12 Some people think of an electron in a molecule as a diffuse cloud of negative charge
   density; others think of it as a particle in rapid, random motion. Which model do you prefer? Why?

13 Discuss the electronic structure and the geometry of the SF_4 molecule.

14 The planar geometry of BF_3 has been rationalized in terms of the tendency of three σ bonding electron pairs to avoid each other. How does the introduction (Section 7–10) of an extra
   electron pair in π bonding affect this electron repulsion argument?

15 Is XeF_5^+ a reasonable species? What geometry would it have?

16 Rationalize the following bond lengths in terms of a molecular orbital model: N_2, 1.10 Å;
   O_2, 1.21 Å; F_2, 1.42 Å.

17 Would you expect the peroxide ion, O_2^{2-}, to have a longer, or shorter, bond length than
   O_2? Why?

18 Compare the electronic structures of NO^+ and NO, using a molecular orbital model.

19 The neutral CN molecule is paramagnetic and absorbs light of very long wavelengths. Attempt an explanation of these two facts, using molecular orbital theory.

20 Would you expect benzene to absorb light of lower, or higher, energy than ethylene? Why?

LIQUIDS AND SOLIDS

In Chapters 6 and 7, we discussed the bonds that hold atoms together in molecules. In this chapter, we will discuss the bonds that hold atoms and molecules together in solids and liquids, the *condensed phases of matter*. Some condensed systems are actually giant molecules that are held together by bonds like those in small molecules. In others, discrete molecules are held together by rather weak, intermolecular forces. We will discuss melting, boiling, and sublimation, the processes that convert matter from one phase to another. We will introduce phase diagrams, which are useful graphical forms for presenting information about the different physical forms of a material. We will consider also the energy required to change matter from one phase to another.

8–1 Phase Changes

The different forms of matter—solid, liquid, and vapor—are known as *phases*. When matter in one phase is converted to another phase, the process is called a *transition,* or *phase change.* In a phase change, energy is either liberated or absorbed by the material undergoing the change. For example, energy is absorbed in the form of heat when a solid melts. When a liquid freezes, it liberates heat energy to its surroundings.

In Chapter 2, we stated that the average kinetic energy of gas molecules is proportional to the absolute (Kelvin) temperature of the gas

$$\text{kinetic energy} = \frac{mv^2}{2} = \text{constant} \times T \tag{8–1}$$

In a sample of gas at any temperature, some molecules are moving slowly and some are traveling at high velocities; however, most of the molecules have speeds close to the average velocity. The fraction of molecules having any particular velocity changes as the temperature changes. This is shown in Figure 8–1. The fraction of molecules having a given velocity is plotted against velocity for three different temperatures. Notice that at the higher temperatures the number of molecules moving faster than the average speed is increased. Therefore, if we cool a sample of gas, the number of very fast molecules decreases faster than the *average* velocity.

What will happen to a sample of gas, such as argon, if we cool it at a constant pressure of 1 atm? As the temperature of the gas decreases, the velocities of the atoms, and their kinetic energies, also decrease. As we discussed in Chapter 2, argon atoms are attracted to each other by weak forces, called van der Waals forces. This attraction causes the volume of the gas to be less than we would predict it to be from the ideal gas law. If we continue to lower the kinetic energy of the atoms by cooling the gas further, eventually we will reach a temperature at which the kinetic energy of most of the atoms is less than the potential energy of the van der Waals forces that tend to hold the atoms together. When this happens, the gas condenses to liquid argon. The temperature at which liquid argon forms (at 1 atm) is the boiling point of argon, $-186°C$ ($87°K$). If we cool liquid argon to $-189°C$ ($84°K$), solid argon will form. The melting point of argon is $-189°C$.

8–2 Phase Diagrams

The different forms of matter (gas, liquid, and crystalline solid) are called phases. Whether a sample of material will be gaseous, liquid, or solid depends on both pressure and temperature. The behavior of a substance under different conditions can be shown by a *phase diagram*. The phase diagram for argon is shown in Figure 8–2. Temperatures are given along the abscissa and pressures along the ordinate.

The lines show conditions for phase changes. Line ABCD is the vapor–liquid curve, which shows the pressure–temperature conditions for boiling liquid argon. Under the conditions indicated by any point on the curve, argon vapor and liquid would exist together, and the liquid would be at its *boiling point*. Notice that as the pressure decreases, the boiling point decreases. The phenomenon is familiar in everyday experience. The boiling point of water is lower on a mountain top than at sea level, because atmospheric pressure decreases as the elevation increases. Point B in Figure 8–2 marks the boiling point ($87°K$) of argon at 1 atm, which is the *normal* boiling point.

The line AFG is the *freezing curve*. Points on the line show conditions for coexistence of liquid and solid. For example, Point G has values of about $105°K$ and 48 atm; this means that at that

Figure 8–1. The distribution of molecular velocities in a gas at different temperatures. $T_1 < T_2 < T_3$.

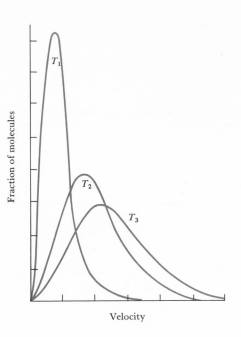

Velocity

pressure argon melts at 105°K. The melting point at 1 atm, called the normal melting point, is 84°K (−189°C). Argon illustrates the general fact that boiling points are much more sensitive than melting points to changes in pressure.

The shaded region between AG and AD is the liquid region. At any pressure–temperature point in this region, such as L, the stable form of argon is the liquid. The solid region is to the left of the liquid region. We can tell at a glance that Point S ($P = 5$ atm, $T = 60$°K) falls in the solid region. Similarly, the region to the right of the figure, where Point V lies, is the pressure–temperature region where the stable form is the vapor.

8–3 Vapor–Liquid Equilibrium

The equilibrium between liquid and vapor at a fixed temperature is a *dynamic equilibrium.* This means that molecules of liquid are constantly leaving the surface of the liquid and becoming vapor, while molecules of the vapor are striking the surface of the liquid and "sticking" to become liquid. We can see why this should happen if we look at the distribution curve in Figure 8–1. Many of the molecules in the vapor have such high energies that when they strike the liquid surface they bounce off. But a large number of molecules have energies less than the energy of the van der Waals forces of attraction in the liquid. When these molecules strike the surface, they are captured by the liquid. Similarly, many molecules on the surface of the liquid have ve-

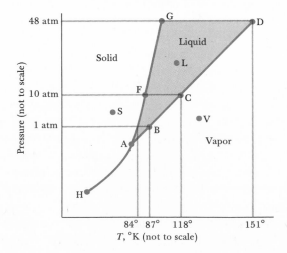

Figure 8–2. Phase diagram for argon.

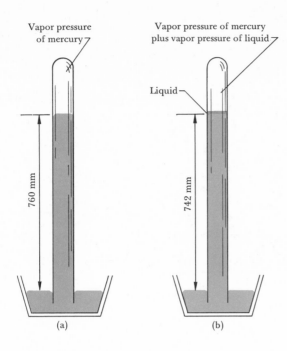

Figure 8–3. A simple experiment for measuring the vapor pressure of a liquid. The vapor pressure of mercury at 20°C is 0.0012 mm Hg and is usually considered a negligible factor; that is, the space above the mercury in (a) may be regarded as a vacuum.

locities sufficiently high that they can escape from the liquid. At equilibrium, the number of molecules escaping from the liquid each second is the same as the number of molecules captured from the vapor by the liquid. Thus, there is no net change in the number of molecules in each phase. The molecules that have escaped into the vapor phase behave as any gaseous substance and exert a given pressure if restricted to a certain volume. A simple experiment can be performed to illustrate this pressure (known as the *equilibrium vapor pressure*). Figure 8–3(a) shows a simple manometer in which the length of the column of mercury is determined by the atmospheric pressure. If we add a few drops of another liquid at the base of the mercury column, the liquid will rise to the top of the mercury column, as shown in Figure 8–3(b). Then the column of mercury will be shorter than it was before the addition of the liquid. If the initial length was 760 mm and the liquid added was water at 20°C, the new column length will be 742 mm. Thus, the equilibrium vapor pressure of water at 20°C is 18 mm Hg.

Now let us consider how changing the temperature changes the position of equilibrium. Increasing the temperature of the system increases the kinetic energy of the molecules in the liquid and in the vapor. Therefore, more molecules in the liquid will be able to escape, and a smaller fraction of the molecules in the vapor will stick when they strike the surface. As a result, more molecules will go into the vapor phase, and if we do not let any of the vapor escape, the vapor pressure will increase. The vapor pressure will continue to increase until the number of molecules of vapor striking the liquid surface is so large that the rate at which molecules are cap-

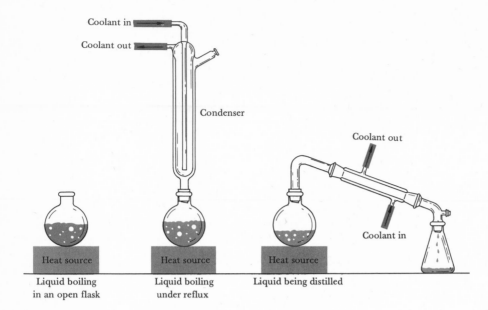

Figure 8–4. Boiling a liquid under various conditions.

tured by the liquid is once again equal to the rate at which they escape from the liquid. This is the new equilibrium vapor pressure at the new temperature.

It is important to notice that the equilibrium vapor pressure is independent of the amount of vapor or liquid present and is independent of the surface area of the liquid. It depends only on the temperature.

So far, we have considered equilibrium situations and what happens when a system goes from one state of equilibrium to another. Now suppose that we heat an open beaker of liquid argon at atmospheric pressure. The boiling point of argon is so low that a laboratory table top is warm enough to use as a hot plate. When the temperature of the liquid argon reaches 87°K, the liquid begins to boil, and the argon vapor produced escapes into the atmosphere. Since the vapor would diffuse away continually, the liquid would continue boiling to try to maintain an argon vapor pressure of 1 atm above the liquid. This process would continue until all the liquid boiled away. If we measure the temperature of the liquid during this process, we find that it is constant at 87°K (−186°C). But all this time we have been putting heat into the liquid argon. Where has the heat gone? The answer is that heat is required to change the liquid to vapor at the boiling temperature. The amount of heat required to vaporize one mole of a liquid at its normal boiling point is called the *heat of vaporization* of the liquid and is represented by the symbol ΔH_{vap}. ΔH is usually expressed in calories per mole; for example, the heat of vaporization of liquid argon is 1500 cal mole^{-1}. Later in this chapter, we shall use the values of ΔH_{vap} to show differences in liquid structures.

Example 8–1. A quantity of 15.3 g of dimethyl ether (CH_3OCH_3) is converted from the liquid phase into the vapor phase at 1 atm. The energy required to do this is 1710 cal. What is ΔH_{vap} in calories per mole?

Solution. Calculate the number of moles of dimethyl ether present

$$\frac{15.3 \text{ g of } CH_3OCH_3}{46.0 \text{ g of } CH_3OCH_3 \text{ mole}^{-1}} = 0.333 \text{ mole of } CH_3OCH_3$$

Since that fraction of a mole requires 1710 cal,

$$\Delta H_{vap} = \frac{1710 \text{ cal}}{0.333 \text{ mole}} = 5130 \text{ cal mole}^{-1}$$

A common laboratory practice consists of passing the vapor from a boiling liquid into a cold region known as a condenser. The gaseous molecules will condense and re-form liquid. Two different setups are shown in Figure 8–4. If the liquid from the condenser is returned to the original container the process is called *refluxing*. If the liquid from the condenser goes to a new container, the process is called *distillation*. Distillation is a very useful method for separating liquid mixtures into components. A mixture is heated to vaporize the more volatile component, leaving the other constituent or constituents behind.

8–4   Solid–Liquid Equilibrium

The line AFG in Figure 8–2 is the melting curve for argon. It separates the solid region from the liquid region. At any point on this line, liquid and solid argon are in equilibrium. If heat is added to a system containing both solid and liquid, the temperature and pressure will stay constant, and the added energy will be used to convert solid to liquid. This phenomenon is very familiar. If ice and water are placed in a beaker in a warm room, the temperature in the beaker will soon reach 0°C and will remain at that temperature until all of the ice has melted. After all of the ice has melted, the temperature of the water will rise until it reaches the temperature of the room. The heat required to convert a solid to a liquid at the melting point is called the *heat of fusion*, ΔH_{fus}. Heats of fusion are usually given in calories per mole.

Example 8–2. The ΔH_{fus} of Ar is 268 cal mole^{-1}. How many calories are required to melt 100 g of solid Ar at its normal melting point, 84°K?

Solution. The number of moles of Ar in 100 g is

$$\text{moles Ar} = \frac{100 \text{ g}}{39.9 \text{ g mole}^{-1}} = 2.51 \text{ moles}$$

The heat required to melt 100 g or 2.51 moles is

$$\Delta H = \Delta H_{fus} = (2.51 \text{ moles})(268 \text{ cal mole}^{-1}) = 673 \text{ cal}$$

In a crystalline solid, atoms and molecules are perfectly ordered. Crystalline argon, shown in an expanded form in Figure 8–5, is an example. In the perfectly ordered structure, each argon atom touches 12 others, and the weak forces that bind these atoms together are, on the average, as large as they can be. However, the atoms have virtually no kinetic energy. When the solid melts, argon atoms attain some freedom to move in the liquid. Their kinetic energies are not nearly as large as those of the atoms in argon vapor, but they are higher than the kinetic energies of atoms in crystals.

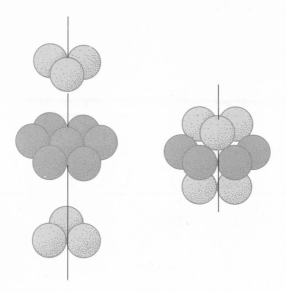

Figure 8–5. Structure of crystalline argon.

When the solid and liquid phases of a substance are maintained at the melting point, the system is in dynamic equilibrium, as is the situation for liquid and vapor phases at the boiling point. Molecules (or atoms) are condensing from the liquid on the crystalline surfaces at the same rate that molecules pass from the crystal into the liquid.

Notice that the liquid region in Figure 8–2 ends at Point A. At pressures below 0.68 atm, liquid argon cannot exist. Along the line AH, solid argon passes directly into the vapor phase, if heat is added. Direct vaporization of a solid without passing through the liquid phase is called *sublimation,* and the solid is said to *sublime.* An example of another solid that sublimes is iodine, I_2. The melting point of I_2 is 114°C, but it sublimes at room temperature. The heat required to sublime a pure substance at 1 atm is called the *heat of sublimation,* ΔH_{sub}.

Example 8–3. A 425.6-g sample of solid iodine at 25°C and 1 atm sublimes to give gaseous I_2, absorbing 24,950 cal of heat. What is ΔH_{sub} for iodine?

Solution. We want to calculate ΔH_{sub} per mole of I_2. Thus,

$$\text{moles } I_2 = \frac{425.6 \text{ g}}{253.8 \text{ g mole}^{-1}} = 1.677 \text{ moles}$$

$$\Delta H_{sub}(I_2) = \frac{24,950 \text{ cal}}{1.677 \text{ moles}} = 14,880 \text{ cal mole}^{-1}$$

At Point A of Figure 8–2, solid, liquid, and vapor are at equilibrium. For any pure substance there is only a single pressure–temperature point at which all three phases can coexist in equilibrium; this is called the *triple point.*

The liquid–vapor equilibrium line of Figure 8–2 cannot be extended to temperatures above 151°K, and it effectively ends at Point D. This is one way of saying that above 151°K (−122°C) argon cannot be liquified, regardless of how high the pressure is raised; 151°K is the *critical temperature* of argon. This phenomenon occurs because, when the kinetic energy of the molecules is sufficiently high, the molecules will not condense to form a liquid, no matter how hard we squeeze them together.

If a liquid is cooled very carefully, a curious phenomenon sometimes can be observed. In Figure 8–6, the dashed line AS passes through the triple point and penetrates the solid region of the phase diagram. Cooling along this line without the appearance of crystals is possible and is called *supercooling*. A kind of equilibrium exists between liquid and vapor, but the system is unstable with respect to crystallization. Crystals may form very slowly, but once crystallization is initiated, crystals grow very rapidly, and the liquid disappears.

8–5   Volume Changes in Transitions

Notice that the solid–liquid line AG in Figures 8–2 and 8–6 is nearly vertical. This indicates that the melting points of most solids change very little with changes in pressure. However, the liquid–vapor line (AD) and the solid–vapor line (AH) show that boiling points and sublimation points are very sensitive to changes in pressure. This sensitivity arises because there is a very large volume change when either a solid or a liquid changes to a gas. At the normal boiling point (boiling point at 1 atm) of argon, the volume of a mole of vapor is more than 100 times greater than that of the liquid.

The increase in melting point with increasing pressure shows that there must be some increase in volume on melting. At the normal melting point of argon, the volume of a mole of solid is 85% of the volume of a mole of liquid. Most materials increase in volume on melting; that is, the den-

Figure 8–6.   Phase diagram showing supercooling of argon.

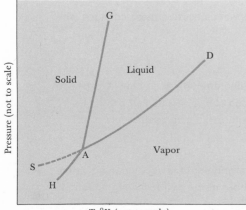

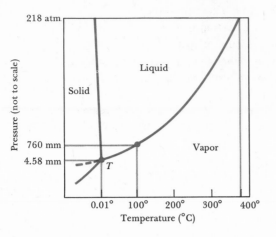

Figure 8–7. Phase diagram for water.

sity of the solid is greater than the density of the liquid. There are exceptions, with water being a familiar example. The density of liquid water is greater than that of ice, so melting causes contraction. This is the reason that ice floats on water. Figure 8–7 shows the phase diagram for water. The freezing point of ice decreases as the pressure is increased, since increasing the pressure tends to squeeze the material into the more dense liquid form.

8–6 Heats of Transition

Figure 8–8 shows all the phase changes that we have discussed. Heats of vaporization, fusion, and sublimation (ΔH_{vap}, ΔH_{fus}, and ΔH_{sub}) are all positive numbers. We decide whether this amount of heat must be added or subtracted to cause one mole of a substance to change from one phase to another by thinking about what process must occur. Condensing and freezing take place when heat is *removed* from the initial phase. Melting, subliming, and evaporating take place when heat is *added* to the initial phase. We can write equations to describe these changes. The equation

$$1500 \text{ cal} + Ar(l) \leftrightarrows Ar(g)$$

means that when one mole of argon liquid at its normal boiling point (87°K, 1 atm) vaporizes to form one mole of argon gas, 1500 cal of heat has to be added ($\Delta H_{vap} = 1500$ cal).

The equation for this process is also commonly written with ΔH_{vap} given separately

$$Ar(l) \leftrightarrows Ar(g) \qquad \Delta H_{vap} = 1500 \text{ cal mole}^{-1}$$

Condensation of one mole of argon gas to one mole of argon liquid is accompanied by the liberation of 1500 cal. This process may be expressed by either of the two following equations

$$Ar(g) \rightarrow Ar(l) + 1500 \text{ cal mole}^{-1}$$
$$Ar(g) \rightarrow Ar(l) \qquad \Delta H = -\Delta H_{vap} = -1500 \text{ cal mole}^{-1}$$

The preceding equations express the fact that the *heat content* (Section 10–4) of one mole of argon gas is 1500 cal greater than the heat content of one mole of liquid argon. It is important to keep the signs (±) of the ΔH quantities straight. The key is to remember that a positive value of ΔH means that heat is absorbed by the substance undergoing change. We say that processes requiring the addition of heat energy ($+\Delta H$) are *endothermic.* A negative ΔH means that energy is liberated as the starting material is transformed. Such a process is called *exothermic.*

Example 8–4. The heat of vaporization (ΔH_{vap}) of water is 9720 cal mole^{-1}. What is ΔH for the condensation of 30.0 g of gaseous H_2O to liquid at 100°C and 1 atm? Is the condensation process exothermic or endothermic?

Solution. The $\Delta H_{vap} = 9720$ cal mole^{-1} refers to the vaporization process

$$H_2O(l) \rightleftarrows H_2O(g) \qquad \Delta H_{vap} = 9720 \text{ cal mole}^{-1}$$

This means that when one mole of $H_2O(g)$ condenses, $\Delta H = -9720$ cal mole^{-1}

$$H_2O(g) \rightleftarrows H_2O(l) \qquad \Delta H = -9720 \text{ cal mole}^{-1}$$

and 30.0 g of H_2O is 30.0 g/18.0 g mole^{-1} = 1.67 moles. The ΔH for the condensation of 1.67 moles H_2O is

$$\Delta H = 1.67 \text{ moles } (-9720 \text{ cal mole}^{-1}) = -16{,}200 \text{ cal}$$

The process is exothermic; 16,200 cal is liberated when 1.67 moles H_2O condense.

8–7   The Structure of Solids

In Chapter 2, we discussed the "structure" of gases. In a gas, the forces between individual molecules are small compared to the kinetic energy of the molecules. We can describe a gas fairly well by ignoring the forces between molecules and concentrating on their kinetic energies. This means that gases consist of molecules in random motion and have no regular structure. In a solid, the forces between individual molecules or ions are relatively large, and the kinetic energy of the particles is small. We can describe a solid by ignoring the motion of the

Figure 8–8.   Phase transition cycle.

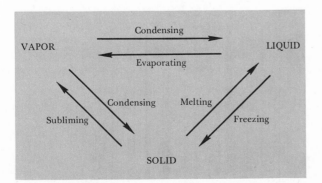

Table 8–1. Classes of Crystalline Solids

Class	Principal binding forces
Molecular	Intermolecular forces
Infinite:	
Covalent	Covalent bonds
Ionic	Electrostatic forces of attraction between oppositely charged ions

particles and concentrating on the forces that hold the particles together in a highly organized pattern.

There are two kinds of solids, crystalline and amorphous. In a crystalline solid, the building blocks are packed in a perfectly ordered arrangement. In an amorphous solid, the constituent molecules are not arranged in an ordered pattern. Rubber, wood, and glass are examples of amorphous solids. In this chapter, we will concentrate on the structures of crystalline solids.

There are several kinds of crystalline structures. The different kinds of crystals can be classified according to the forces that hold the units together. A useful classification is shown in Table 8–1.

Figure 8–9. A model of benzene and its structure in the crystalline state.

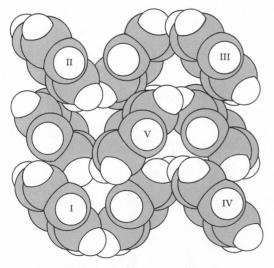

Molecular crystal structure of benzene

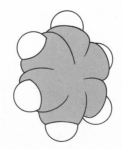

Single molecule of benzene, C_6H_6

MOLECULAR SOLIDS

A molecular crystal is formed by packing molecules together in an orderly, repeating pattern. Figure 8–9 shows the structure of crystals of benzene, C_6H_6 (Section 7–14). The nine molecules shown are packed into an interlocking array with the centers of molecules I through IV forming the corners of a square, or one face of a cube. Molecule V is in the middle of this face. The remaining four molecules are behind this face, and each one is in the center of four other faces of the cube. The important point is that *benzene molecules* are found in the crystal. The shape of the molecules and the distances between atoms within a molecule are not significantly different from those characteristic of gaseous benzene.

The existence of molecular crystals at any temperature above 0°K shows that there must be weak forces of attraction between molecules. These forces do not arise from chemical bonds, such as those that bind the atoms within molecules. The point is well illustrated by the fact that inert elements (the noble gases) form *atomic crystals,* in which the packing units are atoms. Figure 8–5 shows the structure of crystalline argon, in which the argon atoms are packed together like tennis balls in a cubic box.

The forces of attraction between molecules are fundamentally electrical; they arise from the attraction of positive nuclei in one molecule to electrons in another molecule. However, detailed description of intermolecular forces varies considerably from case to case. They can be divided roughly into three classes as follows:

Intermolecular force	*Kinds of molecules*
Van der Waals forces	Most important for symmetrical molecules having no permanent dipole moment
Dipole attractions	Molecules having dipole moments
Hydrogen bonds	Molecules containing groups such as O—H and N—H

DIPOLAR ATTRACTIONS

We will look first at the attractive forces that arise between molecules as a result of their dipole moments (dipole moments were discussed in Section 6–9). Dipolar molecules will attract one another if they are oriented so that the positive end of one dipole is close to the negative end of another. Repulsion occurs if the orientation of dipoles brings like charges together. Figure 8–10 shows examples of attractive and repulsive dipoles.

Extension of these ideas suggests that two hydrogen fluoride molecules should attract each other if they are arranged in one of the following ways:

$$\overset{\delta+}{H}\!\!-\!\!\overset{\delta-}{F}\ \overset{\delta+}{H}\!\!-\!\!\overset{\delta-}{F} \quad \text{or} \quad \overset{\delta+}{H}\!\!-\!\!\overset{\delta-}{F}$$
$$\overset{\delta-}{F}\!\!-\!\!\overset{\delta+}{H}$$

Crystalline hydrogen fluoride melts at −83°C and boils at 19°C. Both transition temperatures are surprisingly high for a molecule having such a low molecular weight. By way of comparison, hydrogen (H_2) melts at −259°C and boils at −253°C, and fluorine (F_2) melts at −223°C and boils at −188°C. Clearly, the unsymmetrical HF molecules are bound together more strongly than either of the symmetrical hydrogen or fluorine molecules in the solid and liquid phases.

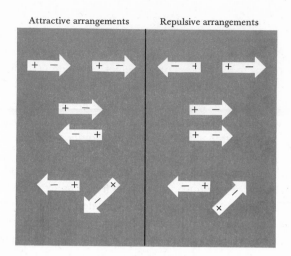

Attractive arrangements   Repulsive arrangements

Figure 8–10.   A schematic representation of the interaction of dipoles.

HYDROGEN BONDS

There is a general tendency for large molecules to have higher melting and boiling points than small molecules have. However, the data in Table 8–2 show that the general rule is not followed by methyl fluoride (CH_3F) and hydrogen fluoride. Both compounds have rather large dipole moments: CH_3F, 1.81 D; HF, 1.92 D.

As discussed in Section 6–9, the dipole moment is a measure of the force tending to make a dipolar molecule align itself with an applied electrical field. The dipole moment of hydrogen fluoride is larger than that of methyl fluoride; however, experience shows that the difference is not nearly large enough to account for the observed differences in the transition temperatures.

Table 8–2.   Transition Temperatures for Methyl Fluoride and Hydrogen Fluoride at One Atmosphere

Compound	Formula	Normal freezing point (°C)	Normal boiling point (°C)
Methyl fluoride	H—C—F (with H above and H below)	−142	−78
Hydrogen fluoride	H—F	−83	19

Apparently, hydrogen fluoride molecules are bound together much more strongly than we would expect on the basis of simple dipole–dipole attraction.

Very strong association forces turn up again and again in the study of compounds containing *polar bonds to hydrogen.* Molecules of such compounds contain F—H, O—H, and N—H groups. To a lesser extent, S—H and Cl—H groups show the same behavior. Interpretation of more data than we give here indicates that molecules containing polar bonds to hydrogen are bound very strongly to the negative ends of other dipolar molecules

$$\overset{\delta-}{X}\!\!-\!\!\overset{\delta+}{H}\cdots\overset{\delta-}{Y}\!\!-\!\!\overset{\delta+}{Z}$$

Bonding of polar X—H molecules to other molecules is called *hydrogen bonding.* A hydrogen bond is fundamentally no different from the bond between any pair of polar molecules. However, the intermolecular forces are unusually large when hydrogen is the bridging atom. This effect probably arises because hydrogen atoms are exposed, and thus they can be approached closely by other atoms. By way of contrast, the carbon atoms of methyl fluoride are shielded from close approach of external atoms by a canopy of nonpolar carbon–hydrogen bonds.

$$\overset{\delta+}{H}\!\!-\!\!\overset{\delta-}{F}$$
hydrogen fluoride

$$
\begin{array}{c}
H \\
\diagdown \\
H\!\!-\!\!\underset{\diagup}{C}\!\!-\!\!\overset{\delta+\ \ \delta-}{F} \\
H
\end{array}
$$
methyl fluoride

VAN DER WAALS FORCES

We have shown that molecular crystals having high melting points usually are held together by dipole–dipole forces, with especially powerful attraction being associated with hydrogen bonding. However, we have not yet come to grips with the fact that benzene, and even argon, form stable crystals. The normal melting point of benzene is 5.5°C, and that of argon is −189°C. The heats of fusion of the two materials are 2360 cal mole^{-1} and 268 cal mole^{-1}, respectively. These binding forces are rather small, but they are important in understanding the physical state of

Figure 8–11. Synchronous fluctuation of the electron clouds around a pair of non-bonded atoms.

Direction of "instantaneous" dipoles

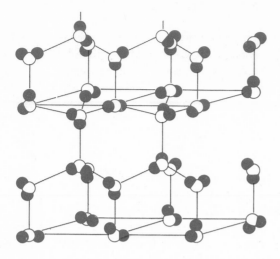

Figure 8–12.   The structure of ice crystals.

many of the materials that we encounter in everyday life. An interesting model to account for these binding forces was first presented by Johannes van der Waals. We will not try to explain his theory in detail, but it depends upon the notion that the electron cloud in any atom is in a constant state of fluctuation. Consequently, even the nucleus of an atom such as argon is not surrounded continuously by a uniform electrical field. If two atoms are close together, the fluctuations in their electron clouds can get into phase so that the little dipoles of the fluctuating electron clouds attract each other. The idea is shown crudely in Figure 8–11.

The concept of van der Waals forces may seem esoteric to students, but the binding of argon atoms in crystalline argon and the binding of benzene molecules in a benzene crystal are facts. Van der Waals forces are largest when large atoms are involved in the interactions, because the outer electrons in heavy atoms are relatively loosely held by their nuclei. Furthermore, molecules containing many atoms are bound together more strongly than small molecules, because many fluctuating charge clouds are in contact when a pair of such molecules are in contact.

Table 8–3.   Comparison of Physical Properties of Acetonitrile and Propane

Property	Acetonitrile	Propane
Molecular weight	41	44
Dipole moment	3.94 D	0
Normal freezing point	−46°C	−190°C
Normal boiling point	82°C	−42°C
Heat of fusion	(Not available)	802 cal mole^{-1}
Heat of vaporization	7530 cal mole^{-1}	4280 cal mole^{-1}

Table 8–4.   Phase Transition Temperatures and Energies for Noble Gases

Substance	Atomic weight	Normal melting point (°C)	ΔH_{fus} (cal mole^{-1})	Normal boiling point (°C)	ΔH_{vap} (cal mole^{-1})
Helium	4.0	—	—	−269	24
Neon	20.2	−249	57	−246	416
Argon	39.9	−189	268	−186	1500
Krypton	83.8	−157	358	−153	2240
Xenon	131.3	−112	490	−107	3200

In the preceding pages, we have discussed the forces that bind molecules together. Crystals in which hydrogen bonds supply most of the intermolecular binding are especially stable. Such compounds have relatively high melting points and heats of fusion. A classic example is ice, in which water molecules are arranged so that the crystal is held together by an infinite network of hydrogen bonds. Figure 8–12 shows the structure of ice crystals.

The heat of fusion of ice is only 1434 cal mole^{-1}. Interpretation of this value must be made cautiously. When ice melts, liquid water forms, and water contains many hydrogen bonds. The heat of vaporization of water is a whopping 9720 cal mole^{-1}. By adding the heats of fusion and the heats of vaporization, we find that 11,154 cal mole^{-1} of heat is required to convert a mole of ice to a mole of isolated, gaseous H_2O molecules.

A good example of the effects of intermolecular attractions due to polar bonds not involving hydrogen is provided by comparison of acetonitrile and propane, two carbon compounds having similar molecular weights. The comparison is shown in Table 8–3.

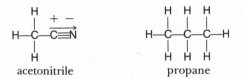
acetonitrile   propane

The influence of increasing atomic size on intermolecular binding forces is shown by the inert elements that form atomic crystals and liquids. The comparison is shown in Table 8–4.

8–8   Heats of Transition and Transition Temperatures

As was implied at the beginning of this chapter, the kinetic energy of molecules is increased when a substance melts or boils. Molecules in a liquid are not nearly as free to move as those in a gas, but they have more freedom than molecules have in a crystal. We can think of the energy put into a melting crystal or a boiling liquid as appearing in the form of increased kinetic energy of the molecules. The increase in the speeds of molecules is greater when they go from liquid to gas than from solid to liquid. Consequently, we expect that heats of vaporization usually will be larger than heats of fusion. This is virtually always the case. The result is reasonable, since

Table 8–5. Comparison of ΔH_{fus} and Molecular Weights of Sets of Compounds with Identical Melting Points

Compound	Melting point (°C)	ΔH_{fus} (cal mole^{-1})	Molecular weight
Methyl amine, CH_3NH_2	−92	1510	31.06
Propionitrile, CH_3CH_2CN	−92	1206	55.08
Sodium, Na	98	623	22.99
Aluminum bromide, $AlBr_3$	98	2792	266.72
1,1,2-Trichloroethane, $CH_2ClCHCl_2$	−37	7960	133.42
2,3-Dimethylhexane, $CH_3CH(CH_3)CH(CH_3)CH_2CH_2CH_3$	114	8020	114.23
2-Methyl-1-hexene $CH_2{=}C(CH_3)CH_2CH_2CH_2CH_3$	91	7450	98.18
5-Methyl-2-hexene, $CH_3CH{=}CHCH_2CH(CH_3)_2$	91	7120	98.18
sec-Butylbromide, $C_2H_5CH(CH_3)Br$	91	7360	137.03

molecules are still in contact with each other in the liquid state, and the average forces of inter-molecular attraction may be nearly as large in the liquid as they are in the solid.

It is also reasonable to expect that, if we compare a large number of substances, we will find those having the highest heats of fusion will also have the highest melting points. A similar correlation is expected between boiling points and heats of vaporization of liquids. Examination of the data in this chapter will show that these generalizations usually are true. However, there is no exact relationship between heat absorbed in transitions and the transition temperatures. Comparison of the data in Table 8–5 for pairs of compounds having matched transition temperatures illustrates the point.

Factors other than the increase in kinetic energy of molecules must be involved in melting and boiling. The most important of these factors is *the increase in disorder* that occurs during the processes. The molecules in a crystal are completely ordered, and those in a liquid are less ordered. However, some of the orderliness of a crystal may be maintained in a liquid. In a gas, molecules are almost completely disordered. In general, matter in nature tends to become dis-ordered, so we say that disorder is more probable than order. For example, if we drop a dozen blocks, they probably will end up in some disorganized array as is shown in Figure 8–13. The disorganized arrangement is more probable than the organized system. Work is required to rearrange the parts of a disordered system in an orderly fashion. The work required to create order in a system is called *entropy*. Since the probability of disorganization is inherently greater

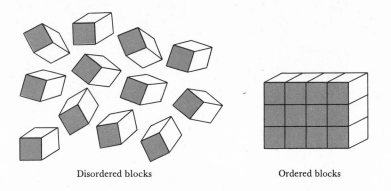

Disordered blocks   Ordered blocks

Figure 8–13. Disorganized and organized arrangements of blocks.

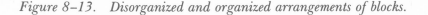

than the probability of order, entropy increases when the parts of a system fall into their natural state of disorder.

Melting and boiling tend to increase disorder; therefore, the processes are accompanied by an increase in entropy. The temperature at which a change from solid to liquid, or liquid to gas, occurs depends upon competition between intermolecular attractive forces, which tend to organize the system, and entropy, the tendency of the system to move toward the most disorganized state. Entropy changes in melting and boiling processes vary from one substance to another; consequently, there is no exact correlation between heats of transitions and transition points. However, we repeat, high melting and boiling points are *usually* associated with high heats of fusion and vaporization, and they reflect relatively large intermolecular attractive forces.

8–9 Ionic Solids

Ionic solids are held together by electrostatic forces between positive and negative ions. These forces are the same as those that hold a molecule of sodium fluoride together in the vapor phase. These ionic bonds are very strong; thus ionic crystals have high melting points and high heats of fusion, as shown in Table 8–6.

Potassium chloride, a representative example, has a structure (Figure 8–14) in which each potassium ion is surrounded by six chloride ions at the corners of an octahedron, an eight-sided

Table 8–6. Some Ionic Substances and Their Transition Temperatures and Energies

Substance	*Melting point (°C)*	ΔH_{fus} *(cal mole^{-1})*
Sodium fluoride (NaF)	992	7000
Barium chloride (BaCl$_2$)	963	5370
Potassium chloride (KCl)	772	6410

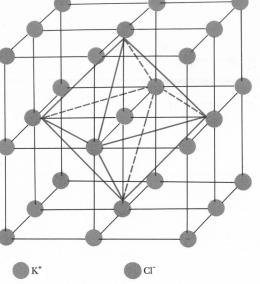

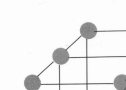

K⁺   Cl⁻

Figure 8–14.   The structure of crystalline potassium chloride.

polygon. Each chloride ion is similarly surrounded by six potassium ions. Remember that potassium follows argon in the periodic table, and chloride precedes it; thus K^+, Ar, and Cl^- all have the same electronic structure. But the properties of solid KCl are entirely different from the properties of solid Ar. Also, notice that there are no discrete molecules of KCl in the KCl crystal. This is in contrast to a molecular solid such as benzene, in which the crystal is composed of individual molecules. An ionic crystal is, in a sense, a single molecule of enormous molecular weight and is said, therefore, to have an infinite structure.

Figure 8–15.   The diamond structure.

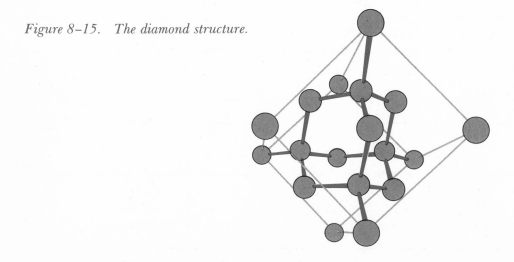

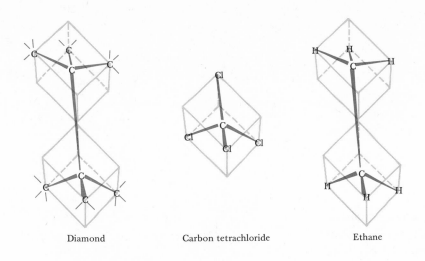

Diamond　　　　　　　　Carbon tetrachloride　　　　　　　Ethane

Figure 8–16.　Some tetrahedral carbon compounds.

8–10　Covalent Solids

Covalent solids are held together by covalent bonds between atoms. A familiar example is diamond, which is composed entirely of carbon atoms. A model of the diamond structure is shown in Figure 8–15. Each carbon atom is bonded to four others at the corners of a tetrahedron. This is the usual shape of molecules containing a four-valent carbon atom in a *sp*³ hybridized form as discussed in Section 7–8. The geometry of this bonded structure is the same as that of carbon in such molecules as carbon tetrachloride and ethane, shown in Figure 8–16.

Graphite, another *allotrope* (from Greek *allos,* meaning other, and *tropos,* meaning way) of elemental carbon is also a covalently bonded solid. The graphite structure is shown in Figure 8–17, along with structural models of two common compounds containing carbon, benzene and naphthalene. The planes in graphite, which contain hexagons of carbon tightly bound together, resemble benzene molecules placed together. The "extra" electrons, which are used to bind hydrogen to carbon in benzene, act like glue to hold the planes of carbon atoms together in graphite. Graphite is an effective lubricant and tends to flake easily because the planes can slip across one another and be pulled apart.

Many solid compounds have infinite structures, in which the atoms are bound together with infinite networks of covalent bonds. An example is aluminum oxide, the principal constituent of common clays. The empirical formula of the oxide is Al_2O_3. If we were to write a formula for a molecule having the simplest composition, it would probably be

$$:\overset{..}{O}=Al—\overset{..}{O}—Al=\overset{..}{O}:$$

However, in this formula the aluminum atoms are electron-poor, having only six electrons (those of the aluminum–oxygen bonds) in their valence shells. It is not surprising that in solid aluminum oxide the atoms form an infinite structure, in which the oxygen electrons (shown as unshared in the formula) are used to bind to aluminum atoms, thereby satisfying the electronic requirements of aluminum.

Figure 8–17.   Skeleton structures of graphite, benzene, and naphthalene (multiple bonding characteristics are not shown).

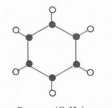

Benzene (C_6H_6)   Naphthalene ($C_{10}H_8$)

Covalent solids generally have very high melting points (diamond melts above 3500°C, and graphite sublimes at about 3950°C) and very high heats of fusion.

INTERMEDIATE TYPES

Just as in the case of single molecules, in which we find ionic bonds (CsF), pure covalent bonds (F_2), and a wide range of intermediate types, many solids show a type of structure that is intermediate. For example, AgF crystallizes to form an ionic solid like KCl. But AgI forms a covalent solid like diamond. Both AgCl and AgBr have the same crystal structure as AgF and KCl have, but the bonding is intermediate between the extremes of ionic and covalent. Solid AgF contains silver *ions* and fluoride *ions*. Solid AgI contains silver *atoms* and iodine *atoms.* We can think of AgCl and AgBr as solids that contain ions with a fractional positive or negative charge because of partial covalent bond formation.

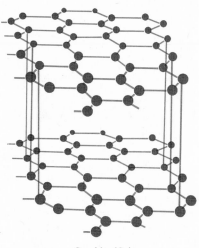

Graphite (C_n)

8–11   Metals

Solid metals have unique and useful properties. They are excellent conductors of electricity and heat, many are rather easily shaped under pressure, whereas others have great strength. The characteristic shininess of a bright metal surface is due to the reflection of light from the surface. Metals are not like molecular solids, because the metal atoms are bound together in infinite structures. However, they are unlike

Figure 8–18.   Structure of crystalline aluminum oxide.

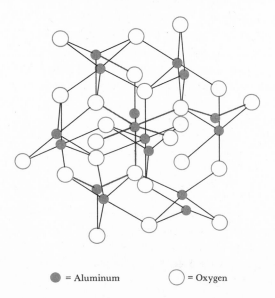

● = Aluminum   ○ = Oxygen

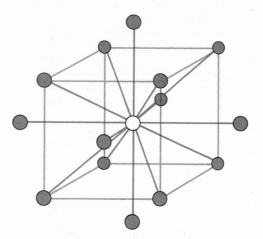

Figure 8–19. Body-centered cubic structure (crystalline potassium).

infinite solids such as aluminum oxide, in that there are not enough valence electrons in a metal to use all of the bonding orbitals of the atoms.

Consider potassium metal as an example. Potassium has a body-centered cubic structure, shown in Figure 8–19. Each atom has eight nearest neighbors and six neighbors only slightly further away with which it can form bonds. Each atom has one valence electron and four orbitals (one $4s$ and three $4p$) that could be used for bonding. Clearly, there are not enough electrons to go around. As a result, electrons in a metal do not form definite bonds between any two atoms, rather they tend to bind atoms to their numerous neighbors. A metal is bound together by many-center bonds, somewhat similar to the three-center bonds in diborane (Section 6–4), an electron-deficient molecule. Metals conduct electric current because an electron can "jump" to a nearby site and find a low-energy metal orbital incompletely filled with electrons. Metals are good conductors of heat for the same reason.

The bonds in potassium metal are not very strong, compared to other metals. Potassium melts at 64°C and boils at 774°C. The heat of fusion is only 574 cal mole^{-1}, and the heat of vaporization of liquid potassium is 18,800 cal mole^{-1}. The sum of the two, 19,400 cal mole^{-1}, is a rough measure of the energy required to convert a mole of solid potassium metal to its vapor. The measure is rough because the heats of fusion and vaporization are measured at quite different temperatures, the melting point and boiling point at atmospheric pressure. Many metals have much higher melting points, boiling points, and heats of transition than potassium has.

Table 8–7 compares the transitions of some metals of the first three families of the periodic table. The data show that as more valence electrons are added, the binding energy of the solids becomes greater. The Group IIIA metals, which have three valence electrons per atom to use in the four low-energy orbitals, have high melting points and large heats of fusion.

8–12 The Structure of Liquids

We know a great deal about liquids. We can measure their boiling points and densities and find how these properties change with pressure. We can determine how well different liquids mix

Table 8–7.   Melting and Boiling Data for Some Metals

Metal	Family	Melting point (°C) at atmospheric pressure	Boiling point (°C) at atmospheric pressure	$\Delta H_{fus} + \Delta H_{vap}$ (cal mole^{-1})
Lithium	IA	186	1367	33,200
Sodium	IA	98	914	24,000
Potassium	IA	62	774	19,400
Beryllium	IIA	1283	1500[a]	87,700[a]
Magnesium	IIA	651	1120	33,700
Calcium	IIA	850	1482	40,800
Aluminum	IIIA	660	2300	70,500
Scandium	IIIB	1400	2700[a]	84,000[a]

[a] *Estimated.*

with each other. We can see if they conduct electricity; we can measure the amount of heat required to change their temperature from one value to another. We can measure the rate of diffusion of molecules in a liquid. We could predict or explain many of these properties, if we knew the properties of the individual molecules of the liquid. But we do not know very much about the *structure* of liquids and the effect liquid structure has on these properties.

We have described gases as being composed of molecules in random motion with very little potential energy of interaction between the molecules. Using this model of a gas, we are able to explain the experimentally determined laws that describe how gases behave. We have also described solids as composed of molecules (or ions) held together in a rigid structure by the potential energy of interaction between the molecules. The difficulty in describing the structure of a liquid is that *both* potential energy and kinetic energy are important. The molecules in a liquid are closely packed and tend to be held together in a regular structure because of the attraction of the molecules for each other. But the molecules in a liquid have so much kinetic energy that they move through the liquid from one position to another. This is why liquids assume the shape of their container.

Liquids probably contain small regions that have a fairly definite arrangement of molecules in a structure that is at least as dense as the corresponding solid, and other regions that are "holes" in the liquid. The regions of definite structure change as the molecules move through the liquid, but the moving molecules stay out of the holes. The holes cause the liquid (usually) to have a lower density than the corresponding solid. Chemists would like to be able to answer questions such as, "How big are the holes?" "Do the holes stay in the same place?" "What is the structure?" "Does the structure change as the molecules move about?" These questions provide a great challenge to chemists working in this field today, and their answers are important to understanding the properties of liquids and the way in which chemical reactions occur. Here we will have to content ourselves with describing some of the properties of typical kinds of liquids.

Molecular liquids, such as argon, are held together by van der Waals forces. The boiling points and heats of vaporization of all the noble gases are low (Table 8–4) because the van der Waals forces holding the liquid together are small.

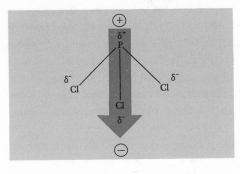

Figure 8–20.   A representation of the dipole associated with the molecule PCl_3.

Polar liquids, such as phosphorus trichloride (PCl_3), are those whose molecules have, or can form, dipoles in which some of the atoms have a partial charge (Section 8–7). PCl_3 can be pictured as a dipolar molecule resulting from the separation of the positive and negative charge, as shown in Figure 8–20. In the liquid, the molecules tend to orient themselves so that positive phosphorus atoms are near the negative chlorine atoms of another molecule. The force of attraction between positive and negative ends of the dipole tends to hold the liquid together.

The most extreme example of this type of interaction is in hydrogen-bonded liquids. We have already discussed hydrogen bonding in solids. The same sort of attraction occurs in liquids, the only difference being that the liquids do not have a completely ordered structure. Figure 8–21 shows how the boiling points of some families of hydrides (Chapter 3) vary. For all of these

Figure 8–21.   Boiling points of several hydrides.

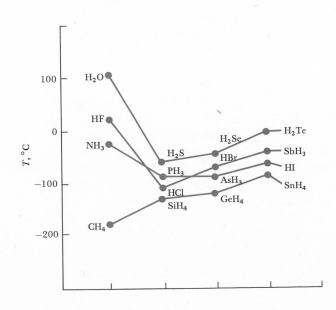

compounds, *except* NH_3, HF, and H_2O, the boiling point increases with molecular weight in a given family. The compounds NH_3, HF, and H_2O are those in which strong hydrogen bonding occurs. Hydrogen bonding holds these liquids together so well that their boiling points are much higher than we might predict, knowing all the other boiling points. Their heats of vaporization are also very high. The heat of vaporization of water is 9720 cal mole^{-1}, whereas the heat of vaporization of hydrogen sulfide (H_2S) is only 4490 cal mole^{-1}. We assume that in liquid water there are large aggregates, sometimes called icebergs, in which groups of molecules are bound together much as in crystalline ice (Figure 8–12).

Just as solid metals have high melting points and large heats of fusion, so *liquid metals* have high boiling points and large heats of vaporization. The only metal that is a liquid under normal conditions is mercury. Its melting point is −38.9°C, and its boiling point is 356.6°C. This has an important practical consequence. Mercury vapor is a dangerous cumulative poison. At room temperature, the equilibrium vapor pressure of mercury is high enough to be toxic. For this reason, it is important to be careful not to spill mercury, and to clean up any spilled mercury immediately. Cesium and gallium are liquids near room temperature, melting at 28.7°C and 29.8°C, respectively.

8–13 Summary

1 The three phases of matter are solid, liquid, and vapor.

2 The important phase transitions and the designations for the heat absorbed when the transition occurs are freezing or melting (solid $\rightleftarrows$ liquid, ΔH_{fus}), boiling or condensation (liquid $\rightleftarrows$ vapor, ΔH_{vap}), and sublimation or condensation (solid $\rightleftarrows$ vapor, ΔH_{sub}).

3 Information about the phase transitions is presented conveniently in graphical form by means of phase diagrams.

4 Crystalline solids contain units packed together in a perfectly ordered way.

5 Crystals can be either molecular solids or infinite; infinite crystals may be ionic or covalent.

6 Metals are a special class of infinite, covalent solids, in which there are too few electrons to use all available bonding orbitals of the atoms.

7 Liquids are more disordered than solids, but molecules in liquids are still in close contact. The energy of a liquid is partly potential energy of attraction between molecules and partly kinetic energy.

8 A vapor is mostly space. The molecules interact very weakly, and nearly all of the energy of a vapor is kinetic energy. (We are ignoring the large chemical energies of compounds, which can be changed only by chemical reactions.)

9 Intermolecular forces are separated into two kinds: van der Waals forces, due to the rapid fluctuations in the electron clouds of atoms and molecules, and dipolar interactions, arising from interaction of molecular dipoles.

10 Hydrogen bonds are especially strong dipolar interactions that occur when a partially positive hydrogen in a molecule forms a bond to an electronegative atom on a neighboring molecule, $\overset{\delta-}{X}\!-\!\overset{\delta+}{H}\!-\!-\!-\!\overset{\delta-}{Y}$.

11 Hydrogen bonds are especially strong when the electronegative atoms involved are oxygen, nitrogen, or fluorine.

12 Van der Waals forces increase as the atomic number of the atoms involved is increased.

8—14   Postscript: Ice Ages and Floods

The water cycle in nature is an example of dynamic equilibrium. Vaporization of water from rivers, lakes, and oceans continuously supplies water vapor to the atmosphere. There, the water condenses and finally falls to earth again as rain, snow, or hail. Without the water cycle, no land creatures could live on the earth. Land plants depend upon water supplied by their root systems for growth, and land animals depend directly upon plants to manufacture the basic food stuffs for animal nourishment.

Many interesting things occur during the cycle. Water vapor condenses to liquid water when the partial pressure exceeds the liquid vapor pressure at the local atmospheric temperature. The first-formed liquid particles do not fall immediately. They are so small that they can be kept suspended by the buffeting action of molecules in the atmosphere. However, the liquid particles in a cloud grow in time. The growth occurs in two ways. First, more water vapor supplied by diffusion can condense so that the total amount of liquid in a cloud increases. Second, in a collection of particles in a state of dynamic equilibrium, the smallest particles tend to disappear, while the larger particles increase in size. When the liquid droplets reach a sufficiently large size, they fall to earth as rain, unless something happens to them first. If the temperature falls low enough, the liquid droplets freeze, forming ice crystals, and fall as snow. Sometimes liquid droplets falling as rain pass through cold zones in the atmosphere where they are frozen rapidly into ice balls, which fall to earth as hail.

The reason for the growth of large droplets at the expense of small ones during the ripening of a cloud is straightforward. Compare the formulas relating the area and volume of a sphere to its radius

$$A = \pi r^2$$

$$V = \frac{4}{3}\pi r^3$$

The ratio of the area to the volume depends upon the size of the sphere

$$\frac{A}{V} = \frac{3}{4r}$$

The weight of material is directly proportional to the volume

$$W = dV \qquad (d = \text{density})$$

The fraction of the total amount of material that is surface will be greater for a small droplet than for a large one. Since surface molecules are responsible for evaporation, the large droplet, which has a greater fraction of its molecules "out of action" in the interior, will slowly grow as molecules are passed back and forth between droplets by evaporation and condensation.

We should stop and ask ourselves if the water in a cloud is really in a state of equilibrium, if the size of the droplets is changing. Clearly, the answer is "No." The most stable state of liquid water is not many small droplets, but one continuous phase. The energy of surface molecules is higher than that of molecules in the interior of the liquid, and the total energy of the system will be least when the water is all gathered in one huge ball. The fact that a water droplet tends to assume a spherical shape illustrates the same fact, since a sphere has less surface area than any other geometric figure having equivalent volume.

The water that percolates through the earth during the cycle dissolves soluble minerals, thus rivers and lakes are not pure water, but dilute solutions. They may also contain solid material in fine suspension, both because of natural erosion and because of the debris added by man. Actually, all of the water-soluble material in the earth's crust is transported slowly to the oceans where most of it remains, because it is nonvolatile and does not reenter the cycle by evaporation. Water itself behaves almost as a dynamic equilibrium system, but the materials carried by water to the oceans do not.

The amount of water involved in the dynamic cycling process varies with time. Some water is removed from the cycle and stored in frozen form as glaciers, the largest of which are the two polar ice caps. We know that the amount of water stored in this way must have varied in the past, because there is strong geological evidence that during the Ice Age the north polar ice cap extended south over much of Europe, all of Canada, and the northern part of the United States. The amount of water stored as ice depends on the average temperature of the earth's atmosphere. As the temperature rises, the glaciers melt, and if the temperature falls, they grow.

Some temperature fluctuations are controlled by conditions on the sun, but others depend upon events occurring on earth. Much of the sunlight reaching the earth is reflected. We can see this by looking at other planets, which appear as bright stars in the night sky because of reflected sunlight. In addition, the earth continuously loses energy by radiating "light" in the infrared region of the spectrum. If the infrared rays are absorbed by materials in the atmosphere, the energy is not lost, and the average temperature of the atmosphere will rise.

A particularly important absorbing species is carbon dioxide, which enters and leaves the atmosphere in its own cycle. Ecologists are beginning to worry about the consequences of man's desire to use more and more energy. Whenever energy is used, including both nuclear energy and chemical energy stored in the form of coal and petroleum, heat is produced. Furthermore, the burning of coal and petroleum produces carbon dioxide. We are both producing more heat and putting into the atmosphere a substance that tends to reduce heat loss by absorbing radiation from the earth's surface. It is estimated that the present rate of increase in the temperature of the atmosphere is at least $0.01°C$ per year, and that the rate of increase may soon be so great as $0.1°C$ per year. This does not seem like a dangerous planetary fever, yet in 100 years the increase would be $10°C$, enough to gradually melt all of the polar ice caps. If this occurred, the level of the oceans would rise more than 200 feet, decreasing the dry land area by 20%, and inundating all of the cities of the world located in coastal regions.

However, the effects of changing the atmosphere because of the activities of a technological society are not simple. Along with carbon dioxide, other materials are being added. Included are small particles of solid matter. Particles absorb some sunlight, but they also reflect and scatter light. A large increase in the number of particles in the upper atmosphere could act as a screen, preventing sunlight from reaching the lower atmosphere and the earth's surface. This would tend to lower the average atmospheric temperature and, perhaps, initiate a new Ice Age!

The prospect of a Flood Age instead of an Ice Age reminds us of the Biblical story of the flood, and of the fact that massive floods are mentioned in the legends of other peoples, including some American Indian tribes. A little reflection indicates that Noah's problems could not have been due to melting of the ice caps. His flood receded in 40 days, whereas the advance and recession of the glacial ice is likely to be measured in terms of thousands of years.

New Terms

Allotrope: Two different forms of the same element are called allotropes. For example, two allotropes of carbon are diamond and graphite.

Boiling curve: The temperature–pressure curve, which gives the boiling point of a liquid as a function of pressure.

Boiling point: The temperature at which the vapor pressure of the liquid is the same as the atmospheric pressure. The normal boiling point is the temperature at which the vapor pressure of the liquid is 1 atm (760 torr).

Condensed phase: A phase in which the atoms are close together; a liquid or a solid.

Covalent solid: A solid consisting of atoms bound together through a network of covalent bonds. Graphite and diamond are covalent solids.

Critical temperature: The temperature above which a substance cannot exist as a liquid no matter how much pressure is applied to it.

Dipole–dipole attraction: The attraction that results from the interaction of molecules that have permanent dipole moments.

Distillation: The process of boiling a liquid and allowing the vapor to be removed from the vessel and condensed to the liquid phase in another vessel. It is a method of separation of a more volatile component of a mixture from the less volatile components.

Endothermic: Any process that consumes heat energy is endothermic and has a positive value of ΔH.

Entropy: The work required to create order in a system is called entropy. Entropy is therefore a measure of the disorder in a system.

Equilibrium vapor pressure: The pressure exerted by the vapor above a pure liquid when the two phases are in equilibrium with each other. The value depends on the temperature of the system, but at any given temperature it is independent of the amount of liquid present.

Exothermic: Any process that releases heat energy is exothermic and has a negative value of ΔH.

Freezing curve: The curve that gives the freezing point of a liquid as a function of pressure.

Heat content: The "amount of heat contained" in a molecule. (See Chapter 10.)

Heat of fusion, ΔH_{fus}: The heat energy required to transform one mole of solid into one mole of liquid at one atmosphere of pressure; ΔH_{fus} is expressed commonly in units of calories mole^{-1}.

Heat of sublimation, ΔH_{sub}: The heat energy required to transform one mole of substance from the solid phase to the vapor phase at one atmosphere pressure; ΔH_{sub} is expressed usually in units of calories mole^{-1}.

Heat of vaporization, ΔH_{vap}: The heat energy required to transform one mole of substance from the liquid phase to the vapor phase at one atmosphere pressure; ΔH_{vap} is expressed usually in units of calories mole^{-1}.

Hydrogen bond: The special name for the dipolar interaction between molecules that have a hydrogen atom attached to a highly electronegative atom such as N, O, or F. The dashed line indicates a bond of this type $-\overset{\delta-}{O}---\overset{\delta+}{H}-\overset{\delta-}{O}-$.

Ionic solid: A solid consisting of discrete positive and negative ions held together by electrostatic attraction; NaCl and KBr are ionic solids.

Melting point: The temperature at which the solid and liquid phases of a substance are in equilibrium with each other. At the normal melting point, pressure is 1 atm.

Molecular solid: A solid consisting of discrete molecules held together by van der Waals or dipole–dipole attractive forces. Solid benzene is an example of a molecular solid bound primarily by van der Waals forces. Solid hydrogen fluoride is held together mainly by dipole–dipole forces.

Phase: The physical state of matter—solid, liquid or gas.

Phase diagram: A graph showing the temperature dependence of the vapor pressure above the solid and liquid phases of a substance, and the phase of that substance under particular temperature and pressure conditions.

Phase transition: The change of a substance from one phase to another.

Reflux: Boiling a liquid and returning the condensed vapor; reflux allows a liquid to boil continuously.

Sublime: A substance sublimes when it passes directly from the solid phase to the vapor phase, without passing through the liquid phase.

Supercooling: If a liquid is cooled slowly and with no mechanical disturbance, the liquid phase may remain well below the normal freezing point. This is an unstable state, and a slight disturbance usually will bring about the phase transition from the supercooled liquid to the stable solid.

Triple point: The temperature at which the liquid, solid, and vapor phases are in equilibrium.

Van der Waals attraction: The intermolecular attraction due to the continuous fluctuation of electron clouds throughout atoms and molecules, which produce transient dipole moments.

Questions and Problems

1   Define the terms phase, phase transition, phase diagram, ΔH_{vap}, ΔH_{sub}, and ΔH_{fus}.

2   Give the stable phase for H_2O under the following conditions (refer to Figure 8–7): (a) 300°C, 760 mm Hg; (b) 0.01°C, 4.58 mm Hg; (c) 150°C, 100 atm; and (d) −50°C, 760 mm Hg.

3   It can be shown experimentally that 755 cal is required to transform 15.5 g of solid methylamine, CH_3NH_2, into liquid methylamine. What is the ΔH_{fus} in cal mole^{-1}? How many grams of propionitrile, CH_3CH_2CN, ($\Delta H_{fus} = 1206$ cal mole^{-1}) could be melted with this same quantity of heat?

4   Write equations for an exothermic and an endothermic process and show how the heat change may be expressed.

5   What is the difference between attractive van der Waals forces and a permanent dipole–dipole interaction? Is a hydrogen bond either of these? Why is a hydrogen bond special?

6   What is the source of bonding in each of the following: (a) ionic attraction, (b) covalent bonding, (c) van der Waals attraction, (d) hydrogen bonding, and (e) metallic bonding. How do the bonds compare in strength? Can you think of examples of each?

7   What effect do hydrogen bonds have on the boiling points of liquids? Explain, and give an example.

8   Which molecule of each pair would you expect to have the higher boiling point? What factor is important in each case? (a) $(CH_3)_2NH$ or $(CH_3)_3CH$; (b) CH_3OCH_3 or CH_3CH_2OH; (c) HF or HBr; and (d) CH_4 or SiH_4.

9   How do we know that some hydrogen bonding in H_2O persists in the liquid phase?

10 What physical effect is responsible for the attraction in van der Waals interactions? What is responsible for the repulsion? Compare the origin of attraction and repulsion in van der Waals interactions and in ionic and covalent bonds.

11 The boiling temperatures and heats of vaporization are given for four compounds. Plot a graph of the boiling point temperatures (ordinates) vs. ΔH_{vap} (abscissa). Write a mathematical expression describing the relationship between T and ΔH_{vap}.

Compound	Boiling temperature, T	ΔH_{vap} (cal mole^{-1})
HCl	188°K	3860
Cl_2	239°K	4878
CS_2	319°K	6400
C_8H_{18}	399°K	8360

12 There are two factors that seem to work in opposite directions in determining the phase of a substance; one is to put the substance into the state of lowest heat content, and the other is to put it in the state of maximum entropy. What is meant by "state of maximum entropy"? Use the melting of a substance to illustrate that this state of maximum entropy is not the same as the state of lowest heat energy.

13 Metals are good conductors of electricity. Explain the conduction properties of metals in terms of the bonding that holds the atoms together.

14 The heats of vaporization for the hydrides of Group VI are given as follows (in kcal mole^{-1}): H_2Te (5.55), H_2Se (4.62), H_2S (4.46), and H_2O (9.72). There appears to be a trend through the first three hydrides listed that suggests that the lower the molecular weight the lower the heat energy required to vaporize the substance. Explain why water deviates from this apparent trend.

15 Refer to the phase diagram for water, Figure 8–7, and explain why the boiling temperature of water is lower than 100°C on a mountain where the atmospheric pressure is 0.80 atm.

16 If van der Waals bonds are extremely weak, why are they discussed at all?

17 Describe the type of solid structure that you would expect the following substances to form: LiH, Si, Br_2, $SrCl_2$, N_2, $CHCl_3$, B_2O_3, CH_3OH, CaO, Kr, SF_6, Ba, $(CH_3)_2CO$, and Cd.

SOLUTIONS

The preceding three chapters have given us a glimpse of the structural and bonding characteristics of many substances. With this chapter, we begin the study of chemical reactions and reactivity relationships. We have learned that chemical reactions can take place in gases, liquids, and solids. Recall that in Chapter 2 we explored the properties of the gaseous state, and in Chapter 8 we discussed the structures and properties of pure liquids and solids. Most common chemical reactions, however, take place in liquid *solutions.* This includes almost all chemical reactions occurring in living things. It is evident, therefore, that basic knowledge of the structures and properties of solutions is vital to an understanding of chemistry. In this chapter, we will discuss some uses of solutions and the interactions between like and unlike molecules that determine the solubilities of substances.

9–1 Homogeneous and Heterogeneous Mixtures

A solution is a homogeneous mixture. Familiar examples are water solutions of salt or sugar. If a spoonful of sugar is added to a tumblerful of water and the contents are stirred, the sugar dissolves. The resulting mixture is almost identical in appearance to pure water. However, the solution has many properties that are different from those of pure water. For example, if you drink some of the solution, you can taste the sugar.

We have already pointed out (Section 1–2) that homogeneous mixtures cannot be separated by mechanical means, whereas heterogeneous mixtures can. However, some heterogeneous mixtures are so difficult to separate into components that this distinction between homogeneous and heterogeneous mixtures is inadequate and impractical. Consider, for example, a pan of dishwater, which contains many suspended particles of food and grease. If the dishwater is filtered, some particles are removed, but the liquid that passes through the filter (called a filtrate) is still heterogeneous. It looks cloudy because light does not pass straight through the mixture, but it is scattered by tiny oil droplets that are too small to be seen with the naked eye. Another example of light scattering by a heterogeneous mixture is provided by the *diffuse* light of the sun passing through atmospheric haze on a foggy morning. A large amount of light gets through the haze, but we can tell that it has been scattered in all directions because objects illuminated by this diffuse light do not cast shadows.

A mixture that can be separated by mechanical means into two or more components is heterogeneous. But even if it cannot be separated, it is not necessarily homogeneous, for it may be separable by some method that was not tried. To avoid confusion, we define a homogeneous mixture as one in which the molecules of one material are distributed uniformly throughout another material. In a heterogeneous mixture, there are islands (or lakes) of one substance dispersed through the other. Examples are grains of sand in a sand–water mixture and the oily droplets in dirty dishwater.

Solutions may be gases, liquids, or solids. The earth's atmosphere is a solution containing approximately 80% nitrogen and 20% oxygen, along with small amounts of other gases such as water vapor, carbon dioxide, and so forth. Table 9–1 shows the average composition of the atmosphere. Salt dissolved in water is an example of a liquid solution. Steel is a solid solution, in which small amounts of other elements are dissolved in iron. Table 9–2 shows the composition of several kinds of steel.

Table 9–1. The Average Composition of the Earth's Atmosphere, a Gaseous Solution

Element	Percentage by volume of dry air
Nitrogen (N_2)	78.03
Oxygen (O_2)	20.99
Argon (Ar)	0.93
Carbon dioxide (CO_2)	0.03
Hydrogen (H_2)	0.01
Neon (Ne)	0.0018
Helium (He)	0.00052
Krypton (Kr)	0.0001
Xenon (Xe)	0.000008

Table 9–2.   The Compositions of Three Kinds of Steel, Solid Solutions

Name	Composition (weight %)
Silicon steel	97.6 Iron (Fe)
	2 Silicon (Si)
	0.4 Carbon (C)
Tungsten steel	94.5 Iron (Fe)
	5 Tungsten (W)
	0.5 Carbon (C)
High-Speed steel	75 Iron (Fe)
	18 Tungsten (W)
	6 Chromium (Cr)
	0.3 Vanadium (V)
	0.7 Carbon (C)

9–2   Definitions of Some Useful Terms

A number of terms are used in discussing solutions. Many solutions consist of a major component and one or more minor components. The major component is called the *solvent* and the minor components are called *solutes.* Seawater is a liquid solution, with water as the solvent and many components as solutes. The most abundant solute is sodium chloride (common table salt), but the oceans also contain many other salts in lesser amounts. Table 9–3 shows the average composition of seawater, excluding dissolved gases. Since the oceans do not mix very rapidly, the amounts of solutes will vary from ocean to ocean, and also with the depth from which a sample is taken.

The amount of a solute in a given amount of solution is called its *concentration.* Concentrations are sometimes given in terms of the amount of the individual components found in a given volume of the mixture, and sometimes in terms of the amount of a component found in a given weight of the mixture. In addition, the amount of a component may be reported in terms of weight, number of moles, or, occasionally, in terms of volume. As a result, there are a large number of measures of concentration that must be kept straight.

The multiplicity of measures of concentration is certain to irritate and frustrate students, as it does the authors. Why should there be so many ways of reporting the concentration of a solution? The reasons are partly logical and partly illogical, and they arise from the fact that solutions are used by many different people for many different reasons. Consider, for instance, the objectives of people interested in iron ore. The ore contains oxides of iron mixed with many other materials. Whatever its composition, the ore will almost certainly be sold for a price quoted in dollars per ton, because the weight can be measured accurately. A steel manufacturer will be interested in the amount of metallic iron that he can get for his dollars. Consequently, the concentration of the mixture will probably be reported as *weight percent of iron,* even though the ore contains iron oxides and not metallic iron. For example, suppose that a sample of iron ore weighs 30.5 g and is composed of a mixture of substances, including iron oxides. Chemical analysis shows that the total amount of pure iron that can be recovered from this particular sample is 10.7 g. This information usually would be reported as weight percent, according to the calculation

$$\frac{10.7 \text{ g iron}}{30.5 \text{ g sample}} \times 100 = 35.1\% \text{ iron}$$

The interest of a chemist or a geologist may center around the magnetic properties of the ore. Then he would wish to know the amount of a particular iron oxide in the sample and the amounts of certain other compounds, which also have magnetic properties. The geochemist (geological chemist) would want to know the number of moles of each compound of interest in a given weight of sample. Since iron ore is a loosely packed, heterogeneous mixture, the moles per weight of sample is a better measure than moles per volume.

A chemist working in an analytical laboratory may have a still different outlook. He may prepare a solution, for use as a chemical reagent, by dissolving iron ore in hydrochloric acid. Since the cost of a small sample of ore would be a negligible part of the entire cost of his work, he would have little interest in the actual composition of the ore. But he would want to know the concentration of the solution that he has prepared. Measurement of solutions by volume is easier than measurement by weight. Thus, the laboratory chemist probably will measure the concentration of iron salts in solution in terms of moles of compound *per liter of solution*. Since measurement of the amount of iron (or iron oxide) is ordinarily done by chemical methods, chemists must be familiar with all common methods for reporting concentrations of the components in mixtures.

MOLARITY

Suppose that a solution is prepared by dissolving 40.0 g (1.00 mole) of NaOH in enough water to make a final solution having a volume of 1000 ml (1 liter). Obviously, if the entire liter of solution

Table 9–3.   The Composition of Seawater

Element[a]	Concentration (g metric ton^{-1})
Chlorine (Cl)	18,980
Sodium (Na)	10,561
Magnesium (Mg)	1,272
Sulfur (S)	884
Calcium (Ca)	400
Potassium (K)	380
Bromine (Br)	65
Carbon (C)	28
Strontium (Sr)	13

[a] *There are at least 45 other elements found in seawater in measurable or detectable amounts.*

is used in some reaction, you will use the entire 40.0 g of NaOH. If 1 ml is used, only 1/1000 of the one mole of NaOH (0.00100 mole) is consumed. Making solutions with a known weight of solute per volume of solution is extremely useful, since the amount of solute present in any volume of the solution can be calculated by multiplying the volume used times the concentration. If the volume is expressed in liters and the concentration of the solution in moles per liter, we have

$$\text{liters} \times \text{moles liter}^{-1} = \text{moles} \tag{9-1}$$

Molarity is the concentration unit defined as the number of moles of solute dissolved in one liter of solution. The symbol for molarity is M. A solution is said to be $0.5M$ (read "0.5 molar"), if one half of a mole of solute is dissolved in one liter of solution, or if the number of moles of solute divided by the total volume of solution (in liters) is 0.5.

Example 9–1. Exactly 25.0 g of sugar is dissolved in water, and the resulting solution is diluted to make exactly one liter of solution. What is the molarity of the solution?

Solution. The molecular formula of sugar is $C_{12}H_{22}O_{11}$ (if the molecular formula of a common substance is not known, it can be found in a chemical handbook or a textbook). One mole of $C_{12}H_{22}O_{11}$ contains 12 moles of C atoms, 22 moles of H atoms, and 11 moles of O atoms. We first calculate the molecular weight of sugar

$$
\begin{array}{lll}
\text{Carbon:} & 12 \text{ moles} \times 12.01 \text{ g mole}^{-1} = & 144.1 \text{ g} \\
\text{Hydrogen:} & 22 \text{ moles} \times \ 1.01 \text{ g mole}^{-1} = & \ 22.2 \text{ g} \\
\text{Oxygen:} & 11 \text{ moles} \times 16.00 \text{ g mole}^{-1} = & \underline{176.0 \text{ g}} \\
\text{Molecular weight of sugar:} & & = 342.3 \text{ g}
\end{array}
$$

A $1M$ (one molar) solution of sugar contains 342 g liter^{-1} of solution. The molar concentration of the solution containing 25.0 g liter^{-1} can now be calculated

$$\frac{25.0 \text{ g}}{\text{liter}} \times \frac{1 \text{ mole}}{342 \text{ g}} = 0.0731 \text{ mole liter}^{-1} = 0.0731M$$

Therefore, the concentration of the solution is $0.0731M$.

An alternative approach to this problem is to recognize that you are trying to calculate the number of moles of sugar in one liter of solution, given that 25.0 g of sugar has been dissolved in enough water to make one liter of solution. The simple conversion of grams to moles will give us moles liter^{-1}. This conversion is accomplished by the calculation

$$\frac{25.0 \text{ g}}{342 \text{ g mole}^{-1}} = 0.0731 \text{ mole of sugar}$$

Since this is the number of moles in a liter of solution, it is the molar concentration, 0.0731 mole liter^{-1}, or $0.0731M$.

Example 9–2. How much sodium chloride is required to prepare 150 ml of a $0.120M$ solution?

Solution. The molar weight of sodium chloride (NaCl) is $23.0 + 35.5 = 58.5$ g. One liter of a $0.120M$ solution of sodium chloride contains

$$0.120 \text{ mole liter}^{-1} \times 58.5 \text{ g mole}^{-1} = 7.02 \text{ g NaCl liter}^{-1}$$

We want to make only 150 ml (or 0.150 liter) of solution, so the weight of sodium chloride required is

$$0.150 \text{ liter} \times 7.02 \text{ g liter}^{-1} = 1.05 \text{ g}$$

This calculation can be done in a more compact form

$$\text{grams of NaCl} = 0.120 \text{ mole liter}^{-1} \times 58.5 \text{ g mole}^{-1} \times 0.150 \text{ liter} = 1.05 \text{ g}$$

Calculations of this type should always be checked to make sure that the answer is reasonable. The first test is to make certain that the units are correct. In the preceding example, a formula is given for calculating the number of grams of NaCl required. This is not the sort of formula that should be memorized; it is simply a guide to thinking about the problem. Notice that the units of moles and liters appear both in the numerator and in the denominator (recall that mole^{-1}, for example, is the same as 1/mole); therefore, these units cancel, leaving the units of grams. Also notice that the molarity of the solution is about 0.1, and the volume of solution is about 0.1 liter, so the answer to the problem must be greater than 0.585 g and less than 10 g. Therefore, the answer makes sense.

It is sometimes necessary in the laboratory to calculate the molarity of a solution when its concentration is known only in terms of weight percent. It is possible to convert this information into molarity, if the density of the solution is known.

Example 9–3. Calculate the molarity of a solution of commercially prepared nitric acid (HNO_3), which is 70.3% nitric acid and has a density of 1.42 g ml^{-1}.

Solution. Since the amount of HNO_3 present is expressed as weight percent, we could select any weight of the solution to work with and calculate what part of it is HNO_3. However, since we want to express the concentration in moles per liter, it is convenient to start with a liter of the solution. The weight of a liter of the solution is determined easily by multiplying the density and the volume

$$1000 \text{ ml liter}^{-1} \times 1.42 \text{ g ml}^{-1} = 1420 \text{ g liter}^{-1}$$

Of this total weight of solution, 70.3% is HNO_3, which, in grams, is

$$0.703 \times 1420 \text{ g liter}^{-1} = 998 \text{ g } HNO_3 \text{ liter}^{-1}$$

Conversion of g HNO_3 liter^{-1} to moles HNO_3 liter^{-1} will give molarity

$$\frac{998 \text{ g liter}^{-1}}{63.0 \text{ g mole}^{-1}} = 15.8 \text{ moles liter}^{-1} = 15.8M$$

FORMALITY

The definition of formal concentrations is similar to that of molar concentration. The formality of a solution is the number of formula weights of solute per liter of solution. The symbol for formality is F. The formula weight is the weight obtained by adding the atomic weights of the atoms in the formula of a compound. In many cases, formula weight is the same as molecular weight.

The most important distinction between formality and molarity is that formality gives a recipe for preparing a solution, whereas molarity tries to describe a solution. Consider, for example, a solution of acetic acid (vinegar) in gasoline. In such a solution, some of the acetic acid exists as single molecules, some as dimers (two acetic acid molecules joined together), and some as higher polymers (several acetic acid molecules joined together). It is misleading to talk about the molarity of such a solution, because when we say the solution is $0.1M$ in acetic acid, we imply that the solute is present as single acetic acid molecules. However, if we refer to a $0.1F$ solution of acetic acid in gasoline, we are saying that the solution was made by diluting 6.005 g (the molecular weight of acetic acid is 60.05) of acetic acid to one liter with gasoline. Reporting solution concentrations in formality units is an intelligent expression of ignorance (we don't know exactly what form the solute has in the solution), or a way of avoiding the complexity of a complete description of the solution (we know, but for our purposes, it doesn't matter). In this book we will use both molar and formal concentrations.

An important application of solutions is their use in carrying out chemical reactions. For example, the reaction of sulfuric acid and sodium hydroxide (a base)

$$H_2SO_4 \; + \; 2NaOH \; \rightarrow \; Na_2SO_4 \; + \; 2H_2O \qquad\qquad (9\text{–}2)$$
$$\text{sulfuric} \quad \text{sodium} \qquad \text{sodium} \quad \text{water}$$
$$\text{acid} \qquad \text{hydroxide} \quad \text{sulfate}$$

would not be performed by pouring sulfuric acid on solid pellets of NaOH. A solution of a known concentration of H_2SO_4 would be prepared and then mixed with a solution of known concentration of sodium hydroxide. Suppose that a 100-ml sample of $0.1M$ H_2SO_4 were used. What volume of $0.1M$ NaOH solution would be required to react with all the H_2SO_4? From Equation 9–2, two moles of NaOH are required for every mole of H_2SO_4. Since the number of moles is the concentration times the volume, we have

$$\frac{\text{moles NaOH}}{\text{moles } H_2SO_4} = 2 = \frac{x \text{ liter NaOH} \times 0.1 \text{ mole liter}^{-1}}{0.1 \text{ liter } H_2SO_4 \times 0.1 \text{ mole liter}^{-1}}$$

So $x = 0.2$ liter, and 200 ml of NaOH solution is required.

Example 9–4. Phosphoric acid reacts with sodium hydroxide as follows:

$$H_3PO_4 \; + \; 3NaOH \; \rightarrow \; Na_3PO_4 \; + \; 3H_2O$$
$$\text{phosphoric} \quad \text{sodium} \qquad \text{sodium} \quad \text{water}$$
$$\text{acid} \qquad \text{hydroxide} \quad \text{phosphate}$$

Calculate the weight of phosphoric acid required to react with 500 ml of $1.00M$ NaOH solution.

Solution. The 500 ml of $1.00M$ NaOH solution contains 0.500 mole of NaOH. From the equation, 1 mole of H_3PO_4 will react with 3 moles of NaOH, or 1/6 mole of H_3PO_4 will react with 0.500 mole of NaOH. The molecular weight of phosphoric acid is 98.0 g mole^{-1}, so the required weight is 98.0 g mole$^{-1} \times$ 1/6 mole $= 16.3$ g H_3PO_4.

Another important application of solution concentrations is the calculation of the *dilution* required to convert a solution of a known concentration to a specific lower concentration. Standard commercial preparations of some of the more common reagents are shown in Table 9–4. Suppose that it is necessary to have a $9M$ H_2SO_4 solution. This can be made by mixing 500 ml of standard $18M$ H_2SO_4 with water to make a total volume of one liter. Doubling the

volume cuts the concentration in half. This is a simple example of the general relationship between a solution and one prepared from it by dilution

$$concentration_1 \times volume_1 = concentration_2 \times volume_2$$

Example 9–5. A solution is $0.245M$ in HNO_3. It is necessary to have 150 ml of a $0.100M$ HNO_3
solution. What volume of the original acid solution would you use to prepare the new
solution?

Solution. The moles of HNO_3 in the final solution can be calculated by multiplying the final
volume ($V_2 = 150$ ml) times the final concentration ($M_2 = 0.100$). This must also equal the
number of moles of HNO_3 in the original volume, V_1, of the more concentrated solution ($M_1 = 0.245$). Therefore,

$$M_1 \times V_1 = M_2 \times V_2$$

$$0.245M \times V_1 = 0.100M \times 150 \text{ ml}$$

$$V_1 = \frac{0.100M \times 150 \text{ ml}}{0.245M}$$

$$V_2 = 61.2 \text{ ml}$$

This volume, 61.2 ml, of the $0.245M$ acid would be diluted to 150 ml, giving the new $0.100M$
solution.

Other measures of concentration — molality, weight percent, mole fraction, and normality —
are discussed in Appendix 6.

9–3 Chemical Calculations with Solutions

Concentrations expressed in molar or formal units are useful because they can be used directly
to calculate the amounts of solutions needed for chemical reactions. Consider the reaction
between bromine (Br_2) and ethylene ($H_2C \!=\! CH_2$) to form 1,2-dibromoethane (Br—CH_2—CH_2—Br).
The first fact we need to know is how many moles of bromine react with one mole of ethylene.
This we find by writing a balanced equation for the reaction

$$Br_2 + H_2C \!=\! CH_2 \rightarrow Br\!-\!CH_2\!-\!CH_2\!-\!Br$$

Table 9–4. Concentrations of Some Commercial Reagents

Reagent	Concentration (M)
Ammonia (NH_3) solution	15
Hydrochloric acid	12
Nitric acid	15
Sulfuric acid	18

Thus, for every mole of ethylene, we need one mole of bromine. Suppose that we have 1.65 g of ethylene and wish to know what volume of 0.125M bromine solution is required to react with all of the ethylene. The key to the calculation is that the number of moles of bromine can be calculated from the volume and the molar concentration of the solution, and that the following equality exists

moles bromine = moles ethylene

We calculate the number of moles of ethylene by dividing the sample weight by the molecular weight of ethylene

$$\text{moles ethylene} = \frac{\text{g ethylene}}{\text{mol wt ethylene}} = \frac{1.65 \text{ g}}{28.0 \text{ g mole}^{-1}} = 0.0589 \text{ mole}$$

Let V stand for the number of liters of bromine solution. Then

$$\text{moles bromine} = V \times M = V \text{ liters} \times 0.125 \text{ mole liter}^{-1}$$

Since moles ethylene = moles bromine, we have

0.0589 mole of bromine = 0.125 mole liter^{-1} × V liters

or

$$V = \frac{0.0589 \text{ mole}}{0.125 \text{ mole liter}^{-1}} = 0.471 \text{ liter or 471 ml bromine solution}$$

In many cases, one of the reactants in a chemical reaction is a gas. This is the case with ethylene. The most convenient way to specify the amount of gaseous ethylene in the sample is to take a sample of a given volume at a certain temperature and pressure.

Example 9–6. Calculate the volume of the same solution of bromine required to react completely with 500 ml of gaseous ethylene at 25°C and 740 torr.

Solution. The most direct way to calculate the number of moles of ethylene present is to use the combined gas law (Section 2–4) $PV = nRT$. Changing temperature, pressure, and volume to the proper units and substituting into the gas law equation gives

$$n \text{ (number of moles)} = \frac{740 \text{ torr}/(760 \text{ torr atm}^{-1}) \times 500 \text{ ml}/(1000 \text{ ml liter}^{-1})}{0.0820 \text{ liter atm mole}^{-1} \text{ °K}^{-1} \times 298°K}$$

$$= 0.0200 \text{ mole ethylene}$$

Since the number of moles of bromine equals the number of moles of ethylene,

0.0200 mole bromine = $V \times 0.125M$

$V = 0.160$ liter

Often, we wish to use a chemical reaction to prepare a given amount of product, as in the following example.

Example 9–7. Calculate the volume of the 0.125M solution of bromine required to produce 20.0 g of 1,2-dibromoethane (Br—CH_2—CH_2—Br) by reaction with excess ethylene.

Solution. From the balanced equation,

moles of bromine = moles of 1,2-dibromoethane

The number of moles of 1,2-dibromoethane can be calculated by dividing the grams of this substance by its molecular weight

$$\frac{20.0 \text{ g}}{188 \text{ g mole}^{-1}} = 0.106 \text{ mole}$$

Therefore,

moles of bromine = 0.106 mole = $V \times 0.125M$

$V = 0.848$ liter of bromine solution

Let us emphasize once again the idea that the product of the volume (in liters) and the concentration will lead to the amount of solute in that volume

$M \times V =$ moles liter^{-1} $\times$ liter = moles

$F \times V =$ formula weights liter^{-1} $\times$ liter = formula weights

TITRATIONS

Chemical reactions often are accomplished by mixing solutions. To avoid waste or contamination of the product of the reaction by unused starting materials, accurately measured volumes of solutions of known concentrations can be used. Graduated cylinders, burettes, volumetric flasks, and pipettes commonly are used to measure liquid volume. Figure 9–1 shows some of these pieces of equipment, which should also be familiar to students from their own laboratory experiments.

Figure 9–1. Some glassware commonly used for making volumetric measurements.

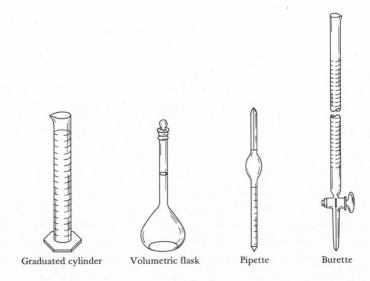

Graduated cylinder   Volumetric flask   Pipette   Burette

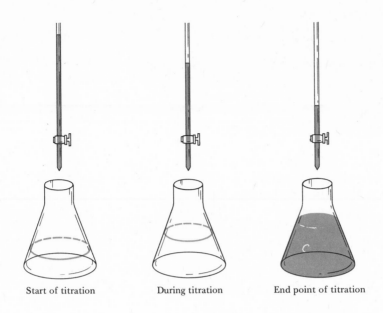

Start of titration	During titration	End point of titration

Figure 9–2.   The titration of an acidic solution with a basic solution to a colored end point.

   Reactions in solution can be used to prepare chemicals or to eliminate chemicals. In addition, they are very important in determining the concentrations of solutions. For example, suppose that we wish to prepare a solution of sodium hydroxide (NaOH) of known concentration. Because solid sodium hydroxide reacts with water vapor and carbon dioxide in the air, it is usually impure. Thus, it is impractical to prepare a solution of known concentration by weighing a sample and dissolving it in water. However, it is very easy to prepare a solution of hydrochloric acid of known concentration. Sodium hydroxide and hydrochloric acid react with each other to produce water

$$HCl + NaOH \rightarrow NaCl + H_2O$$

Suppose that we prepare a solution of sodium hydroxide of unknown concentration by dissolving impure sodium hydroxide in water. Then if we can determine how much hydrochloric acid solution (of known concentration) is required to react completely with a known volume of sodium hydroxide solution, we can determine the concentration of the sodium hydroxide solution. That is,

$$moles\ HCl \rightarrow moles\ NaOH$$
$$V_{HCl} \times M_{HCl} = V_{NaOH} \times M_{NaOH}$$

We know V_{NaOH} and M_{HCl}. If we can find V_{HCl}, we can use the above relationship to calculate M_{NaOH}. The process of allowing the two solutions to react in a way that allows us to determine accurately V_{HCl} is called *titration*. A common way of determining the *equivalence point* (the point at which the number of moles of NaOH added is exactly equivalent to the number of moles of HCl originally present) is to add to the hydrogen chloride solution a small amount of an indicator, a substance that changes color at the equivalence point. There are many different kinds of indicators. For this titration, we would choose an indicator that is one color in basic solution and another color in acidic solution. The point at which the indicator changes color is called the end point. To have a successful titration, we must choose the indicator carefully so that the end point and equivalence point are the same.

Figure 9–2 shows how such a titration is performed. A measured volume of HCl solution is placed in a flask along with a few drops of indicator solution, and the NaOH solution is added slowly from a burette while the flask is swirled to mix the solutions.

Example 9–8. A 25.0-ml sample of a solution of 0.100M hydrochloric acid is titrated with a sodium hydroxide solution. The end point is reached when 42.0 ml of the basic solution has been added. What is the concentration of the solution of base?

Solution. The condition at the equivalence point is

$$\text{moles HCl} = \text{moles NaOH}$$
$$V_{\text{HCl}} \times M_{\text{HCl}} = V_{\text{NaOH}} \times M_{\text{NaOH}}$$
$$25.0 \text{ ml} \times 0.100 \text{ mole liter}^{-1} = 42.0 \text{ ml} \times M_{\text{NaOH}}$$

$$M_{\text{NaOH}} = \frac{25.0 \text{ ml}}{42.0 \text{ ml}} \times 0.100 \text{ mole liter}^{-1}$$

$$M_{\text{NaOH}} = 0.0595 \text{ mole liter}^{-1}$$

Example 9–9. A 0.750-g sample is taken from a mixture of sodium chloride and sodium hydroxide. The sample is dissolved in water and titrated with 0.200M sulfuric acid (H_2SO_4). The amount of the acid solution required is 30.0 ml. What is the weight percent of sodium hydroxide in the mixture?

Solution. Only the NaOH will react with the dilute sulfuric acid; the NaCl does not enter into the reaction at all. Two moles NaOH react with one mole H_2SO_4. Thus, the number of moles of NaOH in the sample will be twice the number of moles of H_2SO_4 consumed in the titration.

$$\text{moles } H_2SO_4 = 0.0300 \text{ liter} \times 0.200 \text{ mole liter}^{-1} = 0.006 \text{ mole}$$
$$\text{moles NaOH in the sample} = 0.012 \text{ mole}$$

Convert moles NaOH to grams of NaOH by the following:

$$\text{grams NaOH} = 0.012 \text{ mole} \times 40 \text{ g mole}^{-1} = 0.480 \text{ g NaOH}$$

$$\text{wt \% NaOH} = \frac{0.480 \text{ g}}{0.750 \text{ g}} \times 100 = 64.0\%$$

9–4 Solubility and Equilibrium

The amount of a solute that can be dissolved in a given amount of a solvent depends on the temperature. When a solution contains the maximum amount of a solute at a specific temperature, it is said to be *saturated*. The concentration of a solute contained in a saturated solution is called the *solubility*. The solubilities of a few common substances in water at 20°C are listed in Table 9–5.

Table 9–5.   Solubilities of Common Substances in Water at 20°C

Substance	Solubility (grams per 100 ml)
Sugar (sucrose)	203.9
Salt (sodium chloride)	36.0
Sodium hydroxide	109
Barium hydroxide	3.89
Silver chloride	0.00015
Iodine	0.029

The terms soluble and insoluble commonly are used by chemists, even though they are not very precise. Sugar and salt both are soluble in water, but the solubility of sugar is much greater than that of salt. Silver chloride usually is said to be insoluble in water, because only a very small amount will dissolve. Although the boundaries are not well defined, we usually say that a substance is soluble if as much as 10 g will dissolve in a liter of water.

EQUILIBRIUM IN SATURATED SOLUTIONS

If 100 g of salt is placed in 100 ml of water at 20°C, salt will go into solution until 36.0 g has dissolved. If the temperature is raised to 100°C, more salt dissolves until the solution contains 39.8 g. If the temperature is then lowered to 20°C, salt will crystallize from the solution until the amount in solution is again 36.0 g. Why does the salt go into solution at the higher temperature and then recrystallize at the lower temperature? Why is the final amount of salt in solution at 20°C the same when the hot solution is cooled to 20°C as when the salt is allowed to dissolve at that temperature? The answer is that the solubility of salt in water at a certain temperature does not depend on how we prepare the solution. It may seem intuitively reasonable that salt should have a fixed solubility that does not depend on how the solution is prepared. However, we can gain a great deal of insight into the properties of a solution by trying to explain why this is true.

Experience has taught us that a saturated solution in contact with undissolved solute is in a state of *dynamic equilibrium.* This means that solute continues to go into solution, but the *rate of dissolution (dissolving) is exactly the same as the rate at which solute comes out of solution.* Let us assume that all soluble molecules are identical, except that we have a way to keep track of individual molecules so that we can watch them go in and out of solution. Figure 9–3 shows the concept. Let us pretend that all of the molecules in the solid phase at the beginning of our imaginary experiment are dark; likewise we pretend that those in solution are light. After a time, if we reexamine the solid and the solution, we will find that the number of molecules in each phase is the same, but that now there are light molecules in the solid and dark molecules in the solution.

The experiment that we have described is imaginary, but real experiments very much like it can be performed. In place of the black and white colored labels, we can use two different isotopes of the same element. For example, chlorine nuclei in naturally occurring chlorine come in two varieties, those having mass 35 and those having mass 37. Consequently, there are two varieties of chloride ions. The isotopic chloride ions have almost the same chemical properties. However, the two isotopes of chlorine can be separated, and mixtures of the two can be analyzed to determine the isotopic composition. To do the experiment, a solution is saturated with ordinary sodium chloride, in which the $^{35}Cl/^{37}Cl$ ratio is 3.36, and the excess undissolved salt removed by filtration. If sodium chloride enriched in $^{35}Cl^-$ is then added to the saturated solution, it may appear not to dissolve; that is, the molar concentration of NaCl in solution does not change. However, isotopic analysis after time has passed will show that the $^{35}Cl^-$ concentration in the solution has increased, and that the amount of $^{37}Cl^-$ in the solid also has increased. This experiment tells us that an exchange of chloride ions between the solid and solution has taken place.

The equilibrium condition for such an exchange is expressed by writing the two states (solid and solution), and placing between them two arrows pointing in opposite directions

solid $\rightleftarrows$ solution

The top arrow indicates molecules are leaving the solid phase and going into solution, whereas the bottom arrow indicates that an equal number of molecules are leaving the solution and becoming solid. This symbolism will become quite useful when we deal in the next chapter with chemical reactions at equilibrium.

HEATS OF SOLUTION

Now let us return to the question of the temperature dependence of solubility. The solubility of most solids in liquids increases as the temperature is raised. However, there are numerous exceptions to this generalization. Figure 9–4 shows how the equilibrium solubilities of some common substances in water vary with temperature. This indicates that heat changes occur when materials dissolve. In many cases, these changes are quite large. When pure sulfuric acid is added to water, so much heat is liberated that mixing must be done slowly to avoid an explosion, caused by the rapid vaporization of water. However, dissolution of potassium nitrate (KNO_3) in water *absorbs* enough heat to make the solution feel cold to the touch. We recommend that students verify the second experiment in their laboratories.

Figure 9–3. A model experiment showing the exchange of molecules in the solid phase with molecules in solution.

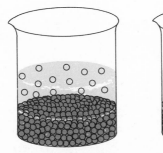

Beginning of experiment   After time passes

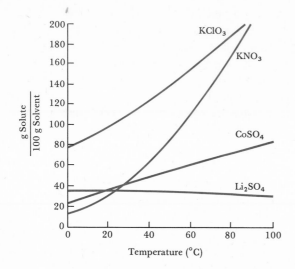

Figure 9–4.   A graph of the solubility of some common salts in water versus the tempera-ture of the solution.

The heat evolved in a solution process can be measured in the same way as are the heats of vaporization and fusion (Section 8–6). Table 9–6 shows the heats of solution of some common substances in water. (Remember that if $\Delta H_{solution}$ is negative, heat is liberated when the substance goes into solution.) Table 9–7 shows the heats of solution of one substance, the element iodine, in several different solvents.

If we examine the values given in Table 9–6, we see several connections between the solubility of a compound and its heat of solution. The solubilities of the sodium salts are in the order NaCl < NaI < NaOH, whereas the heats of solution are in exactly the opposite order, NaCl > NaI > NaOH. That is, the more heat that is produced when the compound is dissolved, the more

Table 9–6.   Heats of Solution of Some Common Substances in Water at 25°C

Substance	Heat of solution ΔH (kcal mole^{-1})	Saturation solubility (moles liter^{-1})
Sodium chloride (NaCl)	0.93	6.18
Sodium iodide (NaI)	−1.8	12.3
Sodium hydroxide (NaOH)	−10.6	38.5
Silver fluoride (AgF)	−3.4	14.1
Silver chloride (AgCl)	15.81	0.000014
Silver iodide (AgI)	26.71	0.00000016
Magnesium bromide (MgBr$_2$)	−43.30	5.62
Ammonium chloride (NH$_4$Cl)	3.53	7.35
Potassium chloride (KCl)	4.12	4.76

Table 9–7. *Heats of Solution of Iodine in Common Solvents at 25°C*

Solvent	Heat of solution, ΔH (kcal mole^{-1})	Saturation solubility (g I$_2$ liter^{-1})
Cyclohexane (C$_6$H$_{12}$)	−5.8	29.1[a]
Benzene (C$_6$H$_6$)	−4.25	144[a]
Toluene (C$_6$H$_5$CH$_3$)	−3.85	3.56
Ethanol (C$_2$H$_5$OH)	−1.65	214[a]
Carbon tetrachloride (CCl$_4$)	−5.8	29.1

[a] *Grams* I$_2$ *per liter of solvent. The other solubilities are given in grams* I$_2$ *per liter of solution.*

soluble it is. The same is true of the silver halides; the solubilities are in the order AgF > AgCl > AgI, whereas the heats of solution are in the order AgF < AgCl < AgI. This is reasonable because, if energy is evolved in the form of heat when a solute is dissolved, it must contain less heat energy in the solution than it had in the solid form. Therefore, as far as the heat energy is concerned, the compound would "prefer" to be in solution, and it should dissolve. This argument can be illustrated by the problem of getting a marble through a plastic maze in the shape of a cube (Figure 9–5). When a marble is dropped into the maze, it has a large amount of potential energy. But as the marble drops to lower levels, it loses energy, and it has the lowest energy on the bottom level. If a marble is on the second level, it looks stable, in the same way that solid

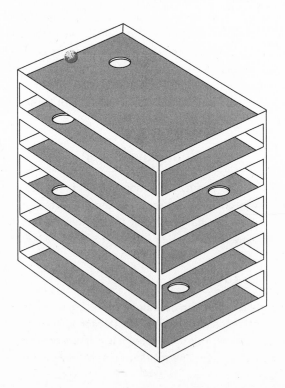

Figure 9–5. A cubic plastic maze in which a marble placed on the top level may fall to the second level, and so on, to the ground level, giving up its potential energy in doing so.

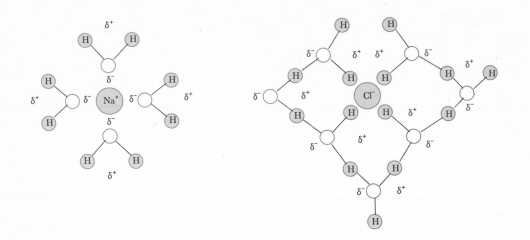

Figure 9–6.   A model of the solvation of the Na$^+$ *and* Cl$^-$ *ions with the solvent water.*

silver fluoride is stable (for instance, if you pound it with a hammer, it doesn't explode). But if you jiggle the box a bit, the marble will roll around and, if it reaches a hole, it will fall through to the next level. In a similar way, solid silver fluoride will dissolve when brought into contact with water, to reach a lower heat energy state.

The heats of solution do not explain entirely the solubilities of substances. In the first place, there is not a strict relationship between the amount of heat evolved and solubility. Referring to Table 9–6, we see that magnesium bromide, which evolves a large amount of heat when it dissolves, is less soluble than sodium chloride, which absorbs heat when it dissolves. Also, ammonium chloride and potassium chloride, which are very similar compounds, have almost the same heats of solution (about 4 kcal mole^{-1}). Yet, the solubility of ammonium chloride is about 7 moles liter^{-1}, whereas that of potassium chloride is about 5 moles liter^{-1}. The data in Table 9–7 further illustrate that heats of solution alone cannot be used to predict solubilities.

9–5   Structure–Solubility Relationships

In Chapter 10, we will discuss factors other than heat changes that help to determine the position of equilibrium in chemical systems. For the present, we can obtain some useful general ideas about solubilities by thinking about the energy required to take a molecule out of one environment and place it into another.

IONIC COMPOUNDS IN POLAR SOLVENTS

Let us think about the dissolution of sodium chloride in water. In sodium chloride crystals, which are similar to potassium chloride (Figure 8–14), each ion is surrounded by six others having the opposite electrical charge. This arrangement is a stable one, and to disturb it requires the input of a large amount of energy. The melting point of sodium chloride is 801°C, and the heat of fusion is 7.24 kcal mole^{-1}. Even in the molten state, the attraction of the positive and negative ions to each other remains very strong; liquid sodium chloride boils at 1439°C with a heat of vaporization of 44.2 kcal mole^{-1}. By adding the heats of fusion and vaporization, we conclude that evaporation of sodium chloride crystals requires an energy input of 51.4 kcal mole^{-1}. Even then, separated ions are not produced, since sodium chloride vapor consists mostly of molecules—NaCl, $(NaCl)_2$, and larger species.

When sodium chloride dissolves in water, the chloride ions and sodium ions become free to move independently of each other in solution. Only a small amount of heat is used during dissolution. Where do we get the energy required to tear apart the firmly bound crystals of sodium chloride? There seems to be only one reasonable answer: The ions in solution must be very strongly attracted by water molecules. Knowledge of the structures of liquid solutions is just about as indefinite as the knowledge of the structures of pure liquids (Chapter 8). However, the evidence available indicates that ions in aqueous solutions are surrounded by clusters of rather rigidly bound water molecules. Figure 9–6 shows models of the way in which we believe sodium ions and chloride ions gather layers of water molecules around them in water solution. The interaction of solutes with solvent molecules is called *solvation,* and the energy required to separate solute molecules from a solvent is called the *solvation energy.*

The layer of water molecules around each ion provides a "friendly" environment. This is because water molecules are strongly attracted to positive ions when the negative (oxygen) end of the molecule is directed toward the positive charge

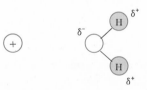

Similarly, water molecules are strongly attracted to negative ions when the positive (hydrogen) end of the molecule is directed toward the negative charge

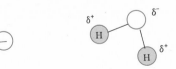

In Figure 9–6, several additional water molecules that form hydrogen bonds (Section 8–7) to each other are shown around the chloride ion. This illustrates the way in which the water molecules attracted to the dissolved ions fit into the overall hydrogen-bonded structure of pure water. This phenomenon raises another factor that affects solubilities. Large amounts of energy are required to separate sodium and chloride ions from crystals of sodium chloride. However, a great deal of stability is gained from the interaction of the dissolved ions with water molecules. An additional energy change is that which occurs in the solvent. The large heat of vaporization of water (10.5 kcal mole^{-1}) shows that energy is required to break up the structure of water. When a salt is dissolved in water, energy must be expended to disrupt the water structure. One of the reasons that sodium chloride is fairly soluble in water is that the sodium and chloride ions do not disturb the water structure very much.

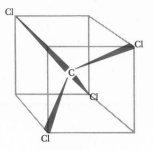

Figure 9–7.   The symmetrical tetrahedral molecule CCl_4.

Although sodium chloride dissolves in water, it has virtually no solubility in many other liquids, such as carbon tetrachloride (CCl_4). In contrast, the solubility of iodine is 29.1 g liter^{-1} in carbon tetrachloride, whereas its solubility is only 0.34 g liter^{-1} in water. This means that there are large differences in the ways these pairs of solutes and solvents interact with each other.

The preferred orientations of the water molecules shown in Figure 9–6 arise because water is a polar molecule (Section 8–7). The hydrogens have a net positive charge and the oxygen a net negative charge. *In general, ionic compounds tend to have high solubilities in polar solvents.*

IONIC COMPOUNDS IN NONPOLAR SOLVENTS

The carbon tetrachloride molecule is nonpolar. Figure 9–7 shows the symmetrical arrangement of the chlorine atoms around the central carbon atom. This is the tetrahedral arrangement discussed in Section 7–8. Because the molecule is symmetrical, it is nonpolar, regardless of the nature of the individual carbon–chlorine bonds. The *individual bonds* are polarized so that the chlorine atoms have a net negative charge

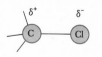

We might expect that salts dissolved in carbon tetrachloride would be more stable if positive ions were placed close to chlorine and negative ions close to carbon. However, as is shown in Figure 9–11, the carbon atoms are effectively shielded from close approach by any atom from outside the molecule. Although we have rationalized the fact that sodium chloride is less soluble in carbon tetrachloride than in water, we should not conclude that the ions are not attracted at all to carbon tetrachloride molecules. On the contrary, some interaction is required to produce the small solubility that is observed. Indeed, carbon tetrachloride is by no means the worst possible solvent for sodium chloride. The solubility of sodium chloride in a liquid such as gasoline is so low that it is essentially unmeasurable. Gasoline is a mixture of hydrocarbons (compounds of carbon and hydrogen), which are soluble in each other, and which are nonpolar.

Table 9–8.   Solubilities of Sodium Chloride in Water and Some Alcohols at 25°C

Solvent	*Solubility (g per 100 ml solvent)*
Water	36.1
Methanol, CH_3OH	1.66
Ethanol, CH_3CH_2OH	0.082
Propanol, $CH_3(CH_2)_2OH$	0.015
Pentanol, $CH_3(CH_2)_4OH$	0.0024

There is a series of compounds, called alcohols, that have the general formula $CH_3(CH_2)_nOH$. The first member of the series ($n = 0$, CH_3OH) is methanol, the second ($n = 1$, CH_3CH_2OH) is ethanol, and so on. Molecular models of the first few members of the alcohol series are shown in Figure 9–8. The —OH part of alcohols is somewhat like the —OH in water, and the rest of the molecule, $CH_3(CH_2)_n$—, is nonpolar like a hydrocarbon. Methanol is slightly similar to water in its solvent properties, but the larger the number of carbons in the alcohol, the more it resembles hydrocarbons in its solvent properties. The data of Table 9–8 show the regular decrease of solubility of sodium chloride in alcohols with increasing numbers of carbon atoms.

NONPOLAR COMPOUNDS AS SOLUTES

Now we turn to the case of iodine to illustrate the solubility of nonpolar solutes. In crystals of solid iodine, pairs of atoms are associated in diatomic molecules. Iodine also exists as diatomic molecules in most solutions. Iodine sublimes (passes directly from the solid to vapor state on heating) with a heat of sublimation of 14.9 kcal mole^{-1}. This rather low value indicates that iodine molecules are not very tightly bound together in crystals. Comparison of the heat of sublimation of iodine with the heats of fusion and vaporization of sodium chloride indicates that iodine might be expected to be considerably more soluble than sodium chloride in any solvent. This is not the case. Why doesn't iodine have a high solubility in water? One important factor is that the nonpolar iodine molecules, unlike sodium and chloride ions, do not bind the

Figure 9–8.   The first members of the series of alcohols having the formula $CH_3(CH_2)_n$ OH; $n = 0, 1, and 2$.

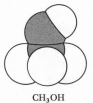   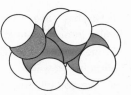

CH₃OH   CH₃CH₂OH   CH₃CH₂CH₂OH

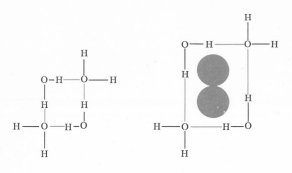

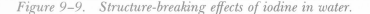

Figure 9–9.   Structure-breaking effects of iodine in water.

dipolar water molecules at all well. Instead, the large molecules break up the water structure, and tend, therefore, to be "squeezed out" of solution. This is illustrated in Figure 9–9.

Carbon tetrachloride molecules do not have the cohesive tendencies that water molecules have. The heat of vaporization of carbon tetrachloride is only 7.14 kcal mole^{-1}. The principal interaction between the nonpolar molecules is a weak binding due to van der Waals forces (Section 8–7). Iodine holds its outer electrons very loosely. Consequently, there is a relatively strong van der Waals interaction between iodine and carbon tetrachloride molecules. The energies involved in all of the possible pairs of interactions, iodine–iodine (in iodine crystals), carbon tetrachloride–carbon tetrachloride (in the pure liquid), and iodine–carbon tetrachloride (in the solution) are similar. Since there is no large gain or loss of energy when iodine dissolves in carbon tetrachloride, the solubility is fairly high. This is because the most probable arrangement for two sets of molecules would be a thoroughly mixed solution (high entropy; Section 8–8), provided no heat energy is required to separate and recombine the individual molecules.

Returning to Table 9–7, we see that there are large differences in the solubility of iodine in various nonpolar solvents. For instance, benzene and toluene are very similar solvents, yet the solubility of iodine is much greater in benzene than in toluene. The main reason for this, and the other solubility differences, is that iodine forms complexes with benzene that strongly affect the solubility. In benzene, some of the iodine exists as single molecules, and some is in loosely bound complexes such as $C_6H_6 \cdot I_2$. When discussing solubilities, it is always important to know what form the solute has in the solution.

SOME GENERAL SOLUBILITY RELATIONSHIPS

One useful rule for solubilities is that "like dissolves like," as far as polarity is concerned. That is, molecules that are similar in polarity tend to mix with each other. For example, water and methanol are miscible (can be mixed) in all proportions. Benzene, a hydrocarbon, is miscible with gasoline in all proportions. However, water and methanol have limited solubility in gasoline. (As we would expect, the solubility of methanol in gasoline is greater than that of water.) Figure 9–10 shows models of molecules similar and dissimilar in polarity.

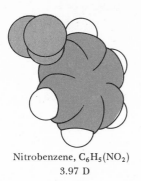

Nitrobenzene, $C_6H_5(NO_2)$
3.97 D

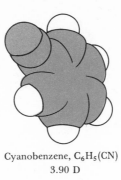

Cyanobenzene, $C_6H_5(CN)$
3.90 D

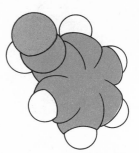

Benzaldehyde, $C_6H_5(COH)$
2.7 D

*Figure 9–10. Like and unlike molecules.
Polarity increases with dipole moment, D.*

Acetone, $(CH_3)_2CO$
2.8 D

Like and unlike polarity characteristics are not always as obvious as those shown in Figure 9–10. Carbon tetrachloride can be mixed with benzene in all proportions. In this case, likeness arises from the fact that molecules of both compounds are nonpolar, even though the atoms on the outsides of the molecules are different, as illustrated in Figure 9–11. The likeness rule does not explain the high solubility of salts in water. A special explanation has already been given for this phenomenon. Additional special explanations are sometimes necessary for other unusual solubility relationships, as for the solubility of iodine in various solvents, but the number of such cases is surprisingly small.

Example 9–10. Oxalic acid melts at 189°C and has the following structure:

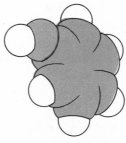

Phenol, $C_6H_5(OH)$
1.6 D

Is the compound likely to be more soluble in water or in benzene? Is it likely to be extremely soluble in either, or both, solvents?

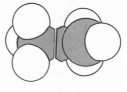

Ethanol, $C_2H_5(OH)$
1.7 D

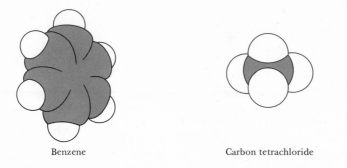

Benzene   Carbon tetrachloride

Figure 9–11. Two nonpolar substances that are miscible in any proportion.

Solution. Oxalic acid has many polar groups and will probably form strong hydrogen bonds to water molecules. Consequently, it will be much more soluble in water than in benzene. However, the high melting point indicates that oxalic acid molecules are tightly bound together in the crystal, probably by hydrogen bonding between the oxalic acid molecules. This bonding energy will be lost when the compound dissolves, so the solubility in water may not be very high. The solubility of oxalic acid in water at 25°C is 9.81 g per 100 g solution. It is virtually insoluble in benzene.

9–6 Summary

1 There are several different ways to express the concentration of a solution. When dealing with solution volumes, use molarity (*M*) or formality (*F*).

2 The product of concentration times solution volume yields the quantity of solute present in that volume of solution; molarity × volume = moles; formality × volume = formula weights.

3 There is a rapid exchange between the solute molecules in the solid phase of a saturated solution and the dissolved solute molecules in the solution phase. When the exchange rate in both directions is equal, the system is in dynamic equilibrium. The condition of equilibrium is represented by two oppositely directed arrows between the two states; solid ⇌ solution.

4 There is not always a simple relationship between the solubility of a given substance in some solvent and the heat evolved when it dissolves in that solvent.

5 A general rule of solubility is that "like dissolves like."

6 The interaction of solute molecules or ions with a solvent is called solvation.

7 Solubility is determined partially by the effect that solute molecules have in breaking up solvent structure. The cohesive forces in the solvent may tend to squeeze solvent molecules out of solution.

9–7 Postscript: Models for Liquid Solutions—A Hippopotamus in a Bog, or a Hummingbird on the Wing?

Model builders like to work with gases and crystalline solids. Such relatively simple systems are delightful because the use of simple models allows us to say wise things about them. Our success with these two phases has been so great that we would probably be content to ignore other types of systems, if it were not for the fact that practically all the chemical reactions of direct importance to us occur in a liquid medium—aqueous solution.

The basic simplicity of gases and crystals is lost in a liquid. Molecules in a liquid are neither very weakly interacting, as they are in a gas, or highly ordered, as they are in a crystal. Liquids and liquid solutions are so difficult to model because they are disordered systems, in which there are strong interactions among the component molecules and ions.

Most attempts to model liquid solutions make them more beautiful than they really are. We pretend that molecules of a solute in a liquid solution behave almost as though they were molecules of a gas. Gaseous molecules move around through their surrounding vacuum, collide with one another, and react or rebound. We often treat molecules in solution in much the same way; molecules diffuse, collide, and react in solution just as they do in the vapor phase. It is true that the diffusion process is much slower in the something that is the liquid than in the nothing that lies between molecules of a gas. It may take hours or days for a solute molecule in solution to traverse a distance that would be covered by a gaseous molecule in a fraction of a second. However, we keep up the pretense, and assume that chemical activity of a solute will vary with concentration in about the same way that the chemical activity of a component of a gaseous mixture varies with its partial pressure. A remarkable thing about such a fantastically oversimplified procedure is the fact that it works rather well. Especially in very dilute solutions, laws patterned after the gas laws do predict the behavior of solute species accurately.

Perhaps the success of oversimplification should recommend it without qualification. However, every time we work with a model less complicated than the real thing, we throw away chances to use available information to create some more complicated, and perhaps more powerful, model. For example, we know that a change of solvent can change the rates of some reactions a thousandfold or more. This inspires attempts to develop models for solutes that take into account the fact that solute molecules are really not small hard balls, but highly structured objects with attendant solvent molecules that are very much a part of the real structure of the system. For example, it is believed that many solutes in water solution are surrounded by "icebergs" of solvents, which are hard shells of rigidly organized water molecules, perhaps having fixed structures like that of ice.

Today, our models for liquids and liquid solutions remain embarrassingly imperfect. However, a number of new techniques, such as low-angle scattering of x rays and nuclear magnetic resonance spectroscopy, are being brought to bear on the problems, and we may soon know a great deal more than at present about the complex beauty of the liquid state. As this occurs, future model builders will no doubt come to speak condescendingly of the cloddish views of chemists in 1970 who really couldn't distinguish a hippopotamus in a bog from a hummingbird on the wing.

New Terms

Concentration: The amount of solute dissolved in a given amount of solution or solvent.

Equivalence point: That point in a titration when equivalent amounts of reactants (e.g., HO^- and H^+) have been added.

Formality: The number of formula weights of solute dissolved in one liter of solution.

Heat of solution ($\Delta H_{solution}$):   The amount of heat released (exothermic) or absorbed (endothermic) when a solute is dissolved. $\Delta H_{solution}$ is usually expressed in cal mole^{-1}, or kcal mole^{-1}.

Molarity: The number of moles of solute dissolved in one liter of solution.

Saturated solution: A solution that has undissolved solute in equilibrium with dissolved solute. In other words, a solution that cannot dissolve any more solute at a given temperature.

Solubility: The amount of solute that can be dissolved to make a saturated solution.

Solubility equilibrium: The condition that exists when the number of molecules dissolving equals the number of molecules coming out of solution. The equilibrium condition is represented with two arrows, solid $\rightleftarrows$ solution.

Solvation: The interaction of a solute molecule with solvent molecules.

Titration: The process of adding a measurable volume of one solution of known concentration to a measured volume of a second solution of unknown concentration for the purpose of determining the unknown concentration.

Weight percent: The ratio of the weight of one component of a mixture or solution to the total weight of the mixture or solution, multiplied by 100.

Questions and Problems

1   How many metric tons of sea water are needed to provide 1 mole of bromine, Br_2? (See Table 9–3.)

2   Calculate the molarity (M) of the following solutions: (a) 10 g of NaOH in 100 ml of aqueous solution, (b) 0.42 mole of sugar dissolved in enough water to make 33 ml of solution.

3   How many grams of sodium bromide are required to prepare 725 ml of 0.50M solution?

4   How many formula weights of nickel chloride are there in 400 ml of 0.25F $NiCl_2$ solution? How many grams of nickel chloride?

5   An amount of 50.0 ml of ether, C_2H_5—O—C_2H_5, which has a density of 0.714 g ml^{-1}, is dissolved in enough ethyl alcohol to make 100 ml of solution. Calculate the molarity of ether in the solution.

6   Explain the difference between formality and molarity.

7   A solution is prepared by dissolving 65 g of acetic acid (CH_3COOH) in sufficient benzene (C_6H_6) to make a final solution volume of 500 ml. The acid is known to associate in benzene solutions to form dimers. Why would it be more correct to use formality than molarity to describe the concentration of this solution? Calculate the formality of the solution.

8   A 5.28-g sample of chromium chloride, $CrCl_3$, is dissolved in water to make 200 ml of solution. Calculate the formality of this solution.

9   Why is water a good solvent for KI, whereas carbon tetrachloride is not? What sorts of bonds are formed between solvent and solute when water dissolves KI?

10   Why is carbon tetrachloride a good solvent for I_2, whereas water is not? What sorts of bonds are formed between solvent and solute when carbon tetrachloride dissolves I_2?

11   If the state of a solute after it is dissolved is much more disordered than before, does this promote, or hinder, solution?

12   If the solute absorbs heat as it dissolves, does this promote, or hinder, solution?

13   A solution is 16.4% H_2SO_4 (by weight) and has a density of 1.12 g ml^{-1}. Calculate the molarity of this solution.

14   A solution is 2.5M in NaCl. How many grams of NaCl would be required to make 750 ml of solution?

15   The bromination of ethylene was discussed in Section 9–3. How many grams of ethylene, $H_2C=CH_2$, could be brominated with 750 ml of a 0.23 M Br_2 solution?

16   Calculate the molarity of a solution that contains 0.0156 g of $Ba(OH)_2$ in 25 ml of solution.

17 The amino acid alanine reacts with sodium hydroxide to give the salt sodium alanate, according to the reaction

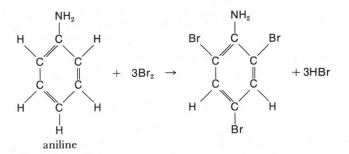

How many milliliters of 0.25M NaOH are required to react with 1.78 g of alanine?

18 What volume of 0.20M HCl is needed to titrate 20 ml of 0.35M NaOH?

19 A solution is 0.500M in HCl. How would you prepare 500 ml of a 0.125M solution of HCl by diluting the original solution?

20 A volume of 40.0 ml of water is added to 90.0 ml of a 1.40M solution of HCl to give a final volume of 130 ml. What is the molarity of the new solution?

21 Bromine in water reacts with an organic compound known as aniline as shown by the equation

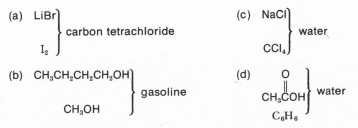

How many grams of Br_2 are required to brominate 3.72 g of aniline? If the Br_2 to be used is in a 0.10M solution of bromine in water, how many milliliters of the solution would be required to obtain the required amount of bromine?

22 A solution of sodium hydroxide is prepared and is approximately 0.1M. It is found that upon titration of 20.0 ml of 0.185M HCl solution, 35.3 ml of the sodium hydroxide solution is needed to reach the equivalence point. What is the exact molarity of the sodium hydroxide solution?

23 Listed below are several groups containing two solutes and a solvent. Select the solute that in your opinion will be most soluble in the given solvent

(a) LiBr ⎫
⎬ carbon tetrachloride
I_2 ⎭

(b) $CH_3CH_2CH_2CH_2OH$ ⎫
⎬ gasoline
CH_3OH ⎭

(c) NaCl ⎫
⎬ water
CCl_4 ⎭

(d) O
‖
CH_3COH ⎫
⎬ water
C_6H_6 ⎭

24 Predict the order of solubility of each of the following solutes in the solvents water, ethanol (C_2H_5OH), and hexane (C_6H_{14}): NaBr, C_6H_6, $CHCl_3$, Na_2SO_4,

HCl, CH_3CN, NH_4NO_3, $C_2H_5NO_2$, and $H_2C{=}C{-}C{-}C{-}H$. Justify your answer.

CHEMICAL EQUILIBRIUM

10

In Chapter 9, we discussed reactions in solution on a go-or-no-go basis. We assumed that if a reaction takes place at all, it goes to completion. In this chapter, we will present evidence that this is not true. That is, reactions *never* use *all* of the reactants, and we will try to explain why this is so. First, we will describe how reactions occur in a sample, even though the amounts of products and reactants do not change. This is called a state of *dynamic chemical equilibrium.* Then we will show how to specify the point of equilibrium (the amounts of products and reactants) in such a state, using a quantity called the *equilibrium constant* for the reaction. Finally, using the ideas of *heat content* and *entropy,* we will describe the condition that must be satisfied for a chemical system to be at equilibrium and will show how this condition is related in a simple way to the equilibrium constant. This condition enables us to predict the tendency for a reaction to occur.

The major ideas of this chapter should not be completely new to you. In Chapter 8, we described the dynamic equilibrium between a liquid and its vapor. We discussed heats of vaporization and entropy (a measure of disorder), which are factors that help to determine equilibrium vapor pressure at any temperature. In Section 9–4, we described the dynamic equilibrium between a solid and its saturated solution, and we discussed solubilities, using heats of solution and entropy. Now we will discuss these ideas in much more detail and will try to come to grips with the question of what determines equilibrium in a chemical system.

10–1 Dynamic Chemical Equilibrium

We are familiar with the idea of equilibrium in mechanical systems. Children on a seesaw easily can find positions for which the seesaw is balanced and does not move about the center support. This is an example of *static equilibrium.*

Let us consider another type of "system" with a different type of equilibrium. Eileen and George arrive at a tennis court, but decide that they won't play tennis because they left their rackets at home. However, they invent a new game, made possible because they happen to have a supply of 1000 practice balls. Each initially takes 500 balls, and they proceed to see who can pick up balls and throw them over the net faster (Figure 10–1). The one who finishes with all of the balls on his side loses the contest.

At first, it seems that George will win, because he tears around the court like a madman, snatching balls and hurling them across the net, while Eileen is much more relaxed about the whole contest. However, it is soon apparent that Eileen knows what she is doing because, as the number of balls on her side increases, she only has to reach down and pick them up to throw them over the net. At the same time, George's supply of balls has diminished to such an extent that he has to run all over his side of the court to get balls to throw back over the net. This makes his rate of throwing balls decrease compared to what it was at the beginning of the contest.

Finally, George realizes that, although he is throwing the balls over the net as fast as he can, Eileen, with much less effort, is throwing them back just as fast. Thus, the number of balls on each side of the court is constant. Although George has both greater prowess and greater enthusiasm for the game than Eileen, he knows that he can't win. Eileen, who didn't really care about winning, knew that from the outset. This is an example of *dynamic equilibrium.* Balls are going across the net in both directions, but the number of balls on each side of the court does not change.

A similar dynamic equilibrium can occur in chemical systems. Consider the simple case of a mixture of two *isomeric* molecules such as the dichloroethylenes

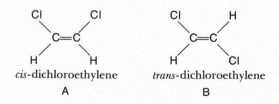

Isomers have the same chemical composition, but the atoms in the molecules are arranged differently. Thus, these two isomers are distinguishable, although their properties are very similar. At room temperature, any mixture of the two compounds is inert; that is, a mixture containing nine moles of A and one mole of B retains that composition almost indefinitely. So will a mixture of nine moles of B and one mole of A, or a mixture containing an equal number of moles of the two compounds. However, if any of the mixtures is heated at 200°C with a trace of iodine, the composition changes. Iodine acts as a *catalyst.* A catalyst is something that makes a system

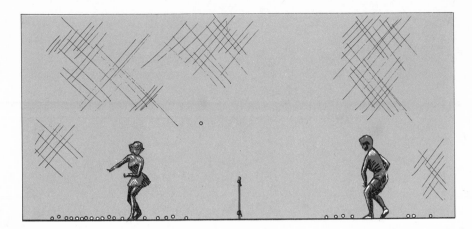

Figure 10-1. The saga of Eileen and George.

reach equilibrium sooner (Chapter 13). In this case, iodine increases the rate of conversion of A to B and of B to A. The reaction will proceed until the composition of the mixture, in moles, is 67% A and 33% B. That is, the mole fraction (Section 2–5) of A is 0.67 and the mole fraction of B is 0.33. Any mixture of the two compounds eventually will reach the same final composition at 200℃. Thus, in mixtures that originally contain more than 67% A, some of A will be converted to B

$$A \rightarrow B$$

In mixtures originally containing more than 33% B, some of B will be converted to A

$$B \rightarrow A$$

These results may seem strange at first glance. Why does the reaction "stop" after some special composition is reached? We know that pure A reacts to form B when heated at 200℃ in the presence of iodine. Why shouldn't A do the same when it is in a mixture containing 33% B?

This seems much like the game that George and Eileen were playing. If their game starts with all of the balls on George's side (pure B), he will be able to get some of the balls over to Eileen's side at a fast rate (B → A). Then Eileen, who sat and watched at the beginning, can begin to throw balls back to George's side (A → B). George cannot maintain his advantage because his rate slows down as the number of balls on his side decreases. And as we described previously, a state of dynamic equilibrium will be reached when a certain number of balls are on George's side (33% B) and a certain number are on Eileen's side (67% A). If the game starts with the equilibrium number of balls on each side (67% A, 33% B), even though George and Eileen throw balls back and forth (B → A and A → B), the total number of balls on each side remains the same (the mixture remains 67% A and 33% B).

When the composition of the reaction mixture remains constant, it is in a state of dynamic chemical equilibrium. Both the reactions still occur, but the rates of the two processes are exactly equal. Compound A continues to react to give B, but the loss of A is replenished continuously by the formation of A from B. We use two arrows in an equation (A ⇌ B) to show that a reaction is reversible; that is, the reaction system can be brought to a condition of dynamic equilibrium.

INERT MIXTURES

Chemical equilibrium between the two dichloroethylene isomers can be attained at 200°C because in the presence of iodine the rates of the reactions A → B and B → A are fast. At room temperature in the absence of iodine, a nonequilibrium mixture of the two isomers can exist because the reactions A → B and B → A are immeasurably slow. This means that during a reasonable time interval so little reaction takes place that no change in the macroscopic sample is detected.

By measuring the rates of these two reactions at higher temperatures in the absence of iodine we can calculate what the rates are at room temperature (Section 13–4). The rate of conversion of B to A at room temperature is less than 10^{-20} mole sec^{-1}. This means that a mole of B would be converted to A in about 10^{20} sec, or about 10^{13} years. Suppose that we observe one mole of B for one year. At the end of that time (about 10^7 seconds), only 10^{-13} mole of A would have formed. The fraction of A molecules in our sample of B would be $1/10^{13}$, a number that is about equal to the reciprocal of the national debt in pennies. No wonder we see no change in the sample.

10–2 Equilibrium Constants

When the dichloroethylene isomers (Section 10–1) are in equilibrium at 200°C, the ratio of the components in the mixture is 1 mole of B for every 2 moles of A, or 1:2. The ratio of 1:2 is determined experimentally; it is independent of the ratio of A to B at the start of the reaction, and depends only on the temperature. This constant ratio is called the *equilibrium constant* (K) for the reaction

$$K = \frac{\text{mole \% B}}{\text{mole \% A}} = \frac{[B]}{[A]} = \frac{33}{67} = 0.50 \tag{10–1}$$

The square brackets in Equation 10–1 signify that we are using the concentrations of the species within the brackets. Thus, we read [B] "the concentration of B." In the equation, we have expressed the concentrations in mole percent. We could also express them in moles per liter, weight percent, or any other concentration units. In this special case, the value of the equilibrium constant is the same no matter what units are used, but this is not true for all reactions. To avoid any confusion about units, we will always express the concentrations used for equilibrium constants in moles per liter (molarity).

THE DISSOCIATION OF IODINE

Now let us examine the experimental results for a slightly more complicated reaction to see what ratio of the concentrations of reactants and products is constant at equilibrium. The purpose of this is to reason from the example to obtain a general method for writing the expression for the equilibrium constant for any reaction.

For the example, we choose the dissociation of iodine, a reaction in which the number of molecules changes. Although the element iodine is solid at room temperature, it can be volatilized readily by heating. Iodine vapor has been studied carefully. Except at rather high temperatures, the vapor consists mostly of diatomic molecules, I_2. However, some iodine atoms are always present. Apparently, molecules and atoms are in dynamic equilibrium

Table 10–1. Equilibrium Concentrations^a of I_2 and I at $1100°C$

$[I]$	$[I_2]$	$[I] + [I_2]$	$[I]/[I_2]$	$[I]^2/[I_2]$
3.42	2.80	6.22	1.22	4.18
4.20	3.97	8.17	1.06	4.44
4.89	5.34	10.23	0.92	4.48
5.44	6.74	12.18	0.81	4.39
				Average: 4.37

^{a}*Concentrations are expressed in (mole liter^{-1}) $\times$ 10^3.*

$$I_2 \rightleftarrows 2I$$

Iodine molecules and atoms absorb light of different wavelengths. Therefore, analysis of light absorption by the mixture allows us to determine the amounts of each species present under any set of experimental conditions. Table 10–1 shows the results of measurements made at 1100°C with varying concentrations of gaseous iodine.

The numbers in the fourth column of Table 10–1 show that the ratio of the two concentrations is not constant. However, the last column shows that the quantity $[I]^2/[I_2]$ is constant within the limits of experimental measurement. Therefore, it makes sense to say that the equilibrium constant is given by

$$K = \frac{[I]^2}{[I_2]} = 4.37 \times 10^{-3} \text{ mole liter}^{-1} \text{ at } 1100°C \tag{10–2}$$

Remember that the balanced chemical equation for the iodine dissociation reaction is

$$I_2 \rightleftarrows 2I$$

Notice that I_2 has a coefficient of one and $[I_2]$ appears in the equilibrium expression with an exponent of one. Similarly, I has a coefficient of two and appears in the equilibrium expression as $[I]^2$ with an exponent of two.

Without presenting any more evidence, we state simply that it is true in general that the numerical coefficients of reactant and product molecules in a balanced equation always appear as exponents in the expression for the equilibrium constant.

HOW TO WRITE EQUILIBRIUM CONSTANT EQUATIONS FOR REACTIONS

In the equilibrium constant equations, the concentrations of reaction products usually appear in the numerator and those of the reactants in the denominator. This is a matter of convention. In a system at equilibrium, the designation of reactants and products is meaningless. The species appearing on the right-hand side of the balanced equation as it is written are arbitrarily called the products, and those on the left-hand side are called the reactants. Obviously, any equation could be written backwards. For example, instead of writing $I_2 \rightleftarrows 2I$, with

$$K = \frac{[I]^2}{[I_2]} = 4.37 \times 10^{-3} \text{ mole liter}^{-1}$$

we could write $2I \rightleftarrows I_2$, with

$$K' = \frac{[I_2]}{[I]^2} = 229 \text{ liters mole}^{-1}$$

Notice that $K = 1/K'$.

These two expressions describe the same reaction and give the same information about the point of equilibrium. The choice of which to use is arbitrary. To avoid confusion about what an equilibrium constant means, it is important to write a specific balanced chemical equation, and then write the equilibrium expression for that reaction with the products in the numerator and the reactants in the denominator. (See Table 10–2.)

Example 10–1. Write the equilibrium constant expression, K_{eq}, for the reaction $N_2 + O_2 \rightleftarrows NO$ and evaluate K_{eq}, if the equilibrium concentrations of N_2 and O_2 are each 0.52 mole liter^{-1} and if NO is 0.055 mole liter^{-1}.

Solution. Before the equilibrium constant can be expressed, the equation must be balanced

$$N_2 + O_2 \rightleftarrows 2NO$$

Then we can write

$$K_{eq} = \frac{[NO]^2}{[N_2][O_2]}$$

Substituting the equilibrium concentrations into this expression yields

$$K_{eq} = \frac{[0.055]^2}{[0.52][0.52]} = 1.1 \times 10^{-2}$$

[*Note:* The units of K vary, depending on the exponents of the concentration terms of the numerator and denominator. In this example, K has no units, whereas K for the dissociation of iodine is expressed in moles liter^{-1}. Can you show why there is this difference?]

Table 10–2. Some Reactions and Their Equilibrium Constant Expressions

Reaction	*Equilibrium constant expressions*
$2N_2O_5 \rightleftarrows 4NO_2 + O_2$	$K = \dfrac{[NO_2]^4[O_2]}{[N_2O_5]^2}$
$H_2 + I_2 = 2HI$	$K = \dfrac{[HI]^2}{[H_2][I_2]}$
$HBr \rightleftarrows H^+ + Br^-$	$K = \dfrac{[H^+][Br^-]}{[HBr]}$

To do calculations involving equilibrium constants, it is important to understand the form in which data are reported. For this reason, equilibrium constants are often given names indicating the sense of the reactions to which they apply. For example, the equilibrium constant for the reaction $I_2 \rightarrow 2I$ is called a *dissociation constant,* indicating that the equation should be written with iodine atoms as reaction products (products of the reaction called dissociation). In a similar way, the equilibrium constants for the formation of ions by dissociation of molecules in solution are called *ionization constants.* For example,

$$AB \rightarrow A^+ + B^-$$

$$K_i = \text{ionization constant of } AB = \frac{[A^+][B^-]}{[AB]} \tag{10-3}$$

There is another way to illustrate that the equilibrium expression depends exactly on how the reaction equation is written. For the reaction $I_2 \rightarrow 2I$

$$K = \frac{[I]^2}{[I_2]} = 4.37 \times 10^{-3} \text{ mole liter}^{-1}$$

An equivalent balanced equation for the dissociation of iodine molecules is $\frac{1}{2}I_2 = I$. The equilibrium constant for this reaction is

$$K'' = \frac{[I]}{[I_2]^{1/2}} = \sqrt{K} = K^{1/2} = 6.61 \times 10^{-2} \text{ mole}^{1/2} \text{ liter}^{-1/2}$$

Example 10–2. Previously, we found that K_{eq} for the reaction $N_2 + O_2 \rightleftarrows 2NO$ is 1.1×10^{-2}. If the reaction begins with 1 mole of N_2 and 1 mole of O_2, and no NO is present, what will the concentrations of the three components be at equilibrium? The reaction vessel has a volume of 1 liter.

Solution. A good approach to this type of problem is to construct a table of the concentrations at the start of the reaction and at equilibrium. The starting concentrations can be put into the table immediately

	Concentrations	
Component	*Start*	*Equilibrium*
N_2	1 mole liter^{-1}	$1 - x$
O_2	1 mole liter^{-1}	$1 - x$
NO	0	$2x$

The equilibrium concentrations can be expressed by letting x be the amount of N_2 that reacts. Since N_2 and O_2 react in a 1:1 ratio, x also will be the amount of O_2 that reacts. The concentration of N_2 and O_2 remaining at equilibrium is the initial amount (1 mole liter^{-1}) minus the amount reacted (x), which is $(1 - x)$. The amount of NO formed must be $2x$, since there are two NO molecules formed for each N_2 molecule that reacts. The equilibrium constant expression for the reaction is

$$K_{eq} = \frac{[NO]^2}{[N_2][O_2]} = 1.10 \times 10^{-2}$$

We can substitute the equilibrium values for the three components from the table to give

$$1.10 \times 10^{-2} = \frac{[2x]^2}{[1-x][1-x]} = \frac{(2x)^2}{(1-x)^2}$$

Then we can take the square root of each side of this equation to give

$$1.05 \times 10^{-1} = \frac{2x}{1-x}$$

$$0.105\,(1-x) = 0.105 - 0.105x = 2x$$

$$2.105x = 0.105$$

$$x = \frac{0.105}{2.105}$$

$$x = 5.0 \times 10^{-2} = 0.050 \text{ mole liter}^{-1}$$

The concentration of NO is $2x$, or 0.10 mole liter^{-1}, and that of N_2 and O_2 is $(1-x)$, or 0.95 mole liter^{-1}.

There are two cases in which the general rules for writing equilibrium expressions are slightly modified. One is the case in which a reaction occurs in solution and involves the solvent. The most common example is the neutralization of a basic solution by an acid, as in the reaction

$$HCl + NaOH \rightleftarrows HOH + NaCl$$

The acid–base neutralization reaction is

$$H^+ + HO^- \rightleftarrows H_2O$$

The opposite reaction, or the water dissociation reaction is

$$H_2O \rightleftarrows H^+ + HO^-$$

The equilibrium constant expression for the dissociation reaction is

$$K = \frac{[H^+][HO^-]}{[H_2O]}$$

In most aqueous solutions, the concentration of water is nearly constant at 56 moles liter^{-1}. This is true because we usually work with dilute solutions. In a liter of a dilute water solution, most of the volume and mass is from the solvent, water. In a liter of water, there are 1000 g of water, or 1000 g/18 g mole$^{-1} \cong 56$ moles. Consequently, we can say that the concentration of water, $[H_2O]$, is the constant $56M$. Therefore, we can combine $[H_2O]$ with K and write

$$K[H_2O] = K_w = [H^+][HO^-]$$

K_w is called the *ion product constant* for the dissociation of water.

Example 10−3. At 25°C, the concentration of both $[H^+]$ and $[HO^-]$ in pure water has been determined experimentally to be 1.0×10^{-7} mole liter^{-1}. Calculate the ion product constant for water at this temperature.

Solution. The expression for the ion product constant of water is

$$K_w = [\text{H}^+][\text{HO}^-] \text{ moles}^2 \text{ liter}^{-2}$$

Substituting the experimentally determined concentration of $[\text{H}^+]$ and $[\text{HO}^-]$ in the expression gives

$$K_w = [1.0 \times 10^{-7}][1.0 \times 10^{-7}] = 1.0 \times 10^{-14} \text{ moles}^2 \text{ liter}^{-2} \text{ at } 25°C$$

Example 10–4. Suppose that a quantity of NaOH is added to pure water that greatly increases the amount of HO^- in equilibrium with H^+. Calculate the amount of H^+ in solution, if the equilibrium concentration of HO^- is 1.0×10^{-3} mole liter^{-1}.

Solution. Since $[\text{H}^+]$ and $[\text{HO}^-]$ must be in equilibrium in the solution, their product is 1.0×10^{-14}, the ion product equilibrium constant. If $[\text{HO}^-]$ increases, $[\text{H}^+]$ must experience a corresponding decrease, if we are to maintain the relationship

$$[\text{H}^+][\text{HO}^-] = K_w$$

We can calculate $[\text{H}^+]$ since

$$[\text{H}^+] = \frac{K_w}{[\text{HO}^-]} = \frac{1.0 \times 10^{-14}}{1.0 \times 10^{-3}}$$

$$[\text{H}^+] = 1.0 \times 10^{-11} \text{ mole liter}^{-1}$$

The second case for which we modify equilibrium expressions involves heterogeneous equilibria. An example is the equilibrium between a solid and its solution, such as solid barium sulfate in an aqueous solution containing barium and sulfate ions

$$\text{BaSO}_4(s) \rightleftharpoons \text{Ba}^{2+} + \text{SO}_4^{2-}$$

The equilibrium expression for this reaction is

$$K = \frac{[\text{Ba}^{2+}][\text{SO}_4^{2-}]}{[\text{BaSO}_4(s)]}$$

If the solution is saturated and the concentrations of Ba^{2+} and SO_4^{2-} are at their maximum levels, the addition of more solid BaSO_4 will not affect the amount of Ba^{2+} and SO_4^{2-} in solution. The addition will simply increase the amount of solid present. Thus, for a saturated solution, we can consider the term $[\text{BaSO}_4(s)]$ a constant, and the equilibrium expression may be rewritten as

$$K[\text{BaSO}_4(s)] = K_{sp} = [\text{Ba}^{2+}][\text{SO}_4^{2-}] \qquad\qquad (10–4)$$

K_{sp} is called the *solubility product constant* of barium sulfate.

The two preceding modifications are convenient, and they are consistent with the general method that we have given for writing equilibrium expressions.

Example 10–5. Calculate the solubility of AgCl at 25°C, given that its solubility product constant, K_{sp}, is 2.8×10^{-10} moles^2 liter^{-2} at that temperature.

Solution. The equation for the dissolution of AgCl to give Ag^+ and Cl^- ions is

$$\text{AgCl}(s) \rightleftharpoons \text{Ag}^+ + \text{Cl}^-$$

For this process,

$$K_{sp} = [Ag^+][Cl^-] = 2.8 \times 10^{-10} \text{ moles}^2 \text{ liter}^{-2}$$

From the chemical equation, we notice that for each Ag^+ ion in solution there will always be one Cl^- ion; that is, the two ions will have equal concentrations. Thus, we can let $[x]$ represent both $[Ag^+]$ and $[Cl^-]$ and write

$$2.8 \times 10^{-10} = [Ag^+][Cl^-] = [x][x] = [x]^2$$
$$[x] = 1.7 \times 10^{-5} \text{ mole liter}^{-1} = [Ag^+] = [Cl^-]$$

Since the formula weight of AgCl is 143 g, this is the same as a solubility of 0.24 mg per 100 ml solution.

THE MAGNITUDE OF THE EQUILIBRIUM CONSTANT

Since the equilibrium constant is a ratio of product and reactant concentrations (raised to appropriate powers) measured when a system is at equilibrium, it serves as a measure of the completeness of a reaction. For example, K_{sp} (25°C) for $BaSO_4$ is 1.08×10^{-10}. This is a very small number because of the large negative exponent. Thus, the K_{sp} tells us that $BaSO_4$ does not dissolve extensively in water to form Ba^{2+} and SO_4^{2-} ions in solution. However, the equilibrium constant for the process

$$\tfrac{3}{2} H_2(g) + \tfrac{1}{2} N_2(g) \rightleftarrows NH_3(g)$$

is 826.1 at 25°C. This value indicates that there is a large amount of NH_3 present at equilibrium.

MULTIPLYING EQUILIBRIUM CONSTANT EXPRESSIONS

Often, the product of one chemical reaction will be the reactant in a subsequent reaction. An example is the decomposition of nitrogen pentoxide, N_2O_5. At moderate temperatures (45°C), N_2O_5 decomposes to form nitrogen dioxide, NO_2

$$2N_2O_5 \rightleftarrows 4NO_2 + O_2 \tag{10-5}$$

This reaction has the equilibrium expression

$$K = \frac{[NO_2]^4[O_2]}{[N_2O_5]^2}$$

But at higher temperatures (600°C), nitrogen dioxide decomposes to give the nitrogen oxide NO

$$4NO_2 \rightleftarrows 4NO + 2O_2 \tag{10-6}$$

For this reaction, the equilibrium expression is

$$K' = \frac{[NO]^4[O_2]^2}{[NO_2]^4}$$

The equation for the decomposition of N_2O_5 to produce NO is

$$2N_2O_5 \rightleftarrows 4NO + 3O_2 \tag{10-7}$$

with

$$K'' = \frac{[NO]^4[O_2]^3}{[N_2O_5]^2}$$

We can obtain Equation 10–7 by adding the two equations for the stepwise decomposition

$$2N_2O_5 \rightleftarrows 4NO_2 + O_2 \qquad (10\text{–}5)$$
$$4NO_2 \rightleftarrows 4NO + 2O_2 \qquad (10\text{–}6)$$
$$\overline{2N_2O_5 + 4NO_2 \rightleftarrows 4NO_2 + 4NO + 3O_2}$$

and canceling the 4NO$_2$ on both sides

$$2N_2O_5 \rightleftarrows 4NO + 3O_2 \qquad (10\text{–}7)$$

The interesting fact to notice is that, if we multiply the equilibrium constant expressions for Equations 10–5 and 10–6, K and K', we obtain

$$KK' = \frac{[NO_2]^4[O_2]}{[N_2O_5]^2} \times \frac{[NO]^4[O_2]^2}{[NO_2]^4}$$

$$= \frac{[NO]^4[O_2]^3}{[N_2O_5]^2} = K''$$

It is generally true that if we add two chemical equations to obtain a third equation, the equilibrium constant expression for the third equation is the product of the equilibrium constant expressions for the first two equations. This illustrates a general principle that can save chemists work. It is easy to write the equilibrium expression for any reaction, but it may be difficult to determine experimentally the value of the corresponding equilibrium constant. In the preceding example, $KK' = K''$. Thus, if we know the values of K and K', we can find the value of K'' by a simple calculation, without having to do an experiment. Similarly, if we know K and K'', we can find K'.

Example 10–6. The dissociation of the acid H$_2$S in aqueous solution proceeds in two steps to give the S^{2-} ion

(1) $H_2S \rightleftarrows H^+ + HS^- \qquad K_1 = 1.1 \times 10^{-7}$
(2) $HS^- \rightleftarrows H^+ + S^{2-} \qquad K_2 = 1.0 \times 10^{-14}$

The complete dissociation can be written in one step

(3) $H_2S \rightleftarrows 2H^+ + S^{2-} \qquad K_3 = ?$

Show that K_3 is the product of K_1 and K_2, and calculate the value of K_3.
Solution. The equilibrium expressions for Reactions 1 and 2 are

$$K_1 = \frac{[H^+]\,[HS^-]}{[H_2S]} \quad \text{and} \quad K_2 = \frac{[H^+]\,[S^{2-}]}{[HS^-]}$$

The product K_1K_2 is

$$K_1K_2 = \frac{[H^+][HS^-]}{[H_2S]} \times \frac{[H^+][S^{2-}]}{[HS^-]} = \frac{[H^+]^2[S^{2-}]}{[H_2S]}$$

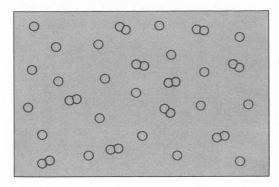

Figure 10–2.   The reaction between A_2 and A in volume V.

which is exactly what would be expected for Reaction 3. The numerical value of K_3 is

$$K_1 \times K_2 = (1.1 \times 10^{-7}) \times (1.0 \times 10^{-14}) = 1.1 \times 10^{-21} \text{ moles}^2 \text{ liter}^{-2}$$

10–3   Le Chatelier's Principle

We have used the example of the iodine dissociation reaction to derive a general method for writing equilibrium expressions. Before giving a justification for this method beyond our assurance that it works, we will present a qualitative argument that shows that the method is at least sensible. This argument comes from *Le Chatelier's principle,* which states: When a stress is applied to a system, the system responds in a way to try to relieve the stress.

Let us return to George and Eileen in their ball-throwing game at the equilibrium condition. If we were to dump two crates of balls on George's side of the court this would apply a stress to the system. George would automatically relieve the stress, however, because now he would have more balls close to him, and his throwing rate would increase. In this section, we discuss some kinds of stress that can be applied to a chemical system at equilibrium and show how the system responds.

CHANGES IN VOLUME

Imagine a box containing ten diatomic molecules (A_2) and twenty atoms (A) in equilibrium (Figure 10–2). Now let us contract the box to half the original size, keeping the temperature constant (Figure 10–3). Halving the volume automatically doubles the concentrations of A and A_2. According to Le Chatelier's principle, the response of the system is for chemical reaction to occur to reduce the total concentration of molecules. Since the reaction we are considering is a dissociation, the total number of molecules, and therefore the concentration, can be reduced by the net reaction of A with itself to form A_2.

Another way of looking at this is to examine the equilibrium constant for the reaction $A_2 \rightleftarrows 2A$

$$K_{eq} = \frac{[A]^2}{[A_2]}$$

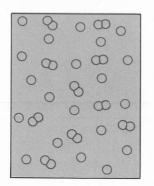

Figure 10–3. The reaction between A_2 and A in volume $V/2$ (before equilibrium is reestablished).

The equilibrium constant depends on concentration, which in turn depends on the volume of the container. If the volume of the container is suddenly changed, the system is no longer in equilibrium, and net chemical reaction must take place until equilibrium is reestablished.

The initial equilibrium condition had 10 A_2 molecules and 20 A atoms in some volume, V liters. From this we can find the equilibrium constant for the reaction

$$K_{eq} = \frac{[A]^2}{[A_2]} = \frac{(20/V)^2}{10/V} = \frac{40}{V}$$

What are the concentrations of A and A_2 when equilibrium is reestablished after the volume is halved? We know that the number of molecules of A_2 will be greater, and the number of A atoms less, than the numbers present under the initial equilibrium conditions, so let us call the new concentrations

$$[A_2] = \frac{(10+x)}{V/2} \qquad [A] = \frac{(20-2x)}{V/2}$$

where x is the increase in the number of molecules of A_2. (Notice that $2x$ atoms of A are required to form x molecules of A_2.) Then the equilibrium constant expression is

$$K_{eq} = \frac{40}{V} = \frac{\left(\dfrac{20-2x}{V/2}\right)^2}{\dfrac{(10+x)}{V/2}} = \frac{8(10-x)^2}{V(10+x)}$$

Rearranging this equation,

$$(10-x)^2 = 5(10+x)$$
$$x^2 - 25x + 50 = 0$$

Using the quadratic formula,

$$x = \frac{25 - \sqrt{625-200}}{2} = 2.2$$

and the numbers of molecules and atoms are

A_2:   12.2 molecules   A: 15.6 atoms

Thus, the equilibrium concentrations after the volume is halved will be

$$[A_2] = \frac{24.4}{V} \qquad [A] = \frac{31.2}{V}$$

and the total concentration of particles is 55.6/V. Comparing this with the nonequilibrium condition when the volume was halved (60/V), we see that the concentration is lower, but not as low as the initial concentration (30/V). This is a general feature of the response of a chemical system to stress. The new equilibrium condition is always between the initial equilibrium condition and the nonequilibrium condition after the stress is applied.

In one way, these numbers, 15.6 atoms and 12.2 molecules, seem silly, since we do not expect to subdivide atoms or molecules. What is the meaning of 15.6 atoms? This question arises only because our imaginary experiment involved a small number of atoms. If we had formulated the problem in terms of moles, there would be no dilemma. Since a mole contains about 6×10^{23} molecules, there is no difficulty in thinking about 12.2 moles or 15.6 g-atoms. If we did an experiment with as few as 40 atoms, we could still interpret the result by admitting that, at any given time, the number of atoms and the number of molecules might not be the same as the *average* numbers. The equilibrium constant is simply a statistical number. It says that if we count the number of A atoms at many different times and calculate the *average* of the results, the *average* number of A atoms will be 15.6. Similarly, the average number of A_2 molecules will be 12.2.

This situation is similar to that of the game played by George and Eileen. The number of balls on each side of the net may not be exactly constant, but if we count the number of balls on each side many different times and calculate the average of the results, we obtain average numbers that tell us better than any single count what the equilibrium constant is. In most chemical systems, the number of atoms or molecules is so large that deviations from the average are not detected readily.

Example 10–7. The gaseous material PCl_5 dissociates into PCl_3 and Cl_2, according to the equation

$$PCl_5 \rightleftarrows PCl_3 + Cl_2$$

If the three substances are in equilibrium with each other, and if the volume of the reaction container is doubled, which way will the equilibrium concentrations change?

Solution. When the volume is doubled, the concentration of each substance is halved. Therefore, net reaction should take place to increase total concentration. This will occur if more PCl_5 dissociates to form PCl_3 and Cl_2. At the new equilibrium condition, the number of molecules of PCl_5 will be less than the initial number, and the total number of molecules in the system will be larger. The new equilibrium concentration of each of the species will be less, however, than the initial equilibrium concentration.

A change in volume of a chemical system does not necessarily produce a change in the equilibrium numbers of molecules. For the interconversion of the dichloroethylene isomers (Section

10–1), we have a reaction of the general type A $\rightleftarrows$ B. Since the total number of molecules in the reaction mixture is constant, a change in volume will not change the equilibrium numbers of molecules of A and B. (Can you verify this?)

Changes in pressure for a gas reaction are similar to changes in volume. For the reaction

$$CaCO_3(s) \rightleftarrows CaO(s) + CO_2(g)$$

an increase in pressure causes the reaction to proceed more toward the left, thereby reducing the number of $CO_2(g)$ molecules and decreasing the pressure. However, for a gas reaction such as

$$H_2(g) + Cl_2(g) \rightleftarrows 2HCl(g)$$

the total number of molecules on each side of the equation is the same, thus a change in pressure does not affect the reaction.

Example 10–8. Predict how the following reactions will change their equilibrium concentrations when the volumes of their reaction containers are reduced:

    (a) $N_2(g) + 3H_2(g) \rightleftarrows 2NH_3(g)$

    (b) $CO_2(g) + H_2(g) \rightleftarrows H_2O(g) + CO(g)$

    (c) $4NH_3(g) + 5O_2(g) \rightleftarrows 4NO(g) + 6H_2O(g)$

Solution.

    (a) There are four moles of reactants and only two moles of products. Decreasing the volume will produce more NH_3.

    (b) There is an equal number of moles of reactants and products. Consequently, volume change has no effect on the equilibrium concentrations in this reaction.

    (c) There are ten moles of product and nine moles of reactant. A decrease in volume will favor the formation of NH_3 and O_2.

10–4   Changes in Equilibrium Constants

Equilibrium constants express relationships among concentrations in a system at chemical equilibrium. If a reaction temperature is changed, its equilibrium constant usually changes. This is also an example of Le Chatelier's principle. A chemical reaction usually produces heat or absorbs heat from the surroundings. If heat is added to a system at equilibrium, the composition will change in the direction favored by heat absorption.

HEAT CONTENT AND STANDARD STATES

Energy changes in chemical reactions usually are described in terms of heat. We frequently indicate the amount of heat produced when a reaction takes place, or the amount of heat required to force a reaction to occur. The same notation is used for chemical reactions as for physical changes (Section 8–3). For example, the heat required to dissociate one mole of water to give one mole of hydrogen and one half mole of oxygen at 1000°K is 65.1 kcal

$$H_2O(g) \xrightarrow{\text{1000°K}} H_2(g) + \tfrac{1}{2}O_2(g)$$

$$\Delta H = 65.1 \text{ kcal mole}^{-1} \text{ of } H_2O(g)$$

Remember that ΔH is the heat change for a reaction, so ΔH is greater than zero for endothermic (heat absorbing) reactions and less than zero for exothermic (heat producing) reactions. If we change the direction in which we consider the reaction to occur, we also change the sign of the heat change. The combustion of one mole of hydrogen gas at 1000°K produces 65.1 kcal of heat

$$H_2(g) + \tfrac{1}{2}O_2(g) \xrightarrow{\ 1000°K\ } H_2O(g)$$
$$\Delta H = -65.1 \text{ kcal mole}^{-1}$$

One way to think about the heat changes in chemical reactions is to imagine that each substance has a certain *heat content, H*.[1] Then the change in heat energy due to chemical reaction can be expressed as the difference between the heat content of the products of the reaction and the heat content of the reactants. Let the symbol $\bar{H}$ stand for the heat content *per mole*. Then, taking the preceding reaction as an example,

$$\Delta H_{\text{reaction}} = \left(\bar{H}_{H_2O(g)}\right) - \left(\bar{H}_{H_2(g)} + \tfrac{1}{2}\bar{H}_{O_2(g)}\right)$$

Notice that the heat content of the products minus the heat content of the reactants is equal to the heat absorbed by the reaction, so ΔH has the proper sign.

This is a very convenient way to formulate heat changes in reactions. The heat change is an important piece of information about a reaction. Although measuring heat changes is often simple, it is time consuming and tedious. We would like to avoid having to measure the heat change for every new reaction we study. The idea of heat content allows us to do this, for if we know the heat contents of the reactants and products, we can *calculate* the heat change instead of measuring it. Since each compound can react in many different ways, a few values of heat contents enable us to calculate the heat changes for a large number of reactions.

There is only one difficulty with this approach. Unfortunately, we can measure only heat *changes.* In other words, we cannot measure the heat "contained" in a molecule. But we can solve this problem by arbitrarily assigning a value of heat content to every element in some well-defined condition. Since we cannot measure heat contents, it does not matter what the absolute values for heat contents of the various elements are; all we care about is that differences in heat content agree with experiment. The arbitrary position of the heat content scale is similar to the arbitrary position of temperature scales. The difference between the freezing point of water and its boiling point is 100°C, if zero is fixed at the freezing point of water (the usual Celsius scale), at absolute zero (the Kelvin scale), or at any other temperature, such as 37°C above the freezing point of water.

The zero point on the heat content scale is defined in the following way: The most stable form of an element at 25°C and 1 atm has, by definition, zero heat content. For example, the stable form of hydrogen under standard conditions is $H_2(g)$. Therefore, the heat content of $H_2(g)$ at 25°C and 1 atm is, by definition, zero.

The *standard heat content* or *standard heat of formation* for any compound is the heat change per mole in the formation of the compound from its elements under standard conditions. For example, the combustion of hydrogen to produce water under standard conditions liberates 57.796 kcal of heat

---

[1] In most textbooks of thermodynamics, *H* is called *enthalpy,* rather than heat content.

$$H_2(g;\ 1\ atm) + \tfrac{1}{2}O_2(g;\ 1\ atm) \xrightarrow{\ 25°C\ } H_2O(g;\ 1\ atm)$$
$$\Delta H^0 = -57.796\ kcal\ mole^{-1}\ H_2O$$

(Notice that the heat change is written ΔH^0. ΔH^0 is called the *standard heat change,* because it is the heat change for a reaction performed under standard conditions.) From the standard heat change for this reaction, we calculate the standard heat content of water vapor

$$\bar{H}^0_{H_2O(g)} - \left(\bar{H}^0_{H_2(g)} + \tfrac{1}{2}\bar{H}^0_{O_2(g)} \right) = \Delta H^0$$

$$\bar{H}^0_{H_2O(g)} - (\ 0\ +\ 0\ ) = -57.796\ kcal\ mole^{-1}$$

$$\bar{H}^0_{H_2O(g)} = -57.796\ kcal\ mole^{-1}$$

Since every compound can be made directly or indirectly from its elements, this procedure allows us to find the heat content of every compound.

Example 10–9. Using the table of heats of formation (Table 10–3), find the standard heat of vaporization of water.

Solution. The standard heat of vaporization of water ΔH^0_{vap}, is the heat change for the reaction

$$H_2O(l) \xrightarrow{\ 25°C\ } H_2O(g;\ 1\ atm)$$

Using the standard heat contents from the table,

$$\Delta H^0_{vap} = \bar{H}^0_{H_2O(g)} - \bar{H}^0_{H_2O(l)}$$
$$= -57.79 - (-68.32)$$
$$\Delta H^0_{vap} = 10.53\ kcal\ mole^{-1}$$

Often we are interested in the heat content of a substance in solution. The standard state for a solute is usually chosen to be a **1M** solution at 25°C and 1 atm.

Example 10–10. The standard heat of solution (in water) of ammonia gas is $-8.28\ kcal\ mole^{-1}$. What is the standard heat content of ammonia in water?

Table 10–3.   Heats of Formation from the Elements at 25°C and 1 atm

Reaction	$\Delta H^0\ (kcal\ mole^{-1})$
$H_2(g) + \tfrac{1}{2}O_2(g) \rightarrow H_2O(g)$	-57.79
$H_2(g) + \tfrac{1}{2}O_2(g) \rightarrow H_2O(l)$	-68.32
$S(s) + O_2(g) \rightarrow SO_2(g)$	-70.96
$\tfrac{1}{2}N_2(g) + \tfrac{1}{2}O_2(g) \rightarrow NO(g)$	$+21.60$
$\tfrac{1}{2}N_2(g) + \tfrac{3}{2}H_2(g) \rightarrow NH_3(g)$	-11.04
$Zn(s) + \tfrac{1}{2}O_2(g) \rightarrow ZnO(s)$	-83.2
$C(s) + 2H_2(g) \rightarrow CH_4(g)$	-17.89
$\tfrac{1}{2}N_2(g) \rightarrow N(g)$	$+112.5$

Solution. Referring to Table 10–3, $\overline{H}^0_{NH_3(g)} = -11.04$ kcal. The standard heat of solution is the standard heat change for the reaction

$$NH_3(g;\ 1\ atm) \xrightarrow{25°C} NH_3(aq;\ 1M)$$

$$\Delta H^0 = -8.28\ kcal$$

$$\Delta H^0 = \overline{H}^0_{NH_3(aq)} - \overline{H}^0_{NH_3(g)}$$

$$\overline{H}^0_{NH_3(aq)} = \Delta H^0 + \overline{H}^0_{NH_3(g)}$$

$$= -8.28 + (-11.04)$$

$$\overline{H}^0_{NH_3(aq)} = -19.32\ kcal$$

TEMPERATURE DEPENDENCE OF EQUILIBRIUM CONSTANTS

Table 10–4 shows how the equilibrium constant for the water dissociation reaction depends on the temperature. As the temperature is increased, the dissociation constant becomes larger, indicating that raising the temperature increases the equilibrium amount of hydrogen and oxygen. This is what we expect from our observation that the dissociation of water requires heat. Also, this is consistent with what we would predict from Le Chatelier's principle. The dissociation of water requires energy, and if we take a reaction mixture in equilibrium ($H_2O \rightleftarrows H_2 + \frac{1}{2}O_2$) and apply heat to raise the temperature, the reaction will move in the direction that will relieve the stress. To consume the added heat, more water will dissociate, and a new equilibrium will be established at the new temperature. At the new equilibrium, there will be more H_2 and O_2 present and less H_2O. Therefore, the value of K at the new temperature will be larger than K at the lower temperature. This is confirmed in Table 10–4.

THE LAW OF CONSERVATION OF ENERGY

When a reaction system responds to the stress of added heat by displacement of the equilibrium position, some of the heat energy is converted into chemical energy. One of the most important

Table 10–4. The Equilibrium Constant for the Reaction $H_2O(g) \rightleftarrows H_2(g) + \frac{1}{2}O_2(g)$ at Several Temperatures[a]

$T(°K)$	$K = [H_2][O_2]^{1/2}/[H_2O]$
300	1.6×10^{-40}
400	6.3×10^{-30}
600	2.5×10^{-19}
1000	1.0×10^{-10}
2000	2.94×10^{-4}
3000	4.78×10^{-2}
4000	6.02×10^{-1}
5000	2.82

[a] Concentrations are expressed as partial pressures in atmospheres.

principles of science is the *law of conservation of energy.* This law states: Energy never disappears, although it may be converted from one form to another. For example, electrical energy can be converted to heat energy in an electric stove. Heat energy can be converted to energy of motion, as occurs when a balloon is heated until it expands and bursts. But in each of these processes, no energy is lost. The principle to remember is that the heat changes in a reaction must be balanced by changes in chemical energy. We implicitly used this fact when defining heat content. We assumed that the heat energy produced (or absorbed) in a reaction is exactly the change in heat content of the chemical system.

10–5   Why Are Reactions Reversible?

We have seen that when chemical systems are in a state of dynamic equilibrium, the amounts of reactants and products present can be calculated from the equilibrium constant for the reaction. We also have seen that the equilibrium constant for a reaction changes when the temperature is changed, and we have used Le Chatelier's principle to predict the direction of that change. But we have said nothing about how to predict what the value of the equilibrium constant will be, or exactly how much the equilibrium constant will change when the temperature is changed.

   We would like to have some condition that will tell us when equilibrium is attained. Such a condition must have something to do with the value of the equilibrium constant, because the equilibrium constant describes what the system is like at equilibrium. Let us see if we can find some way of describing the balance that is maintained at equilibrium. Energy is obviously involved, since we can change the equilibrium point by increasing the temperature, and raising the temperature always increases the energy content of a system.

ENTROPY

Since energy is released or absorbed in the form of heat in chemical reactions, we might think that equilibrium in a chemical reaction involves some kind of balance of heat changes. However, this is only partly the case. In our discussion of solubilities (Section 9–4) we pointed out that heats of solution are not enough to explain the relative solubilities of various compounds in water. In particular, many common water-soluble salts have positive heats of solution (the reactions are endothermic). This would be impossible if heat changes were the only driving force for chemical reactions.

   Consider the reaction of hydrogen and oxygen to produce water

$$H_2(g) + \tfrac{1}{2}O_2(g) \xrightarrow{\ 5000°K\ } H_2O(g) \qquad \Delta H = -67.3 \text{ kcal mole}^{-1}$$

The equation states that one half mole of oxygen reacts with one mole of hydrogen with the release of 67.3 kcal of heat energy. The data given in Table 10–4 allow us to calculate that, if a mole of hydrogen and one half mole of oxygen are mixed at 5000°K in the presence of a catalyst (platinum metal), a reaction will occur, producing 0.16 mole of water. The reaction will evolve 10.8 kcal (67.3 kcal mole^{-1} × 0.16 mole) of heat energy.

   We now face one of the most important questions in chemistry. Why doesn't the reaction occur until 100% of the reactants have been converted to water, producing 67.3 kcal of heat energy? What makes the reverse reaction

$$H_2O(g) \rightarrow H_2(g) + \tfrac{1}{2}O_2(g)$$

"fight back," and what determines the equilibrium point between the two reactions? If we believe that equilibrium represents a state of balance, we must try to find out *what* is balanced. Certainly the heat content of the chemicals in the system is not an accurate measure of balance.

One way of answering this question comes from an intuitive consideration of *probability*. Probability is important in all problems of energy exchange. Common experience tells us so. For example, if we pour cold water into a hot pan, we expect that the temperature of the pan will fall, and the temperature of the water will rise, until the two temperatures are equal. Why does the change occur in this direction? After all, heat might pass from the water to the pan, making the water still colder and the pan hotter. Experience tells us that such a change is improbable, so we do not think very seriously about it.

A similar situation exists in the game played by George and Eileen. Their original agreement was that the person with no tennis balls left on his side would win. But, as George finally realized, this did not make any sense, because no matter how hard he worked to clear balls from his side of the court, it was extremely improbable that Eileen would not get a ball back across the net while he was throwing the last one from his side. The more balls in the game, the more improbable it is. In the same way, it is highly improbable for the hydrogen–oxygen mixture to deliver all of its chemical energy to the surroundings by reacting completely to form water.

The chemical measure of probability is called *entropy*. Entropy is related to the disorder in a system. The less structure, or order, a system has, the more probable it will be, and the higher its entropy will be. This is true because there are more ways of achieving disorder than order. Let us consider a few examples of physical systems in which the equilibrium state is one of maximum entropy, or maximum disorder.

Consider a gas in a container, Figure 10–4(a), which has a partition confining the gas to one half the total volume of the box. If the partition is removed, we would expect from experience that the gas would flow spontaneously to fill the new larger volume, Figure 10–4(b). We cannot explain this flow as being due to the gas moving to a lower energy state, since the energy of the gas is not dependent on the volume it occupies. It must be that the greater disorder the gas has in the larger volume represents a more probable state than that if the gas stayed confined in the original half of the container after the partition is removed, Figure 10–4(c).

Figure 10–4.   The behavior of a gas in containers having different volumes.

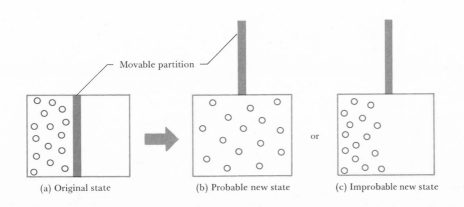

(a) Original state   (b) Probable new state   (c) Improbable new state

Now, ponder the following question: Why doesn't this book jump off the table and hit you in the head? We could imagine it happening in such a way that it would not violate the law of conservation of energy. Suppose that we took the energy to move the book from the molecules that compose it. If we did, the energy loss would be apparent as a decrease in the temperature of the book. But this would mean that for the book to jump spontaneously, its molecules would be required to go from a more probable and disordered state (molecules moving in random directions) to a less probable and more ordered state (all molecules moving in the same direction). The probability of this happening is so remote that we can say with some certainty that you needn't wear safety glasses while reading this book. The tendency to keep the system in, or move it to, the most probable state is a strong factor in determining the likelihood of any process.

Just as every chemical reaction has associated with it a heat change, ΔH, so it has an entropy change, ΔS. For some processes, the entropy change can be determined in a simple way. When water is vaporized at constant pressure, the heat put into the system to accomplish the transformation is the heat of vaporization, ΔH_{vap}. Since this is an endothermic process, ΔH_{vap} is positive. The change in entropy for this process can be calculated from the relationship

$$\frac{\Delta H_{vap}}{T} = \Delta S \qquad\qquad\qquad\qquad (10\text{--}8)$$

ΔS should be positive for this process, since the probability of a highly disordered vapor phase is greater than the probability of the more orderly liquid phase. We cannot justify this equation in detail, but it should seem reasonable that temperature enters in the way it does. Since the temperature is related directly to molecular speeds, the higher the temperature, the higher should be the entropy of a system. At very low temperatures, systems have little randomness, and a phase change, such as vaporization, should cause a large *change* in entropy. But at high temperatures, systems have so much randomness due to thermal motion that a change of phase produces much less change in randomness or entropy.

By analogy with our definition of heat content, we can say that each substance has a certain entropy, S, and that the entropy change of a reaction is equal to the entropy of the products

Table 10–5. Entropies of Some Substances in Their Standard States at 25°C

Substance	$\bar{S}^0$ *(cal mole^{-1} deg^{-1})*
C (diamond)	0.6
C (graphite)	1.4
$H_2(g)$	31.2
$O_2(g)$	49.0
$H_2O(l)$	16.8
$CH_3OH(l)$	30.3
$H_2O(g)$	45.2
$CO_2(g)$	51.1
$C_2H_4(g)$	52.5

minus the entropy of the reactants. For example, the standard entropy change for the dissociation of water vapor is given by

$$H_2O(g; 1 \text{ atm}) \xrightarrow{25°C} H_2(g; 1 \text{ atm}) + \tfrac{1}{2}O_2(g; 1 \text{ atm})$$

$$\Delta S^0 = \left(\bar{S}^0_{H_2(g)} + \tfrac{1}{2}\bar{S}^0_{O_2(g)} \right) - \bar{S}^0_{H_2O(g)}$$

The problem of obtaining entropy values is somewhat different from that of obtaining heat contents of substances. We will simply state the hypothesis that the entropy of any pure crystal is zero at 0°K, the absolute zero temperature. This entropy postulate can be supported by experimental evidence, but it is intuitively reasonable. In a perfect crystal at absolute zero, we expect that there will be as perfect order as can be imagined, and the entropy would be zero for such a system.

The entropies of some substances at atmospheric pressure and 25°C are given in Table 10–5. Notice that the entropy is higher the greater the disorder in the substance.

Example 10–11. What change in entropy will accompany the following reaction under standard conditions?

$$H_2(g) + \tfrac{1}{2}O_2(g) \rightarrow H_2O(g)$$

Solution. The entropy change, ΔS^0, for this process is

$$S^0_{products} - S^0_{reactants}$$

From Table 10–5, we find

$$S^0_{products} = 45.2 \text{ cal mole}^{-1} \text{ deg}^{-1}$$
$$S^0_{reactants} = \bar{S}^0_{H_2(g)} + \tfrac{1}{2}\bar{S}^0_{O_2(g)}$$
$$= 31.2 + \tfrac{1}{2}(49.0)$$
$$= 31.2 + 24.5 = 55.7 \text{ cal mole}^{-1} \text{ deg}^{-1}$$
$$\Delta S^0 = 45.2 - 55.7$$
$$\Delta S^0 = -10.5 \text{ cal mole}^{-1} \text{ deg}^{-1}$$

Notice that this is not a large decrease in entropy, but it is a decrease as expected, since the atoms in the H_2O molecule are more ordered than in the separate elements.

The entropy changes for some gas reactions at atmospheric pressure and 25°C are given in Table 10–6. For the first two reactions, the number of molecules (and therefore the amount of

Table 10–6.   Standard Entropy Changes for Some Gas Reactions

Reaction	ΔS^0 *(cal mole^{-1} deg^{-1})*
$2NO_2 \rightleftarrows 2NO + O_2$	34.74
$2CH_3OH + 3O_2 \rightleftarrows 4H_2O + 2CO_2$	21.95
$3F_2 + 2B \rightleftarrows 2BF_3$	−97.7
$N_2 + 3H_2 \rightleftarrows 2NH_3$	−47.38

disorder) increases, so the reactions have a positive entropy change. For the second two reactions, the number of molecules decreases, and the entropy change is negative.

Example 10–12. Processes that involve an increase in entropy ($\Delta S > 0$) are favored, or are more likely than processes that have a decrease in entropy ($\Delta S < 0$). In which of the following reactions will entropy increase?

(a) $H_2O(g) \rightarrow H_2(g) + \frac{1}{2}O_2(g)$
(b) $CO_2(s) \rightarrow CO_2(g)$
(c) $2H_2S(g) + 3O_2(g) \rightarrow 2H_2O(l) + 2SO_2(g)$

Solution.

(a) $\Delta S > 0$ for this process, since there are more product molecules than reactant molecules.
(b) $\Delta S > 0$, because the gas phase is much more disordered than the solid phase.
(c) $\Delta S < 0$, since there are fewer product molecules than reactant molecules, and, therefore, less disorder. Also, one of the products is in the liquid phase.

FREE ENERGY

We have presented some convincing arguments that both heat changes and entropy changes are important in determining to what extent a chemical reaction can occur. The tendency to produce maximum heat and the tendency toward maximum increase in entropy are both driving forces for chemical reactions to occur. It must be the case, then, that some combination of heat change and entropy change determines the course of a chemical reaction. An experimentally convenient combination of entropy and heat change that we use to predict the point of equilibrium of a chemical system is the *free energy* change, $\Delta G = \Delta H - T\,\Delta S$. If we had to guess a suitable combination of ΔH and ΔS, this would be a reasonable choice. From the earlier discussion of heats and entropies of vaporization, it is clear that the combination $T\,\Delta S$ has the units of heat. Since exothermic reactions have negative values of ΔH, and entropy-increasing reactions have positive values of ΔS, ΔG is the (negative) sum of the driving forces for a reaction. There are many other ways that we might express the balance between heat and entropy changes required for chemical equilibrium. It simply turns out that for systems at constant pressure, which is the usual case in a chemical reaction, the free energy function is the most useful.

Since ΔG is just the sum of two functions that we have already discussed, we can say immediately that each reaction has a value of ΔG determined by $\Delta G = \Delta H - T\,\Delta S$, that standard free energy changes, ΔG^0, describe reactions occurring in the standard state, and that each chemical substance has an intrinsic free energy, G, such that for any reaction $\Delta G = G_{\text{products}} - G_{\text{reactants}}$.

We have defined H, S, and ΔH, ΔS. Therefore, ΔG is determined completely by $\Delta G = \Delta H - T\,\Delta S$. By analogy with heat content and standard heats, we have ΔG^0 and $\bar{G}^0$. It is important to point out that values of free energy content, G, are *not* given by $G = H - TS$. The values of G are determined in such a way that $\Delta G_{\text{reaction}} = G_{\text{products}} - G_{\text{reactants}}$, but the method of calculating values of G is too complicated to explain here. One result of the system for defining G is that $\bar{G}^0$ for each element in its most stable form in the standard state is zero, as is the case with standard heat content, $\bar{H}^0$.

Because ΔG is the negative sum of the net driving force for a reaction, any reaction should occur if $\Delta G < 0$. Consider the conversion of A to B in the system A $\rightleftarrows$ B; what is ΔG? If $\Delta G < 0$, more A will be converted to B. When $\Delta G = 0$, the reaction no longer has a driving force, and there is no further net conversion of A to B. Suppose that $\Delta G > 0$. Then, because changing the direction of a reaction changes the sign of ΔG, we have $B \rightleftarrows A$ and $\Delta G < 0$. Thus, there will be net conversion of B to A until the point at which $\Delta G = 0$. Clearly, the condition for chemical equilibrium in a system at constant pressure must be $\Delta G_{\text{reaction}} = 0$.

In this book, we will deal only with equilibrium relationships at fixed pressures. This does not mean that we will never refer to pressure changes, but that the value of an equilibrium constant will be such that $\Delta G = 0$ for the equilibrium system at the pressure of the experiment.

10–6   Equilibrium Constants and Free Energy Changes

Consider the reaction of $A \rightleftarrows B$. If we know the standard free energies of formation of A and B, we can write

$$\Delta \bar{G}^0 = \bar{G}_B^0 - \bar{G}_A^0 \tag{10–9}$$

$\Delta \bar{G}^0$ is the free energy change when one mole of A in its standard state is changed to one mole of B in its standard state.

When a mole of A is in a system that allows the reversible reaction to occur, some B will always form, but complete conversion of A to B will never occur. The extent of reaction when equilibrium is established depends upon the difference between the standard free energies of A and B. If $\bar{G}_A^0$ is greater than $\bar{G}_B^0$, there will be more B than A at equilibrium. If $\bar{G}_B^0$ is greater than $\bar{G}_A^0$, A will be present in larger amount at equilibrium. Let us see if we can find a quantitative relationship between the free energy of reactants and products and the equilibrium composition.

Consider what change actually occurs when one mole of pure A reacts to produce the equilibrium mixture:

Starting state	*Equilibrium state*
1 mole A	x moles A
0 moles B	$(1 - x)$ moles B

During the reaction, the free energy of A decreases as the amount of A decreases to x moles, and that of B increases as the amount of B increases from 0 moles to $1 - x$ mole. The equilibrium balance will be established when the free energy change for conversion of A to B becomes zero. At equilibrium

$$A_{\text{equil}} \rightleftarrows B_{\text{equil}} \qquad \Delta G = 0 = \Delta H - T \Delta S$$

Note carefully that the G is no longer a standard free energy, since the conditions are not those corresponding to the standard state. Let us assume that A and B are gaseous compounds and that the reaction is carried out at 25°C and 1 atm pressure. At the beginning, A was in its standard state. When equilibrium is established, A can no longer be in its standard state, since the partial pressure of A will have been reduced to x atm from 1 atm.

We have to calculate the free energy per mole of A at x atm and of B at $(1 - x)$ atm. We have seen that both ΔG and K_{eq} are related to the extent a reaction will proceed from reactants to products. Using laws of physics to analyze the relationships between heat and the work involved in expansion and contraction of gases, we can show that the free energy of a substance not in the standard state, G, is equal to the free energy in the standard state, G^0, plus an additional factor, which depends on concentration. If we apply this relationship to the reaction $A \rightarrow B$ and let the nonstandard state be the state of equilibrium, we can write

$$G_{A(equil)} = \bar{G}_A^0 + 2.303 \, RT \log(x)$$

where the second expression on the right is the additional factor mentioned previously. Here, x is the concentration, or partial pressure, of reactant A at equilibrium, R is the gas constant expressed as 1.99 cal mole^{-1} deg^{-1}, T is the absolute temperature, and 2.303 is a factor for converting from natural logarithms to the base e, to common logarithms to the base ten. Similarly, for product B the following relationship holds

$$G_B = \bar{G}_B^0 + 2.303 \, RT \log(1 - x)$$

where $(1 - x)$ is the concentration, or partial pressure, of B. Since the system is at equilibrium, the free energy of the reactants and products are equal, so that

$$G_{A(equil)} = G_{B(equil)}, \text{ or } \Delta G_{(equil)} = 0$$

and

$$\bar{G}_A^0 + 2.303 \, RT \log(x) = \bar{G}_B^0 + 2.303 \, RT \log(1 - x)$$

$$\bar{G}_B^0 - \bar{G}_A^0 = \Delta \bar{G}^0 = -2.303 \, RT \log\left(\frac{1 - x}{x}\right) \tag{10--10}$$

Notice that $(1 - x)/x$ is the *equilibrium constant* for the reaction $A \rightleftarrows B$. We can rewrite Equation 10–10 in the general form

$$\Delta G^0 = -2.303 \, RT \log K_{eq} \tag{10--11}$$

This means that equilibrium constants for reactions can be calculated, if the standard free energies of the compounds are known. Conversely, the equilibrium constant is a measure of the standard free energy of a reaction. Many calculations concerning reactions at equilibrium can be made by simply adding and subtracting standard free energies.

The choice of standard states determines the choice of concentration units for the equilibrium expression. In the preceding case, in which the standard state is 1 atm, the unit of concentration is partial pressure. If the standard state were a one molar solution at 25°C, then the appropriate units of concentration would be moles liter^{-1}.

The negative sign in Equation 10–11 is very significant, and a brief discussion of it may help our understanding of the meaning of ΔG^0. If the value of K_{eq} is greater than one, we know that there are appreciable quantities of product present at equilibrium. A value of $K_{eq} > 1$ would produce a positive value for log K_{eq}. This means that the right-hand side of Equation 10–11 would be negative. This requires that the larger the value of K_{eq} the more negative is the value of ΔG^0 for that process. This is consistent with our earlier statement that if the free energy of A, $\bar{G}_A^0$, is greater than the free energy of B, $\bar{G}_B^0$, there will be more B than A at equilibrium. That is, if $\bar{G}_B^0 < \bar{G}_A^0$ then $K = [B]/[A] > 1$, and $\bar{G}_B^0 - \bar{G}_A^0$ is negative.

If K_{eq} is less than one, its logarithm becomes negative, and the right-hand side of Equation 10–11 becomes positive. The smaller K_{eq} becomes (therefore, the less favorable the process it describes), the more positive ΔG^0 becomes. We may conclude that any process having a large negative value of ΔG^0 will have a strong tendency to make products from reactants.

Table 10–7.   Standard Free Energies for Some Substances

Substance	Standard free energy per mole, $\overline{G}^0$ (kcal mole $^{-1}$)
$CO_2(g)$	−94.26
$CH_3OH(g)$	−38.70
$NO_2(g)$	12.39
$H_2O(g)$	−54.64
$H_2O(l)$	−56.69
$HO^-(aq)$	−37.60
$H^+(aq)$	0.0
$H_3O^+(aq)$	−56.69

Often we say that reactions with $K > 1$ or negative ΔG^0 values are spontaneous; that is, the reactants when mixed in their standard states can react to produce substantial amounts of product. From Table 10–7, we can predict which chemical reactions proceed to give a large percentage of products and which ones do not.

Example 10–13. Which of the following chemical reactions are spontaneous?

(a) $C(s) + O_2(g) \rightarrow CO_2(g)$

(b) $CO_2(g) + 3H_2(g) \rightarrow CH_3OH(g) + H_2O(g)$

(c) $\frac{1}{2}N_2(g) + O_2(g) \rightarrow NO_2(g)$

Solution. Using Table 10–7 and remembering that $\overline{G}^0$ for the most stable elemental form in the standard state is zero, we have

(a) $C(s) + O_2(g) \rightarrow CO_2(g)$

$\Delta G^0 = \overline{G}^0_{CO_2(g)} - (\overline{G}^0_{C(s)} + \overline{G}^0_{O_2(g)})$

$\quad = -94.26 - 0$

$\Delta G^0 = -94.26$ kcal

$\Delta G^0 < 0$, so the reaction is spontaneous.

(b) $CO_2(g) + 3H_2(g) \rightarrow CH_3OH(g) + H_2O(g)$

$\Delta G^0 = (\overline{G}^0_{CH_3OH(g)} + \overline{G}^0_{H_2O(g)}) - (\overline{G}^0_{CO_2(g)} + 3\overline{G}^0_{H_2(g)})$

$\quad = (-38.70 - 54.64) - (-94.26 + 0)$

$\quad = -93.34 + 94.26$

$\Delta G^0 = +0.92$ kcal

$\Delta G^0 > 0$, so the reaction is not spontaneous.

(c) The reaction is not spontaneous. Can you show why it is not?

Example 10–14. The conversion of one mole of sulfur trioxide to produce one mole of sulfur dioxide and one half mole of oxygen has a standard free energy change at 25°C of 16.73 kcal. What is the equilibrium constant K_{eq} for the reaction at 25°C? From the value of K_{eq}, what do you conclude about the spontaneity of the reaction?

Solution. The reaction is

$$SO_3(g) \rightleftarrows SO_2(g) + \tfrac{1}{2}O_2(g) \qquad \Delta G^0 = 16.73 \text{ kcal}$$

$$K_{eq} = \frac{[SO_2][O_2]^{1/2}}{[SO_3]}$$

$$\Delta G^0 = -2.303 \ RT \log K_{eq}$$

In this equation, we take ΔG^0 in cal mole^{-1}, R in cal mole^{-1} deg^{-1}, and T in °K. Thus we have

$$\Delta G^0 = 16{,}730 \text{ cal mole}^{-1}$$
$$R = 1.99 \text{ cal mole}^{-1} \text{ deg}^{-1}$$
$$T = 273° + 25° = 298°K$$

$$\log K_{eq} = \frac{16{,}730}{-2.303(1.99)(298)}$$

$$\log K_{eq} = -12.25$$
$$K_{eq} = 5.62 \times 10^{-13}$$

The value of K_{eq} is extremely small, and the reaction will not proceed spontaneously.

Finally, it is important to emphasize the fact that the free energy criterion allows us to predict whether a reaction *can* occur, but does not tell us if it *will* occur. Many spontaneous reactions occur at such slow rates that no measurable change takes place in hundreds of years. The oxygen–hydrogen reaction is an example. From Table 10–4, it is clear that at room temperature the driving force for the reaction $H_2(g) + \tfrac{1}{2}O_2(g) \rightarrow H_2O(g)$ is *very* large. However, in the earth's atmosphere, the concentration of $H_2(g)$ is many times larger than the equilibrium value predicted from the atmospheric concentration of $H_2O(g)$ and $O_2(g)$. This nonequilibrium situation persists because, at the temperatures in the atmosphere, the rate of the reaction is infinitesimally small.

We began this chapter with a problem to solve: How can we predict whether a chemical reaction can occur? On our way to the solution we have discussed many new ideas and explored the notion of chemical equilibrium. With the foundation of knowing what chemical changes are allowed, now we can go on to learn more about the nature of chemical reactions and how they occur.

10–7 Summary

1 A reacting chemical system in which there is no net change in the number of moles of products and reactants is in a state of dynamic equilibrium.

2 The equilibrium constant expression for a reaction is equal to the product of the concentrations of the products divided by the product of concentrations of the reactants, each concentration raised to the power of the coefficient of that substance in the balanced chemical equation.

$$aA + bB \rightleftarrows cC + dD$$

$$K = \frac{[C]^c[D]^d}{[A]^a[B]^b}$$

3 The Le Chatelier principle states that, if a stress is applied to a chemical system, the system responds in a way to remove the stress.

4 A chemical reaction tends to occur if the reaction can convert chemical energy to heat energy. The heat of a reaction, ΔH, is the difference between the heat content of the products and the reactants ($H_{prod} - H_{react}$). Reactions that produce heat (exothermic) have negative values of ΔH, whereas reactions that absorb heat (endothermic) have positive values of ΔH.

5 A chemical reaction also tends to occur if it increases the disorder of the system. The measurement of disorder is a quantity called entropy, S. The entropy change of a reaction, ΔS, is the difference between the entropy of the products and the reactants ($S_{prod} - S_{react}$). Reactions that increase entropy have positive values of ΔS, whereas reactions that decrease entropy have negative values of ΔS.

6 A balance between changes in heat content and the changes in entropy, which can be used to determine the position of equilibrium for a reaction at a fixed pressure, is the change in free energy, ΔG. The change in free energy for a reaction is the difference between the free energies of the products and the reactants ($G_{prod} - G_{react}$). The relationship among free energy, heat, and entropy is $\Delta G = \Delta H - T \Delta S$.

7 The condition for equilibrium in a chemical reaction is that the change in free energy for the formation of products from reactants is equal to the change in free energy for the formation of reactants from products. That is, $\Delta G_{prod \rightarrow react} = \Delta G_{react \rightarrow prod}$.

8 The absolute values of G and H for any substance cannot be measured, but changes involved in any chemical or physical process, ΔG, ΔH, and ΔS can be determined. For convenience, we define the values of H for all elements in their most stable forms at 25°C (298°K) and 1 atm to be zero, and we use the hypothesis that the values of S for all pure solids at 0°K are zero. The values of the changes in heat content, entropy, and free energy for reactants and products in their standard states are designated ΔH^0, ΔS^0, ΔG^0. The quantities $\Delta \bar{H}^0$, $\Delta \bar{S}^0$, and $\Delta \bar{G}^0$ refer to the changes for one mole of reactant in the standard state.

9 The important relationship between the standard free energy change for a reaction (ΔG^0) and the equilibrium constant for the reaction (K_{eq}) is $\Delta G^0 = -2.303 \, RT \log K_{eq}$.

10-8 Postscript: Phase Equilibria and Juvenile Hormones

Equilibria in solutions are important and relatively easy to discuss. A more complicated kind of equilibrium involves two phases. Because life tends to be more complicated than textbooks, equilibria involving two or more phases are literally vitally important. The absorption of nutrients by intestines and the solution of oxygen gas in blood are two examples of processes involving the distribution of a substance between two different phases. The equilibrium constant that fixes the ratio of concentrations of a substance in two different phases is usually called a partition coefficient, or a distribution coefficient.

Let us give an example. Some metal ions such as Pb^{2+} can react with a variety of complex anions, L^-, to form PbL_2, a neutral compound. With the right choice of the ligand (L^-), PbL_2 becomes much less soluble in a polar solvent, such as water, than it is in a nonpolar solvent, such as carbon tetrachloride (CCl_4). Water and CCl_4 do not mix, so if we put water, CCl_4, and PbL_2 in a beaker and stirred the mixture, we would find, after letting it settle, a carbon tetrachloride layer on the bottom (CCl_4 is more dense than H_2O) and a water layer on top. The lead would be distributed between the two layers, according to the partition coefficient for PbL_2 between the two phases.

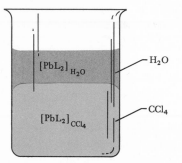

The partition coefficient is

$$P = \frac{[PbL_2]_{CCl_4}}{[PbL_2]_{H_2O}}$$

If we started with an aqueous solution of Pb^{2+} and added enough L^- to convert all the Pb^{2+} to PbL_2, and then went through the *extraction procedure,* we could pour off the water and be left with a solution of PbL_2 in CCl_4.

Many people are interested in the analysis of samples for lead, because common paint pigments contain lead and leaded gasolines pollute the atmosphere with this deadly cumulative poison. Although there is dispute over what concentrations of lead are "safe" in water, food, and other sources, there is no question that finding just how much lead is present in a very dilute solution is so difficult that often the "safe level" is lower than the lowest level that can be detected reliably. This is an example of the inability of chemical technology to respond rapidly with the solution to a new technical problem that has profound social and public health implications.

One approach to this problem has been to *concentrate* samples by *extraction.* For example, if 500 ml of a Pb^{2+} aqueous solution were treated with L^- to form PbL_2 and then extracted with 50 ml of CCl_4, nearly all the lead could be removed to the CCl_4 phase, and the CCl_4 solution would be ten times as concentrated as the original water solution. Suppose that the lowest concentration we can detect is $5 \times 10^{-7}M$, but the "safe" concentration is $5 \times 10^{-8}M$. By concentrating by a factor of ten, we have effectively moved the detection limit down to the "safe" limit. Notice that we are talking about very small amounts of material; 500 ml of $5 \times 10^{-7}M$ Pb^{2+} is about 0.05 mg of Pb, so the solution is about 10^{-5}% Pb. Chemists have powerful tools for *trace analysis,* but even so, finding such a small trace of one substance in a solution is an art that relies on the skill and intuition of individuals as heavily as on scientific principles.

The extraction that we have discussed is a simple basis for a wide variety of separation and identification techniques known collectively as chromatography. Chromatography is the process of separating the components of a mixture by passing the mixture through a column, or sheet, of material in which the components have different affinities. A particular type is called gas–liquid chromatography (glc) or vapor-phase chromatography (vpc). A glc column consists of a metal tube filled with some inert packing material (such as glass beads), which is covered by a film of liquid. The liquid stays in place in the column as the sample is forced through the column in the gas phase by a stream of carrier gas, such as nitrogen or helium. The gas phase and the stationary liquid phase correspond to the water and CCl_4 phases in the extraction process. Passing a sample through the column is like doing thousands of extractions of the same sample. The partition coefficient for a substance on the column is

$$P = \frac{\text{concentration in the stationary liquid}}{\text{concentration in the gas phase}}$$

A substance with a very small partition coefficient will be carried through the column by the carrier gas, but a substance with a large partition coefficient will "get stuck" in the liquid, and will take longer to reach the end of the column. Thus, mixtures can be separated into pure substances by using these phase equilibria, which cause each substance to reach the end of the column at a different time.

An interesting example of this technique is the discovery of a whole class of compounds called insect juvenile hormones. These are compounds responsible for controlling the stages of insect development from egg to mature adult. They were first found and identified by grinding bugs, extracting the mess with a suitable solvent, and passing the resulting solution through a glc column to separate it into individual compounds.

The identification of the composition and function of these compounds is interesting chemistry and fascinating biology. It also opens an entirely new field of insect pest control, for insects in developing stages that are fed juvenile hormones of the right type will fail to mature properly, and therefore will not reproduce. This is the kind of example that supports the argument that the cure for problems caused by science and technology (pollution due to pesticides) is more science and more technology (the development of an entirely new method of insect control using juvenile hormones).

Problem. A compound passed through a glc column in 5 min when the column was at 35°C. Would it take more, or less, time if the column temperature were changed to 60°C?

New Terms

Catalyst: A substance that accelerates the approach of a chemical system to equilibrium (Chapter 13).

Dissociation constant: Equilibrium constant for a dissociation reaction.

Dynamic chemical equilibrium: That state of a reacting chemical system in which there is no net change in the number of moles of products or reactants.

Entropy: A measure of the probability, or disorder, of a state for any chemical system. The higher the disorder (or probability) of a state the higher is the entropy of that state. The entropy, S, of a pure crystal is zero at 0°K, and the entropy change in a reaction is $\Delta S = S_{products} - S_{reactants}$.

Equilibrium constant: An experimentally determined constant that is equal to the product of the concentrations of the products divided by the product of the concentrations of the reactants, with each concentration raised to the power of the coefficient of that substance in the balanced equation. Specifically, we have

$$aA + bB \rightleftarrows cC + dD$$

$$K_{eq} = \frac{[C]^c[D]^d}{[A]^a[B]^b}$$

Extraction: The process of drawing a substance from one phase into another.

Free energy: The free energies, *G*, of substances are defined so that for a chemical reaction $\Delta G = G_{\mathrm{products}} - G_{\mathrm{reactants}}$, where $\Delta G = \Delta H - T \Delta S$. The free energies of the most stable forms of the elements are all zero under standard conditions.

Heat content (enthalpy): The heat contents, *H*, of all elements in their most stable forms under standard conditions are defined to be zero. The standard heat content of any compound is the heat change of the reaction for forming the compound from its elements under standard conditions. For any reaction, $\Delta H = H_{\mathrm{products}} - H_{\mathrm{reactants}}$.

Heterogeneous equilibrium: An equilibrium involving substances in different phases.

Inert: Resistant to chemical change. The term is applied especially to substances that can undergo spontaneous chemical reaction, but for which the reaction proceeds at an immeasureably slow rate.

Ion product constant of water: The equilibrium constant, K_w, for the dissociation of water to H^+ and HO^- ions

$$K_w = [H^+][HO^-]$$

Ionization constant: Equilibrium constant for an ionization reaction.

Isomers: Molecules with the same chemical composition, but different arrangements of the atoms. Example: Urea $[(H_2N)_2C{=}O]$ and ammonium cyanate $(NH_4^+ OCN^-)$ both have the chemical composition CON_2H_4.

Law of conservation of energy: Energy can neither be created nor destroyed, but it can be transformed from one form into another. For example, chemical energy can be transformed to heat energy.

Le Chatelier's principle: A system tends to respond to an imposed stress in such a way as to relieve the stress.

Partition coefficient (distribution coefficient): The equilibrium constant that describes the distribution of a substance between two phases.

Reversible reaction: A reaction that rapidly attains the equilibrium state.

Solubility product constant: The equilibrium constant, K_{sp}, describing a reaction of the type $AB(s) \rightleftarrows A^+ + B^-$ where $K_{sp} = [A^+][B^-]$.

Spontaneous reaction: A reaction capable of producing a substantial percentage of products; that is, a reaction with $K_{eq} > 1$ or $\Delta G^0 < 0$.

Standard state: The set of conditions 25°C, 1 atm (1 atm partial pressure for a gas), and for solutions, the concentration 1*M*.

Trace analysis: Finding the amount of a trace of one substance in a sample.

Questions and Problems

1 Write expressions for the equilibrium constant for the following reactions. Pay particular attention to the phases of the reactants and products.

(a) $2SO_2(g) + O_2(g) \rightleftarrows 2SO_3(g)$

(b) $PCl_3(g) + Cl_2(g) \rightleftarrows PCl_5(g)$

(c) $3H_2(g) + N_2(g) \rightleftarrows 2NH_3(g)$

(d) $CO(g) + H_2O(g) \rightleftarrows CO_2(g) + H_2(g)$

(e) $4NH_3(g) + 5O_2(g) \rightleftarrows 4NO(g) + 6H_2O(g)$

(f) $BaSO_4(s) \rightleftarrows Ba^{2+}(aq) + SO_4^{2-}(aq)$

(g) $H_2O(l) \rightleftarrows H^+(aq) + HO^-(aq)$

(h) $CaCO_3(s) \rightleftarrows CaO(s) + CO_2(g)$

2 The gas N_2O_4 decomposes into NO_2 according to the equation

$$N_2O_4 \rightleftarrows 2NO_2$$

If the concentrations of N_2O_4 and NO_2 at equilibrium at 298°K are, respectively, 1.53×10^{-2} mole liter^{-1} and 6.15×10^{-3} mole liter^{-1}, what is the value of the equilibrium constant?

3 The dissociation of phosphoric acid, H_3PO_4, may be described in a single step by the equation

$$H_3PO_4 \rightleftarrows 3H^+ + PO_4^{3-}$$

or in three separate steps, each producing one new H^+ ion. Write equations for these three equilibrium steps, and show that the product of the equilibrium constants for these three steps is equal to the equilibrium constant for the reaction written as if it were a one-step process.

4 When silver chromate, Ag_2CrO_4, is placed in water, a small amount will dissolve. The ions in solution are then in equilibrium with the undissolved Ag_2CrO_4. If it is determined experimentally that the concentration of the CrO_4^{2-} in solution is $7.8 \times 10^{-5}M$, what will be the concentration of the Ag^+ ion? Calculate the value of the solubility product for Ag_2CrO_4.

5 The solubility product constant of CuCl is 1.0×10^{-6}. Calculate the solubility of CuCl in moles liter^{-1}.

6 What is the solubility of $Mn(OH)_2$, if its solubility product constant is 4×10^{-14}?

7 Explain the following observations in terms of the change in entropy (disorder) of the system and the need for conservation of energy:

(a) Water does not run uphill.

(b) Water in the solid phase at 0°C is in a lower energy state than it is in the liquid phase at 0°C; yet, water melts at this temperature.

(c) Even though the reaction $H_2 + \frac{1}{2}O_2 \rightarrow H_2O$ is highly exothermic, there will always be some H_2 and O_2 molecules with H_2O molecules in an equilibrium system.

8 For the reaction $H_2(g) + Cl_2(g) \rightleftarrows 2HCl(g)$, ΔH^0 (kcal mole^{-1}) is -22.06 and ΔG^0 (kcal mole^{-1}) is -22.77. Calculate the value of ΔS^0 for this process at 25°C.

9 Calculate the change in entropy (ΔS^0) for the following reaction at 25°C:

$$C_2H_4(g) + 3O_2(g) \rightleftarrows 2CO_2(g) + 2H_2O(g)$$

Use the entropy values given in Table 10–5 for this calculation.

10 The free energy, ΔG^0, for the reaction

$$N_2 + 3H_2 \rightleftarrows 2NH_3$$

is -3976 cal mole^{-1} at 25°C. Calculate the value of the equilibrium constant for this reaction.

11 When a chemical reaction reaches the equilibrium state, describe the relationship among ΔH, ΔS, ΔG, and K_{eq}.

12 Consider the reaction

$$A + B \rightleftarrows C + D$$

which has an equilibrium constant of 25.0. If the reaction begins with A and B present in concentrations of 1.00 mole liter^{-1} and no C or D, what will be the final equilibrium concentrations of all four components?

13 The heat of fusion (melting) of ice is 1436 cal mole^{-1} and the increase in entropy in going from the solid to the liquid phase is 5.26 cal mole^{-1} deg^{-1}. Recall what value ΔG has when a system is in equilibrium. What value of temperature will achieve this equilibrium condition for the process

$$H_2O(s) \rightarrow H_2O(l)$$

14 Interpret the meaning of an equilibrium state in terms of the two tendencies described by ΔH and ΔS.

15 What effect would there be on the position of equilibrium of the following reactions, if the volume of the reaction container is reduced to half its original size?

(a) $2Cl_2(g) + 2H_2O(g) \rightleftarrows 4HCl(g) + O_2(g)$

(b) $NH_4CO_2NH_2(s) \rightleftarrows 2NH_3(g) + CO_2(g)$

(c) $2NO_2 \rightleftarrows N_2O_4$

16 Explain why entropy is expressed in energy units (calories) per degree. (*Hint:* What effect on the entropy of a system would you expect an increase in temperature to produce?)

17 A solution of hydrochloric acid is prepared that has a hydrogen ion concentration, $[H^+]$, of $2.5 \times 10^{-3}M$. In aqueous solution, the H^+ and HO^- ions are in equilibrium with undissociated water. Using the value of the ion product constant for water, $K_w = 1 \times 10^{-14}$, calculate the concentration of HO^- ions in equilibrium with the H^+ ion.

18 How will a temperature increase affect the following reactions in their equilibrium states? What effect will the temperature change produce in the value of K_{eq}?

(a) $2NO_2(g) + \frac{1}{2}O_2(g) \leftrightarrows N_2O_5(s)$ $\Delta H = -26.2$ kcal
(b) $CaCO_3(s) \leftrightarrows CaO(s) + CO_2(g)$ $\Delta H = +42.5$ kcal

19 What form does the equilibrium expression for Problem 2 take if the equilibrium quantities of N_2O_4 and NO_2 are expressed as partial pressures? Calculate the standard free energy change, ΔG^0, for the decomposition reaction from the equilibrium constant for this expression. Why is it necessary to use this value of K_{eq} and not the value from Problem 2?

20 The reaction

$$N_2(g) + O_2(g) \rightleftarrows 2NO(g)$$

is endothermic when it goes to the right. Write the expression for K_{eq}. Will K_{eq} increase, or decrease, if the temperature is raised? Why?

CHEMICAL REACTIONS

There are millions of known chemical reactions; no one really knows how many. During a lifetime, a chemist has direct experience with only a small fraction of the known reactions. Yet, a proficient chemist usually can make a reasonably accurate prediction concerning the feasibility of a particular reaction, even though he has not encountered it previously in his experience. Furthermore, people in the field of chemical synthesis often perform routinely reactions that no one has investigated before. This is not true of all reactions used in synthesis. A major challenge of the field is to devise rational plans for carrying through the nonroutine parts of a synthetic program. A principal difference between an expert and a neophyte in chemical synthesis is that the expert will be able to predict whether a particular step is likely to be easy or troublesome.

Our understanding of chemical reactions has not advanced to the same state of refinement as has our knowledge of structural chemistry. However, we do have a large amount of experimental data and many models that are of considerable help in making predictions systematically. Some of us believe that the field of chemical dynamics is poised for a great leap forward, like that made in structural chemistry during the past half century. Other competent chemists feel that such predictions are wishful dreaming. The case will be settled by the work of people who are now students entering the field.

Much of our ability to make useful models of chemical reactions comes from three sources:
1   Reactions can be classified in large groups of very similar processes.
2   Much is known about the sequences of steps, or mechanisms, involved in complex reactions.
3   We have a useful, although primitive, general theory of reaction rates.

In this chapter, we will present a simple scheme for classification of reactions and will discuss the details of writing balanced chemical equations, an absolute necessity for working with reactions. Chapter 12 will be devoted to the study in detail of perhaps the most important group of reactions, those involving proton transfer. In Chapter 13, we will present some theories and ideas about reaction mechanisms, and there will be further examples of mechanisms in Chapter 14.

11–1   Chemical Equations, the Basis for Classification

We have written equations for a few chemical reactions, and we have seen many examples of the critical importance of keeping accurate balance of the materials involved in chemical change. Priestley's experiments with oxygen (Chapter 1), Avogadro's hypothesis (Chapter 2), Faraday's law of electrochemical equivalence (Chapter 4), the formulation of relationships in chemical equilibrium (Chapter 10), and several other topics show that a balanced equation is indispensable to any detailed discussion of reactions. Since examination of a balanced equation provides the principal basis for classifying reactions, we will comment on techniques for balancing equations while we discuss classification.

To classify a reaction, we look at the equation to see the changes that occur in bonding. This means that we must know not only the composition of the molecules, but also their structures.

Consider the reaction of hydrogen with fluorine to form hydrogen fluoride. Balancing the equation can be done very quickly by inspection. The valence number of both hydrogen and fluorine is one, so the formula of the product is HF. We must either remember or look up the fact that elemental hydrogen and fluorine both exist as diatomic molecules. We know that

$$F_2 + H_2 \quad \text{gives} \quad HF$$

A balanced chemical equation requires that the same number of atoms of each kind appear on both sides of the equation. Thus, we must get two molecules of HF from one molecule each of F_2 and H_2

$$F_2 + H_2 \rightarrow 2HF \tag{11-1}$$

In the reactants, hydrogen atoms are bonded to each other, and fluorine atoms are bonded together. In the product, hydrogen atoms are bonded to fluorine atoms. We can think of various ways of describing such a reaction. It is an *exchange* reaction, since the atoms trade partners. It is also a *substitution* or *replacement* reaction, since, if we focus on any atom in the reactant molecules, we find that its partner is replaced by another kind of atom in the product. Such a reaction is sometimes called an *oxidation–reduction* reaction. As we will discuss later in the chapter, oxidation and reduction are defined in terms of the electronic nature of bonds that are made and broken in a reaction. Equation 11–1 could be called an oxidation–reduction reaction, because the bonds in the reactants are electronically symmetrical, whereas the bonds in the product are highly polar. We can also specify some features of the reactants that have not

Table 11–1. Examples of Substitution Reactions

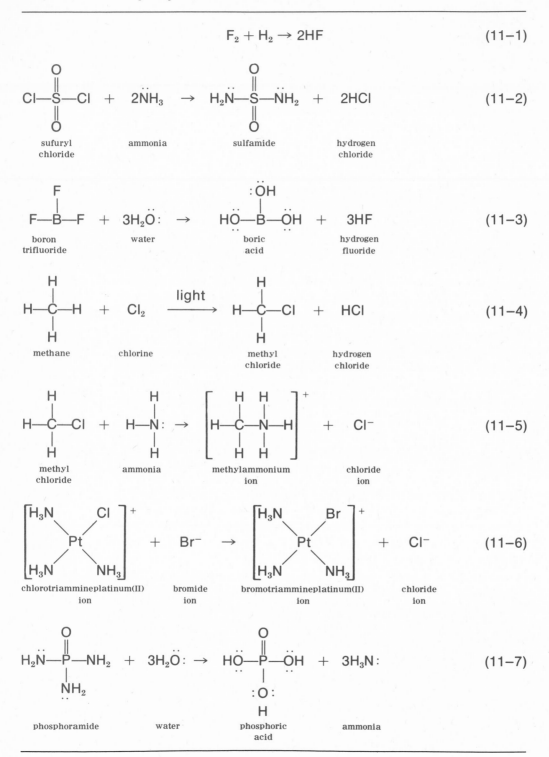

$$F_2 + H_2 \rightarrow 2HF \qquad (11\text{--}1)$$

(11–2)

sufuryl chloride ammonia sulfamide hydrogen chloride

(11–3)

boron trifluoride water boric acid hydrogen fluoride

(11–4)

methane chlorine methyl chloride hydrogen chloride

(11–5)

methyl chloride ammonia methylammonium ion chloride ion

(11–6)

chlorotriammineplatinum(II) ion bromide ion bromotriammineplatinum(II) ion chloride ion

(11–7)

phosphoramide water phosphoric acid ammonia

changed during the hydrogen–fluorine reaction. For instance, there is no change in the number of bonds to any atom; both reactants and products contain only single bonds. Reactions of this type are usually called *substitution* reactions.

No real problem is created when a particular reaction can be classified in more than one way, since the classifications are merely descriptions of chemical changes. The most useful classification will be determined by the context of the discussion. In one conversation, Equation 11–1 might be described most usefully as an oxidation–reduction reaction; in another, it might make more sense to call it a substitution reaction. By analogy, a baseball bat may be called sports equipment, or a deadly weapon, depending on how it is used.

The categories of reaction that we will use are as follows:

1 Substitution
2 Addition
3 Elimination
4 Isomerization
5 Oxidation–reduction

11–2 Substitution Reactions

Substitution reactions always involve replacement of an atom or group of atoms by another atom or group. A number of substitution reactions are listed in Table 11–1.

Balancing most of the equations in the table offers little difficulty, since inspection of the formulas of the reactants and products shows that, in most cases, one group simply trades places with another. The reaction of boron trifluoride with water might present a problem, because more than one group exchange occurs. Look first at the formulas of the reactants and products

reactants: BF_3, H_2O products: $B(OH)_3$, HF

It is obvious that three fluorine atoms are removed from boron and appear as HF. Clearly, three protons from water molecules will be needed to make three HF molecules. Since each oxygen keeps one proton in boric acid, we will need three water molecules to provide the three protons in HF and the three OH groups in $B(OH)_3$.

The reactions in Table 11–1 illustrate several points:

1 Substitution may involve the replacement of one atom by another (Reactions 11–1 and 11–4), one ion for another (Reaction 11–6), the exchange of more complex groups for atoms or ions (Reactions 11–2, 11–3, 11–5), or the exchange of one group of atoms for other groups of atoms (Reaction 11–7).

2 The reactants may be either molecules or ions.

3 In many substitution reactions, one of the reactants has an unshared pair of electrons that can be used for the formation of a new bond in the products. This is done by the ammonia molecule in Reaction 11–5 and by the bromide ion in Reaction 11–6. In other cases, one of the reactants has an unshared pair of electrons that seems to remain intact throughout the reaction (Reactions 11–2, 11–3, and 11–7). In these examples, it is possible that in some intermediate stages of the reaction unshared electron pairs play an important role.

REACTIONS INVOLVING IONS

When dissolved in some solvents, especially water, many substances dissociate into ions. Reactions in such solutions usually involve ions, rather than molecules. The nature of the chemical changes that occur is usually clearer if equations are written using only the ions that actually react in solution, as was done in the case of Reaction 11–6. Consider another example

$$H_3C—Br + NaOH \xrightarrow[\text{solution}]{\text{water}} H_3C—OH + NaBr \qquad (11–8)$$

methyl   sodium   methyl   sodium
bromide   hydroxide   alcohol   bromide

Both sodium hydroxide and sodium bromide are dissociated completely into ions in water solution, and the sodium ions are believed to play no part in the chemical reaction. A much simpler and more accurate representation of this reaction is

$$H_3C—Br + HO^- \rightarrow H_3C—OH + Br^- \qquad (11–9)$$

methyl   hydroxide   methyl   bromide
bromide   ion   alcohol   ion

Reaction 11–9 is clearly a substitution reaction, in which HO^- substitutes for Br^- in CH_3Br. Ionic equations (such as 11–9) generally are used to formulate reactions of ionized substances. In balanced ionic equations, the total charge of all the reactants must equal the total charge of all the products. You should not forget that if the total charge on the reactants is not zero, there are other ions in the solution to balance this charge; thus, the total charge in the solution is zero. Reaction 11–9 can be accomplished with solutions made from any one of several metallic hydroxides, such as sodium hydroxide, potassium hydroxide (KOH), or calcium hydroxide [$Ca(OH)_2$]. The only need to specify the particular hydroxide used is to be able to weigh the correct amount of hydroxide required to perform the reaction.

Example 11–1. Calculate the amounts of sodium hydroxide, potassium hydroxide, and calcium hydroxide required to react with 25.0 g of methyl bromide, according to Equation 11–9.

Solution. According to Equation 11–9, one mole of hydroxide ion is required for each mole of methyl bromide. The number of moles of methyl bromide is

$$\text{moles of } CH_3Br = \frac{\text{grams of } CH_3Br}{\text{g mol wt } CH_3Br} = \frac{25.0 \text{ g}}{94.9 \text{ g mole}^{-1}} = 0.263 \text{ mole}$$

NaOH and KOH provide one mole of HO^- per mole of compound, whereas each mole of $Ca(OH)_2$ provides two moles of HO^-. Therefore, we would need 0.263 mole of NaOH or KOH, or $0.263/2 = 0.132$ mole of $Ca(OH)_2$. The amounts required are

$$\text{grams of NaOH} = \text{moles NaOH} \times \text{g mol wt NaOH}$$
$$= 0.263 \text{ mole} \times 40.0 \text{ g mole}^{-1}$$
$$= 10.5 \text{ g}$$
$$\text{grams of KOH} = \text{moles KOH} \times \text{g mol wt KOH}$$
$$= 0.263 \text{ mole} \times 56.1 \text{ g mole}^{-1}$$
$$= 14.8 \text{ g}$$
$$\text{grams of } Ca(OH)_2 = \text{moles } Ca(OH)_2 \times \text{g mol wt } Ca(OH)_2$$
$$= 0.132 \text{ mole} \times 74.1 \text{ g mole}^{-1}$$
$$= 9.78 \text{ g}$$

ACID–BASE REACTIONS; PROTON TRANSFER

The word "acid" is familiar to nearly everyone; most people know that acids are corrosive materials because of their chemical reactivity. Also, "base" is a familiar name for a group of substances that *neutralize* acids. A preliminary discussion of the reactions of acids and bases was included in Chapter 9 as examples of reactions in solution.

The first formal definition of acids and bases was suggested in the 1880's by a young Swedish student named Svante Arrhenius. Arrhenius was also the first to recognize that many substances dissociate into ions when dissolved in water. His definition of acids and bases arose naturally from the fact that most acids and bases are used in water solution, and that water dissociates to form both hydrogen ions and hydroxide ions (Section 10–2). He designated a material that dissociates in solution to give a hydrogen ion (H^+) as an acid, and called substances that dissociate to give hydroxide ions (HO^-) bases. Arrhenius's definition can be illustrated by the equations

$$HX \rightleftarrows H^+ + X^- \quad \text{acid}$$
$$MOH \rightleftarrows M^+ + HO^- \quad \text{base}$$

A broader definition of acids and bases was suggested in 1923 by Johannes Brønsted and Thomas M. Lowry. Among other matters, they were concerned about the reactions of acids and bases in solvents other than water. In the Brønsted–Lowry scheme, an acid is a proton (H^+) donor and a base is a proton acceptor. An equation that illustrates the Brønsted–Lowry definition is

$$HB_1 + :B_2^- \rightarrow HB_2 + :B_1^- \tag{11–10}$$
$$\text{acid}_1 + \text{base}_2 \rightarrow \text{acid}_2 + \text{base}_1$$

From Equation 11–10, it is clear that an acid–base reaction can be viewed as a substitution reaction in which a base ($:B_2^-$) replaces another base ($:B_1^-$) bound to a proton. We call Reaction 11–10 simply a *proton transfer reaction*. Applying this definition to water solutions, we see that it becomes identical to the Arrhenius definition, if we realize that bare protons do not exist in water solution but are always bound to one or more water molecules. The Brønsted–Lowry equation for the ionization of an acid in water solution is

$$HX + H_2\overset{..}{O}: \rightarrow H_3O:^+ + :X^- \tag{11–11}$$
$$\text{acid} \quad \text{base} \quad \text{hydronium} \\ \text{ion}$$

Then the equation for the neutralization of a water solution of an acid is

$$H_3O:^+ + :B^- \rightarrow H_2\overset{..}{O}: + HB \tag{11–12}$$

G. N. Lewis further generalized the definition of an acid–base reaction by suggesting that a base is any molecule or ion capable of donating an electron pair in forming a bond to an acceptor molecule or ion. The acceptor could be a proton, as in Reaction 11–10, or it could be any other substance with an atom capable of adding an electron pair to its valence shell. Electron-pair donors are called *Lewis bases,* or *ligands,* whereas electron-pair acceptors are called *Lewis acids.* Examples of Lewis acids are BF_3, $AlCl_3$, and metal ions such as Pt^{2+}. Compounds formed between Lewis acids and bases (ligands) are called *addition* compounds, *coordination* compounds, or *complex* compounds. We have already discussed the bonding in one such compound, $H_3N:BF_3$ (Section 6–5). The general Lewis acid–base reaction is a substitution reaction, in which a Lewis base in a complex is replaced by a second Lewis base, as in Reaction 11–13

$$A:B_1 \quad + \quad :B_2 \quad \rightarrow \quad A:B_2 \quad + \quad :B_1 \tag{11-13}$$

Lewis acid–base complex$_1$ Lewis base$_2$ Lewis acid–base complex$_2$ Lewis base$_1$

The substitutions in Table 11–1 that can be described as reactions of Lewis bases with Lewis acid–base complexes are Reactions 11–5, 11–6, and 11–7.

11–3   Addition Reactions

In an addition reaction, molecules are made by the combination of simpler substances, as in the following examples:

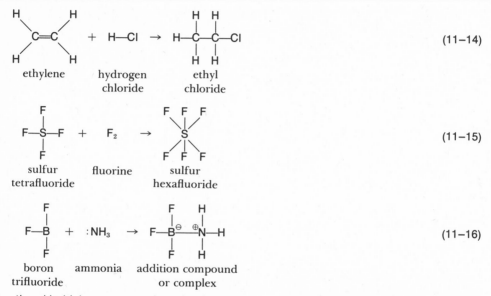

$$(11-14)$$

ethylene   hydrogen chloride   ethyl chloride

$$(11-15)$$

sulfur tetrafluoride   fluorine   sulfur hexafluoride

$$(11-16)$$

boron trifluoride   ammonia   addition compound or complex

Reaction 11–14 is representative of a large group of addition reactions, in which various reagents are added to substances containing double or triple bonds. Another important group consists of reactions such as Reaction 11–15, in which the addition reaction increases the total number of bonds. Reactions of this type always can be classified as oxidation–reduction reactions, as well as addition reactions. Reaction 11–16 of a Lewis base with a Lewis acid to give a complex, or addition compound, is another example of an addition reaction.

11–4   Elimination Reactions

Elimination is the opposite of addition. In an elimination reaction, some molecule is always broken into smaller units. Some examples are

$$H-\overset{\displaystyle H}{\underset{\displaystyle H}{C}}-\overset{\displaystyle H}{\underset{\displaystyle H}{C}}-Cl \quad + \quad KOH \quad \rightarrow \quad H_2C{=}CH_2 \quad + \quad H_2O \quad + \quad KCl \tag{11-17}$$

ethyl chloride   potassium hydroxide   ethylene   water   potassium chloride

$$XeF_6 \quad \rightarrow \quad Xe \quad + \quad 3F_2 \tag{11-18}$$
xenon   xenon   fluorine
hexafluoride

$$PCl_5 \quad \rightarrow \quad PCl_3 \quad + \quad Cl_2 \tag{11-19}$$
phosphorus   phosphorus   chlorine
pentachloride   trichloride

The definition of this reaction type is straightforward, although it is evident that some elimination reactions also could be classified in other ways. For example, Reaction 11–17 could be called an acid–base reaction.

11–5   Isomerization

In an isomerization reaction, a compound is converted to an isomer (from the Greek *isomeres*, which means having equal parts). Isomers have the same parts (atoms), but they have different structures. Examples of isomerization reactions are

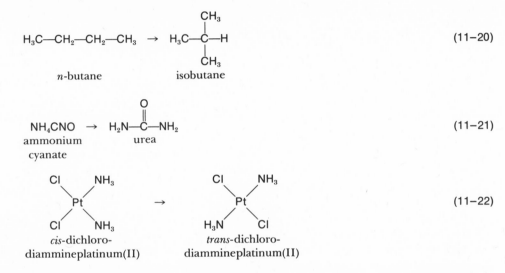

Reaction 11–20 is an example of isomerization involving *structural isomers,* compounds in which the *sequences* of atoms are different. Reaction 11–22 is the isomerization of a compound to a *geometric isomer.* The two materials have the same generalized structural formula, but, because the molecule has a square planar configuration (Section 7–9), there are two geometric arrangements of the ligands around the central platinum atom.

Reaction 11–21 is interesting for two reasons. Ammonium cyanate is an ionic compound, $NH_4^+CNO^-$. So, in a sense, the reaction is not an isomerization, but a complicated addition reaction of ammonium ion to a cyanate ion. The reaction has special historical significance. It was first reported by Friedrich Wöhler, in 1828, and came to have considerable bearing on the relationship between organic and inorganic compounds. In the early days of chemistry, it was believed that organic compounds such as urea derived only from living organisms containing a "vital force" not present in inorganic materials, such as ammonium cyanate, which came from minerals. This model now seems absurd, because the interconversion of materials from living and mineral sources has been accomplished countless times. However, the notion died slowly. Even Wöhler was almost apologetic in his first report, and he mentioned that his sample of ammonium cyanate was obtained by heating bones and might, therefore, be different in some undetectable way from ammonium cyanate of mineral origin.

Isomerization reactions appear deceptively simple. However, the actual mechanisms of changes such as those shown in Equations 11–20 and 11–22 are usually rather complicated. In these two reactions, the conversion from one form to the other involves two or more steps in which the molecules are actually taken apart and reassembled.

11–6   Oxidation–Reduction

Historically, the term "oxidation" meant "reaction with oxygen." It was applied to the process of combustion, both fast and slow, of various materials exposed to atmospheric oxygen. However, most of the changes caused by reactions with oxygen can be accomplished by treatment with other reagents, which came to be called *oxidizing agents.* Development of the electronic theory of molecular structure (Chapters 6 and 7) brought further insight into the structural changes that occur in oxidation reactions. In this section, we will present a set of formal rules that can be used to study oxidation–reduction processes.

First, we need a term to describe what happens to the oxidizing agent. We define *reduction* as the reverse of oxidation. When an oxidizing agent reacts with another substance, it is itself *reduced.* All reactions of the class *oxidation–reduction* have the general formulation

oxidized form of A + reduced form of B → reduced form of A + oxidized form of B

The oxidized form of A is called an *oxidizing agent* and the reduced form of B is called a *reducing agent.* Reactions of this class often are called *redox* reactions.

OXIDATION NUMBERS

Atoms in compounds are assigned *oxidation numbers,* which form the basis for describing the changes that occur in redox reactions. Oxidation numbers are often referred to as *oxidation states.* The following rules cover most situations:

1   The sum of the oxidation numbers of the elements in any neutral compound is 0, and the sum of the oxidation numbers of the elements in an ion is equal to the charge on the ion.

2   The oxidation number of any pure element is zero. For example, hydrogen in H_2 is 0, and fluorine in F_2 is 0.

3   Hydrogen in compounds is usually assigned the oxidation number +1. For example, hydrogen in HF, H_2O, or NH_3 is +1. Exceptions are metallic hydrides, in which hydrogen is assigned the oxidation number −1. For example, hydrogen in LiH is −1.

4   Metallic elements are assigned positive oxidation numbers, except in intermetallic compounds, in which metals are combined with each other. (In intermetallic compounds, elements have the oxidation number 0.) The magnitude of the positive oxidation number is the same as the valence number. For example, sodium in NaF is +1, magnesium in $MgCl_2$ is +2, and aluminum in $AlCl_3$ is +3.

5   With the exception of thallium, all of the metals of Groups IA, IIA, and IIIA have oxidation numbers that are the same as their group numbers. This is reasonable, since the group number is the same as the number of valence electrons in each atom in the group. The most common oxidation number of thallium is +1, although it also can be +3.

6   Oxygen is nearly always assigned a negative oxidation number. That number is −2, except in peroxides (compounds containing O—O bonds), for which it is −1.

7   Other nonmetallic elements (N, P, As, S, Se, Te, F, Cl, Br, I) usually are given negative oxidation numbers corresponding to those assigned to their simplest hydrides. For example, nitrogen in NH_3 is -3, and bromine in HBr is -1.

8   When the elements mentioned in Rule 7 are combined with oxygen, they are assigned positive oxidation numbers. For example, in nitric acid (HNO_3) hydrogen is $+1$, and each oxygen is -2. The sum of these is $(+1) + 3(-2) = -5$, so to make the sum of the oxidation numbers in the compound equal to 0, nitrogen must be $+5$.

9   When the nonmetals mentioned in Rule 7 are combined with each other, the one located further to the left in the periodic table is given a positive oxidation number; if the elements are in the same family, the heavier one is given a positive oxidation number. The following examples illustrate the rule:

$$NCl_3, \quad 3Cl \quad 3(-1) = -3$$
$$N \qquad \quad \underline{+3}$$
$$\text{Total} \quad 0$$

$$SF_6, \quad 6F \quad 6(-1) = -6$$
$$S \qquad \quad \underline{+6}$$
$$\text{Total} \quad 0$$

$$IF_5, \quad 5F \quad 5(-1) = -5$$
$$I \qquad \quad \underline{+5}$$
$$\text{Total} \quad 0$$

Rules 1–7 are easy to remember and to use. Rules 8 and 9 are especially important, because they are needed to treat many of the most important oxidizing agents. Rules 5 and 9 take precedence over Rule 6 (e.g., in OF_2, O is $+2$ and F is -1).

Some compounds contain complex networks of atoms, in which the same element occurs in different bonding situations. This is especially true of carbon compounds. In such cases, the compounds are divided into parts that are joined together by like atoms, and oxidation numbers are assigned to the atoms in each part according to the rules above. A few examples will make this clear

Notice that different oxidation numbers are sometimes assigned to the same element in a single compound

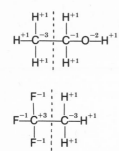

BALANCING REDOX EQUATIONS

If the formulas of reactants and products are known, any equation can be balanced by some combination of inspection and trial and error. However, in some reactions, the relationships are complex enough to make trial and error a tedious process. Redox reactions often involve this level of complexity, so it is useful to have a routine that we can use to generate a balanced equation. We will present a method, one of several, for doing this. However, we wish to make it clear that we are discussing time-saving methodology, not great scientific concepts.

A change in the oxidation number of an element may be thought of as involving a change in the number of electrons in the valence shell of the atom. Consider the reaction

$$\overset{+3}{2FeCl_3} + \overset{+2}{SnCl_2} \rightarrow \overset{+2}{2FeCl_2} + \overset{+4}{SnCl_4} \qquad (11\text{--}23)$$

$$\underset{\substack{\text{oxidant or} \\ \text{oxidizing} \\ \text{agent}}}{} \quad \underset{\substack{\text{reductant or} \\ \text{reducing} \\ \text{agent}}}{}$$

The change can be thought of as the transfer of electrons from $SnCl_2$, the reductant, to $FeCl_3$, the oxidant. Since the oxidation number of tin changes by two units, and that of iron by one, it is clear that we will need two $FeCl_3$ molecules to accept the electrons provided by one $SnCl_2$ molecule. This relationship in itself is enough to allow balancing of Equation 11–23. The chlorine atoms are used to keep the formulas of reactants and products neutral.

We realize that chloride ions play no part in the net reaction, since the oxidation number of chlorine remains −1 throughout the process. We will arrive at the final equation more quickly, if we write the equation in ionic form

$$2Fe^{3+} + Sn^{2+} \rightarrow 2Fe^{2+} + Sn^{4+} \qquad (11\text{--}24)$$

Now, it is evident immediately that we need two Fe^{3+} to react with one Sn^{2+}, just to keep the electrical charges balanced in the equation. Note that the charges on the cations are just the oxidation numbers shown in Equation 11–23. Furthermore, if for any reason we wish to write the equation in molecular form, as in Equation 11–23, all we need to do is attach enough chloride ions to the formulas in Equation 11–24 to make neutral molecules. For our present purposes, we do not care whether the compounds are ionized or not in the reacting solution; all we want is an efficient method for writing balanced equations.

The balancing is made even easier and more automatic if we break Equation 11–24 into two parts, one showing electron gain and the other showing electron loss

$$e^- + Fe^{3+} \rightarrow Fe^{2+} \tag{11-25}$$
$$Sn^{2+} \rightarrow Sn^{4+} + 2e^- \tag{11-26}$$

Equations such as 11–25 and 11–26 are called *half-reactions*. Half-reactions are valuable because they allow us to see the stoichiometric relationships immediately. If we take two units of Equation 11–25 and add them to Equation 11–26, we obtain Equation 11–24

$$
\begin{aligned}
2 \times (11\text{--}25) = 2e^- &+ 2Fe^{3+} \rightarrow 2Fe^{2+} \\
1 \times (11\text{--}26) = \quad\quad &Sn^{2+} \rightarrow Sn^{4+} + 2e^- \\
\hline
\text{Add:} \quad 2Fe^{3+} + Sn^{2+} &\rightarrow 2Fe^{2+} + Sn^{4+}
\end{aligned}
$$

No chemist wants to devote this much labor to balancing the equation for Reaction 11–23. However, the method becomes useful when applied to other systems. This is especially true when we deal with oxidants and reductants containing varying numbers of oxygen atoms. Consider the oxidation of arsenious oxide (As_2O_3) by potassium permanganate ($KMnO_4$), a powerful and useful oxidizing agent. The formulas of reactants and products are given by the unbalanced expression

$$As_2O_3 + MnO_4^- \rightarrow As_2O_5 + Mn^{2+} \tag{11-27}$$

The reaction is carried out in water solution, so we can use the components of the solution (H_2O, H^+ and HO^-), both in writing half-reactions and in formulating the overall balanced equation. If the reaction is actually performed in acidic solution, balance is accomplished using H^+ and H_2O. If the reaction is done in basic solution, we use HO^- and H_2O. This choice is a matter of convention and convenience. Any half-reaction balanced with H^+ can be converted quickly to the equivalent equation balanced with HO^-.

Let us write a half-reaction for the oxidation of As_2O_3 to As_2O_5 in acidic solution. The steps are

1 Write formulas for reduced and oxidized forms

$$As_2O_3 \rightarrow As_2O_5 \tag{11-28}$$

2 Balance oxygen atoms by adding water molecules to the left side of the equation

$$2H_2O + As_2O_3 \rightarrow As_2O_5 \tag{11-29}$$

3 Add hydrogen ions on the right side to balance hydrogen atoms

$$2H_2O + As_2O_3 \rightarrow As_2O_5 + 4H^+ \tag{11-30}$$

4 Add electrons on the right side to balance charges

$$2H_2O + As_2O_3 \rightarrow As_2O_5 + 4H^+ + 4e^- \tag{11-31}$$

Equation 11–31 is now a balanced half-reaction.

Let us do the same for the reduction of MnO_4^- to Mn^{2+}

$$
\begin{aligned}
&\text{(1)} \quad MnO_4^- && \rightarrow Mn^{2+} && \tag{11-32} \\
&\text{(2)} \quad MnO_4^- && \rightarrow Mn^{2+} + 4H_2O && \tag{11-33} \\
&\text{(3)} \quad 8H^+ + MnO_4^- && \rightarrow Mn^{2+} + 4H_2O && \tag{11-34} \\
&\text{(4)} \quad 5e^- + 8H^+ + MnO_4^- && \rightarrow Mn^{2+} + 4H_2O && \tag{11-35}
\end{aligned}
$$

The second half-reaction is shown by Equation 11–35. Looking at the electron balance in the two half-reactions, we can see that five units of Equation 11–31 will supply electrons required by four units of Equation 11–35. Accordingly, we combine the half-reactions to obtain the balanced equation

$$5 \times (11-31) = 10H_2O + 5As_2O_3 \rightarrow 5As_2O_5 + 20H^+ + 20e^-$$

$$\underline{4 \times (11-35) = 20e^- + 32H^+ + 4MnO_4^- \rightarrow 4Mn^{2+} + 16H_2O}$$

$$10H_2O + 5As_2O_3 + 20e^- + 32H^+ + 4MnO_4^-$$

$$\rightarrow 5As_2O_5 + 20H^+ + 20e^- + 4Mn^{2+}$$

$$+ 16H_2O \qquad\qquad (11-36)$$

Equation 11–36 is balanced, but we can simplify it by subtracting items (H_2O, e^-, and H^+) that occur on both sides of the equation. This will eliminate the electrons entirely, since the method makes the gains and losses of electrons equal. Simplification gives

$$5As_2O_3 + 12H^+ + 4MnO_4^- \rightarrow 5As_2O_5 + 4Mn^{2+} + 6H_2O \qquad\qquad (11-37)$$

Equation 11–37 is more simple than Equation 11–36, but it is still complex enough to convince us that the systematic method of balancing equations has something to offer. Arriving at Equation 11–37 by trial and error requires more time and drudgery than should be spent on such problems!

Equations for half-reactions have even more value than we have indicated. A hard-earned item such as Equations 11–31 and 11–35 can be used over and over with various combinations of oxidants and reductants. For example, permanganate ion (MnO_4^-) will oxidize Sn^{2+} to Sn^{4+}. Since we have half-reaction equations already available, we can write a balanced equation for the reaction very quickly by combining five units of Equation 11–26 with two units of Equation 11–35

$$5 \times (11-26) = 5Sn^{2+} \rightarrow 5Sn^{4+} + 10e^-$$

$$2 \times (11-35) = 10e^- + 16H^+ + 2MnO_4^- \rightarrow 2Mn^{2+} + 8H_2O$$

Adding and simplifying gives

$$5Sn^{2+} + 16H^+ + 2MnO_4^- \rightarrow 5Sn^{4+} + 2Mn^{2+} + 8H_2O \qquad\qquad (11-38)$$

While the method is fresh in our minds, we should balance an equation for a reaction in a basic solution, using HO^- instead of H^+. Lead in the $+2$ oxidation state is reduced easily to metallic lead by a number of reducing agents. Sulfites (SO_3^{2-}) are oxidized easily to sulfates (SO_4^{2-}). Since the principal species present in basic solutions of $+2$ lead is $Pb(OH)_3^-$, we will work with that formula, although we could use Pb^{2+} and add HO^- ions later

$$\underset{\text{reactants}}{Pb(OH)_3^- + SO_3^{2-}} \rightarrow \underset{\text{products}}{Pb + SO_4^{2-}}$$

First, obtain the lead half-reaction

$$(1) \quad Pb(OH)_3^- \rightarrow Pb + 3HO^-$$

(Note that there is no problem here in O and H balance.)

$$(2) \quad 2e^- + Pb(OH)_3^- \rightarrow Pb + 3HO^- \qquad\qquad (11-39)$$

Then obtain the sulfite half-reaction

$$(3) \quad SO_3^{2-} \rightarrow SO_4^{2-}$$

Use HO^- to supply oxygen atoms

$$(4) \quad 2HO^- + SO_3^{2-} \rightarrow SO_4^{2-} + H_2O$$

$$(5) \quad 2HO^- + SO_3^{2-} \rightarrow SO_4^{2-} + H_2O + 2e^- \qquad (11\text{--}40)$$

Combination of one unit each of Equations 11–39 and 11–40 gives a balanced equation, since each involves a two-electron change

$$Pb(OH)_3^- + SO_3^{2-} \rightarrow Pb + SO_4^{2-} + HO^- + H_2O \qquad (11\text{--}41)$$

Equations for half-reactions are useful for a variety of chemical purposes. For instance, half-reactions are used continuously by electrochemists in the interpretation of their work. In many textbooks of general or physical chemistry, and in all texts dealing with electrochemistry, tables of standard oxidation or reduction potentials are given. These tables contain not only balanced equations for half-reactions (called half-*cell* reactions in electrochemistry), but also information about the electrical potential required to take electrons out of, or put them into, chemical systems.

We can summarize a procedure for balancing redox equations as follows:

1   Write the formulas of reactants and products; without these, no procedure can work.

2   Check the oxidation numbers of atoms in reactants and products. If these make balancing obvious, do so; otherwise, continue with Steps 3, 4, and 5.

3   Write equations for the half-reactions.

4   Choose the number of units of the half-reactions to make electrons supplied by the reductant equal to the demand for electrons by the oxidant; then combine the equations.

5   Simplify by canceling species that appear on both sides of the equation.

11–7   Summary

1   Reactions can be described by dividing them into classes according to the structural changes that take place.

2   We classify reactions according to the following types: substitution, addition, elimination, isomerization, and oxidation–reduction (redox).

3   Acid–base reactions are an important class of substitution reactions. Most acid–base reactions can be defined in terms of proton transfer. A proton acid is a proton donor, and a base is a proton acceptor. A Lewis acid is an electron-pair acceptor; a Lewis base, or ligand, is an electron-pair donor.

4   Redox reactions can be discussed in terms of oxidation numbers or electron transfer. Systematic methods for balancing redox equations can be developed.

11–8   Postscript: Chemical Gears?

Chemical reactions are fascinating, because they are matter in action. Structural chemistry has its own appeal, and those who agree with Edna St. Vincent Millay that, "Euclid alone hath looked on beauty bare," are captivated by the geometric precision with which atoms organize themselves in molecules and infinite crystalline arrays. However, activists (and they are found in science) see the substance of chemical science in the study of chemical change. Along with the sense of action, chemical dynamics also offers potential for power and control. Reactions can be turned off and turned on, or slowed down and accelerated. In many cases, we can change the rates of a chemical process as delicately as we can tune an electronic circuit, by varying the temperature, the reaction medium, or the amount of a catalyst (Section 13–5).

After the eulogy, what next? Is it worthwhile to control physical phenomena called chemical reactions, or is it just another of the games that scientists play? What can we do with a reaction once we have it under control? One use of reactions is the production of new chemicals (see Postscript to Chapter 14). Manufacture of medicinal drugs, steel, synthetic films and fibers, and even the pigments in paints that ornament our lives, is possible only because chemical reactions can be carried out in a predictable way. Chemical synthesis is also of great value in the development of chemical science itself. Structural chemistry has profited enormously from the ingenuity and persistence of experts in synthesis. During the past 15 years, several dozen compounds have been prepared and studied that many theoreticians had said could not be made. The same is true of the systematic study of the relationship between chemical structure and reactivity (Sections 12–9 and 13–6). Models to explain and predict reactivity relationships are constructed by careful comparison of the reactivities of structurally similar compounds. Frequently, key points in the comparison schemes have been established only because someone did the work required to synthesize the "right" compounds to test the predictive value of models in a definitive way.

We can still ask, "What else?" Production of synthetic fibers and films has changed our lives, but we may not need to go on creating new variants indefinitely. In a similar way, using chemical reactions to produce new compounds to test theories of structure and dynamics may not be an endless mission. Can we use chemical reactions in still other ways? Obviously we can, and do, make reactions work for us in many ways. Burning gasoline in an automobile engine to convert chemical energy into mechanical energy to run the machine is an example. We also use chemical reactions to tell us how manufacturing plants are operating, how badly we are polluting the atmosphere and water resources (see Postscript to Chapter 10), and how sick we are and why.

Although we already have put chemical reactions to work in many ways, we have hardly begun to exploit the full potential for using controlled chemical changes. We should be able to use chemistry in engineering design, just as electrical and mechanical properties are used. We only have to look at the family car to realize that we are far from that goal. The mechanical design of an automobile is remarkably sophisticated, but chemical engineering in the machine remains about as crude as lighting a fire. We see a few areas in which the design of a whole working system has depended on adjustment of the chemical part along with the electrical and mechanical parts. Development of modern rockets has been possible only because chemistry has been considered as a design variable. However, such examples are relatively rare.

New Terms

Addition reaction: A reaction in which simpler molecules combine to form a more complex molecule.

Arrhenius acid: A proton donor.

Arrhenius base: A hydroxide donor.

Brønsted–Lowry acid: A proton donor.

Brønsted–Lowry base: A proton acceptor.

Elimination reaction: A reaction in which a molecule breaks apart into two or more simpler molecules; the opposite of an addition reaction.

Exchange reaction: A reaction in which a molecule gives up an atom or group of atoms in exchange for another.

Geometric isomers: Molecules having the same atoms linked together in the same sequence, but having different orientations in space.

Half-reaction (half-cell reaction):   The reaction of either the oxidant or the reductant in a redox reaction that shows the number of electrons gained or lost.

Isomerization reaction: The conversion of a molecule into an isomer.

Lewis acid: Electron-pair acceptor.

Lewis base: Electron-pair donor.

Oxidation: The increase in the oxidation number of an atom; loss of electrons.

Oxidizing agent (oxidant):   A substance that can remove electrons from another substance (oxidize it) and is itself reduced (gains electrons).

Oxidation number: A somewhat arbitrary number assigned to an atom according to definite rules; it is useful in determining whether a substance is oxidized or reduced in a reaction and in balancing redox equations. A change in oxidation number in a reaction corresponds to loss or gain of electrons.

Oxidation–reduction (redox) reaction: A reaction in which oxidation numbers of one or more atoms increase and decrease by an equivalent amount.

Reducing agent (reductant):   A substance that can reduce another and is itself oxidized.

Reduction: The decrease in oxidation number of an atom; gain of electrons.

Structural isomers: Molecules having the same atoms linked together in different ways.

Substitution (replacement) reaction: An exchange reaction in which there is no change in the number of bonds to each atom.

Questions and Problems

1   Classify the following reactions according to the categories listed in Section 11–1. If a reaction will fit into more than one category, show it in each.

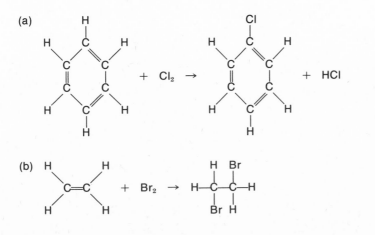

(c) $CH_3OH + HO\overset{O}{\overset{\|}{C}}CH_3 \rightarrow CH_3O\overset{O}{\overset{\|}{C}}CH_3 + H_2O$

(d) $2Na + 2H_2O \rightarrow 2NaOH + H_2$

(e)

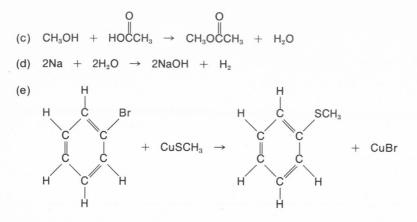

2 Explain the difference between the Arrhenius and Brønsted–Lowry definitions of a base, using the following reactions:

(a) $HCl + NaOH \rightarrow H_2O + Na^+ + Cl^-$
    acid   base

(b) $HCl + NH_3 \rightarrow NH_4^+ + Cl^-$
    acid   base

3 A very important reaction in biochemistry is the formation of a peptide bond from an amine and a carboxylic acid, as shown in the equation

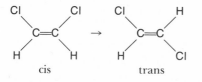

In what category would you place this reaction?

4 It is possible to carry out the following conversion under suitable conditions

$$\underset{\text{cis}}{\overset{Cl}{\underset{H}{>}}C=C\overset{Cl}{\underset{H}{<}}} \rightarrow \underset{\text{trans}}{\overset{Cl}{\underset{H}{>}}C=C\overset{H}{\underset{Cl}{<}}}$$

How would you categorize this reaction?

5 Assign oxidation numbers to each element in $KMnO_4$, SO_4^{2-}, $BrCl$, $Na_2C_2O_4$, CH_3OCH_3, and F_2.

6 Balance the following expressions:

(a) $KMnO_4 + HCl \rightarrow KCl + MnCl_2 + H_2O + Cl_2$
(b) $Cu + HNO_3 \rightarrow Cu(NO_3)_2 + H_2O + NO_2$
(c) $Sb + HNO_3 \rightarrow Sb_2O_5 + NO + H_2O$
(d) $Fe_2(SO_4)_3 + NaI \rightarrow FeSO_4 + Na_2SO_4 + I_2$
(e) $CrO_2^- + ClO^- + HO^- \rightarrow Cl^- + CrO_4^{2-} + H_2O$
(f) $Br_2 + HO^- \rightarrow Br^- + BrO_3^- + H_2O$

(g) $Bi(OH)_3 + SnO_2^{2-} \rightarrow Bi + SnO_3^{2-}$

(h) $I_2 + HNO_3 \rightarrow HIO_3 + NO_2 + H_2O$

(i) $Zn + NO_3^- \rightarrow Zn^{2+} + NH_4^+$

(j) $P_4 + HO^- + H_2O \rightarrow H_2PO_4^- + PH_3$

7   State which of the substances in the following redox reactions have been oxidized and which have been reduced. How many electrons are involved per atom in the reduction and oxidation steps in each reaction?

(a) $ClO_2 + HO^- \rightarrow ClO_2^- + ClO_3^-$

(b) $I_2 + Na_2S_2O_3 \rightarrow Na_2S_4O_6 + NaI$

8   A common oxidizing agent is potassium dichromate ($K_2Cr_2O_7$). It undergoes the following reduction in acidic solution

$$6e^- + 14H^+ + Cr_2O_7^{2-} \rightarrow 2Cr^{3+} + 7H_2O$$

If 20.4 g of $K_2Cr_2O_7$ is reduced, how many electrons must be donated to the $K_2Cr_2O_7$ to complete this reduction?

9   Complete the following half-reactions with electrons, H^+ ions, HO^- ions, water, or whatever might be necessary to achieve the proper balance of charge and matter:

(a) $Cl_2 \rightarrow Cl^-$

(b) $H_2O_2 + H^+ \rightarrow H_2O$

(c) $C_2O_4^{2-} \rightarrow CO_2$

(d) $P_4 \rightarrow P_2H_4$

10   Explain the difference between a proton acid and a Lewis acid. How is the Lewis definition of a base more general than the Brønsted–Lowry one? Give an example of a Lewis acid–base reaction.

11   Explain why acid–base reactions are placed in the substitution reaction category.

12   Classify the following reactions as either substitution or redox:

(a) $2Cu^{2+} + 2I^- \rightarrow 2Cu^+ + I_2$

(b) $AlCl_4^- + 6F^- \rightarrow AlF_6^{3-} + 4Cl^-$

(c) $5H^+ + Cr^{2+} + Co(NH_3)_5Cl^{2+} \rightarrow CrCl^{2+} + Co^{2+} + 5NH_4^+$

(d) $PtCl_2^{2-} + 4I^- \rightarrow PtI_4^{2-} + 4Cl^-$

13   In the redox reactions of Problem 12, label the oxidants and reductants. Show how the substitution reactions can be formulated as Lewis acid–base reactions. Which reactants are complexes? Which are ligands? Try to justify calling these reactions "ligand substitutions."

14   Show by one or more examples the type or types of reactions you could have if:

(a) A Lewis acid reacts with a Lewis base.

(b) A Lewis acid reacts with a ligand.

(c) A complex compound reacts with a ligand.

(d) A halogen reacts with an unsaturated compound.

(e) A proton acid reacts with a ligand.

PROTON ACIDS AND BASES

An important application of chemical equilibrium is the development of a quantitative understanding of the reactions of acids and bases. These substances play an important role in a wide variety of chemical studies and are of direct practical importance in all phases of chemical industry. For example, the manufacture of sulfuric acid (a proton acid) and ammonia (a proton base) are two of the largest volume chemical production processes in the world. Closer to the personal side of life, the reaction of carbon dioxide (CO_2) in the air with water to produce carbonic acid, and the various reactions of carbonic acid, are vital processes that help maintain blood at the proper pH. In this chapter, we will examine the important topic of proton acid–base equilibria in detail. After discussing some of the properties of solutions of proton acids and bases, we will show how the relative strengths of proton acids are related to their structures.

12–1 Conjugate Acids and Conjugate Bases

The Brønsted–Lowry definitions of an acid as a proton donor and a base as a proton acceptor (Section 11–2) are illustrated by the reaction of the acid HCl with the base water to form a new acid, H_3O^+, and a new base, Cl^-

$$HCl + H_2O \rightleftarrows H_3O^+ + Cl^- \tag{12-1}$$
$$\text{acid}_1 \quad \text{base}_2 \quad \text{acid}_2 \quad \text{base}_1$$

The general Brønsted–Lowry equation is

$$\text{acid}_1 + \text{base}_2 \rightleftarrows \text{acid}_2 + \text{base}_1 \tag{12-2}$$

The acid–base pairs are called *conjugates*. We refer to chloride ion in Equation 12–1 as the conjugate base of the acid hydrogen chloride. We would say that hydrogen chloride is the conjugate acid of the (weak) base chloride ion. Finally, the two substances, chloride ion and hydrogen chloride, are called an acid–base conjugate pair.

The position of the equilibrium in the general Reaction 12–2 depends on the relative strengths of acids 1 and 2. The stronger acid (the better proton donor) is more successful in giving up its protons. If acid_1 is stronger than acid_2, then the equilibrium favors the products on the right side of the equation. If acid_2 is the stronger, the equilibrium favors the substances on the left. An equivalent way of stating this is to say that the stronger base is more successful in capturing protons. If base$_2$ is stronger than base$_1$, the equilibrium position lies further to the right. If an acid is "strong" it has a "weak" conjugate base; a "weak" acid is the conjugate of a "strong" base. "Strong" and "weak" are relative terms, which we will define more carefully in Section 12–2.

Notice that electrical charge does not determine whether a substance is an acid, a base, or both. Most common acids are either neutral or positively charged, although there are exceptions. Some acids, such as sulfuric acid (H_2SO_4), contain two removable protons. Looking at the two stages of ionization,

$$H_2SO_4 + H_2O \rightleftarrows HSO_4^- + H_3O^+ \tag{12–3}$$
$$HSO_4^- + H_2O \rightleftarrows SO_4^{2-} + H_3O^+ \tag{12–4}$$

we realize that the hydrogen sulfate ion (HSO_4^-) must be classified as both an acid and a base. It is the conjugate base of H_2SO_4 and the conjugate acid of the sulfate ion (SO_4^{2-}).

Bases are usually anions or neutral molecules. As was pointed out in Section 11–2, the common structural feature of bases is the availability of unshared pairs of electrons to serve as binding sites for protons, which are Lewis acids. Some bases are

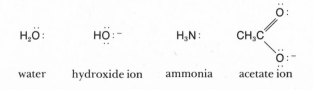

water   hydroxide ion   ammonia   acetate ion

Example 12–1. Write equations for the reactions of the acid HCN (hydrogen cyanide) and the base $(CH_3)_3N$: (trimethyl amine) with water. Label the reactants and products as conjugate acid–base pairs.

Solution.

$$\underset{\text{acid}_1}{HCN} + \underset{\text{base}_2}{H_2O} \rightleftarrows \underset{\text{acid}_2}{H_3O^+} + \underset{\text{base}_1}{CN^-}$$

$$\underset{\text{base}_1}{(CH_3)_3N} + \underset{\text{acid}_2}{H_2O} \rightleftarrows \underset{\text{acid}_1}{(CH_3)_3NH^+} + \underset{\text{base}_2}{HO^-}$$

Table 12–1. Some Strong Acids

Acid	Formula	K_a(moles $liter^{-1}$)[a]
Perchloric	$HClO_4$	10^{10}
Hydroiodic	HI	10^9
Hydrochloric	HCl	10^6
Sulfuric	H_2SO_4	$10^{3\ b}$
Nitric	HNO_3	10^2
Hydrogen sulfate ion	HSO_4^-	1.2×10^{-2}

[a]*The larger values of K_a cannot be measured directly; they are estimated.*
[b]*The K_a value is for the first ionization, $H_2SO_4 \rightarrow H^+ + HSO_4^-$.*

12–2 Strong and Weak Acids

Most references to acidic (or basic) strength relate to water solutions. These strengths can be expressed quantitatively in terms of equilibrium constants. Consider the ionization of an acid, HA, in water

$$HA + H_2O \rightleftarrows H_3O^+ + A^- \tag{12–5}$$

$$K = \frac{[H_3O^+][A^-]}{[HA][H_2O]} \tag{12–6}$$

As we mentioned in Section 10–2, the concentration of water ordinarily is considered a constant in the definition of K_a, the acidity equilibrium constant of the acid

$$K_a = K[H_2O] = \frac{[H_3O^+][A^-]}{[HA]} \tag{12–7}$$

Since K_a is a measure of the tendency of HA to ionize in water, it defines acid strength quantitatively. "Strong" and "weak" are qualitative terms, so there is no clear line of demarcation between them. Usually, acids having values of K_a of 1 or greater are referred to as strong acids. Some common strong acids are listed in Table 12–1. The hydrogen sulfate ion is included, although it is marginally strong.

Using the values of K_a, we can calculate the extent to which an acid is ionized in a solution of any given concentration. Consider a 0.1F solution of nitric acid

$$HNO_3 \quad + \quad H_2O \quad \rightleftarrows \quad H_3O^+ \quad + \quad NO_3^- \tag{12–8}$$

equilibrium
concentrations: $0.1 - x$ $\qquad\qquad\qquad x \qquad\qquad x$

The equation indicates that ionization of HNO_3 produces H_3O^+ and NO_3^- in equal amounts, so $[H_3O^+] = [NO_3^-]$. We have neglected a tiny additional amount of H_3O^+ coming from the self-ionization of water, $2H_2O \rightleftarrows H_3O^+ + HO^-$. The original concentration of HNO_3 is the sum of the concentrations of ionized and un-ionized forms so,

$$0.1F = [H_3O^+] + [HNO_3] = [NO_3^-] + [HNO_3]$$
$$\text{If } x = [H_3O^+] = [NO_3^-]$$
$$\text{Then } [HNO_3] = 0.1 - x$$

Using $K_a = 10^2$ (from Table 12–1) we can calculate x

$$10^2 = \frac{[H_3O^+][NO_3^-]}{[HNO_3]} = \frac{x^2}{0.1 - x}$$

$$10^2(0.1 - x) = x^2$$
$$x^2 + 10^2 x - 10 = 0 \qquad\qquad (12\text{–}9)$$

This quadratic equation can be solved easily, using the standard formula, and gives a value of $x = 0.09990$. From this answer, we can calculate the fraction of the total acid dissociated

$$\text{fraction dissociated} = \frac{x}{F} = \frac{0.09990}{0.1} = 0.9990$$
$$\% \text{ dissociated} = 99.90\%$$

The solution of the quadratic equation is a little tedious, and entirely unnecessary to obtain a reasonable answer. We can calculate an approximate value of x simply. We know that x must be less than 0.1, so it is obvious that $10^2 x$ must be much larger than x^2. If we make that approximation, Equation 12–9 becomes

$$10^2 x \cong 10$$
$$x \cong 0.1 \text{ mole liter}^{-1}$$

This states simply that with a value of K_a as large as 10^2, there will be a very small amount of molecular nitric acid in solution. We can calculate the amount of molecular HNO_3 approximately by setting $x \cong 0.1$.

$$10^2 \cong \frac{(0.1)^2}{[HNO_3]}$$

$$[HNO_3] \cong \frac{10^{-2}}{10^2} = 10^{-4} \text{ mole liter}^{-1}$$

The value calculated using the quadratic solution is 1×10^{-4} mole liter^{-1}, so the approximate answer is close to the real value. Both answers reveal that the ionization of HNO_3 is nearly complete in water.

Example 12–2. Calculate the fraction of hydrogen sulfate that will be ionized in a 0.1F sodium hydrogen sulfate ($NaHSO_4$) water solution.

Solution. Consider the sodium hydrogen sulfate to be ionized completely to Na^+ and HSO_4^-. This is a reasonable approximation for most salts in water solution. The equation for ionization of HSO_4^- is

$$HSO_4^- + H_2O \rightleftarrows SO_4^{2-} + H_3O^+$$

From Table 12–1, $K_a = 1.2 \times 10^{-2}$. If we make the trial approximation that $[H_3O^+] = [SO_4^{2-}] \cong 0.1$, we find

$$1.2 \times 10^{-2} = \frac{[0.1]^2}{[HSO_4^-]} = \frac{10^{-2}}{[HSO_4^-]}$$

$$[HSO_4^-] \quad = \frac{10^{-2}}{1.2 \times 10^{-2}} = 0.83 \text{ mole liter}^{-1}$$

The answer is ridiculous, since it states that $[HSO_4^-]$ is greater than the formality of the solution. Obviously, the values of $[H_3O^+]$ and $[SO_4^{2-}]$ must be much lower than 0.1 mole liter^{-1}. Let us turn to the original equation and do the problem

$$1.2 \times 10^{-2} = \frac{x^2}{(0.1 - x)}$$

$$x^2 + (1.2 \times 10^{-2})x - (0.12 \times 10^{-2}) = 0$$

Solution of the quadratic equation gives x = 0.029 mole liter^{-1}

$$\text{fraction dissociated} = \frac{0.029}{0.10} = 0.29$$

Since only 29% of the hydrogen sulfate is dissociated, we can get a sense of what is meant by the statement that HSO_4^- is a marginally strong acid. The fraction dissociated will be greater in more dilute solutions and smaller in more concentrated solutions. We recommend that students verify this statement.

Acids having values of K_a of 10^{-4} or less are only slightly ionized, except in very dilute solutions, and they are usually called weak acids. Table 12–2 lists some common weak acids. As an example, we will calculate the fraction of the acetic acid molecules ionized in $0.1F$ solution. The equilibrium concentrations, when x is the concentration of acetic acid that ionizes, are as follows:

$$CH_3C\!\!\overset{\displaystyle O}{\underset{\displaystyle OH}{\Big\langle}} \quad + H_2O \;\rightleftarrows\; CH_3CO_2^- \;+\; H_3O^+$$

[HOAc]	[AcO⁻]	[H₃O⁺]
$0.1 - x$	x	x

[*Note:* Ac represents the $CH_3\overset{|}{C}\!\!=\!\!O$ group.] Substituting these expressions in the equilibrium constant equation, we can write

$$1.85 \times 10^{-5} = \frac{x^2}{(0.1 - x)} = K_a \tag{12–10}$$

Before solving Equation 12–10 exactly as a quadratic, let's see what answer we would get if we assume that x is negligible in comparison with 0.1

$$0.1 - x \cong 0.1$$

$$1.85 \times 10^{-5} \cong \frac{x^2}{0.1}$$

$$x^2 \cong 1.85 \times 10^{-6}$$

$$x \cong 1.36 \times 10^{-3} \text{ mole liter}^{-1}$$

Table 12–2. Ionization Constants for Some Weak Acids

Acid	Formula		K_a *(moles liter^{-1})^a*
Hydrofluoric acid	HF		6.9×10^{-4}
Acetic acid	$CH_3C\overset{O}{\underset{OH}{\big/}}$	(HOAc)	1.85×10^{-5}
Telluric acid	$Te(OH)_6$		2×10^{-8}
Boric acid	$B(OH)_3$		6.0×10^{-10}
Aluminum hydroxide	$Al(OH)_3$		4×10^{-13}
Nickel hydroxide	$Ni(OH)_2$		6×10^{-19}

^{a}The K_a values are for the first ionizations of those acids that have more than one proton to lose.

This approximate value of x is larger than the real value, since in the quadratic equation

$$x^2 = 1.85 \times 10^{-6} - (1.85 \times 10^{-5})x$$

and, obviously,

$$1.85 \times 10^{-6} > 1.85 \times 10^{-6} - (1.85 \times 10^{-5})x$$

However, we will not make a large error if we assume x is small in comparison with 0.1, for the answer, $x = 1.36 \times 10^{-3}$, is almost 100 times smaller than 0.1. If we solve the full quadratic equation, we find $x = 1.35 \times 10^{-3}$, a difference hardly worth quibbling about. The fraction of the acetic acid ionized, using the approximate answer, is

$$\text{fraction ionized} \cong \frac{1.36 \times 10^{-3}}{0.1} = 1.36 \times 10^{-2}$$

$$\text{percent ionized} \cong 1.36\%$$

Example 12–3. A 0.10F solution of hydrofluoric acid, HF, is 7.9% dissociated. Calculate the value of K_a for this acid.

Solution. The amount of H_3O^+ and F^- present at equilibrium can be calculated from the original concentration of the acid (0.10F) and the fraction (0.079) of it that has dissociated, according to the equations

$$HF + H_2O \rightleftarrows H_3O^+ + F^-$$

$$K_a = \frac{[H_3O^+][F^-]}{[HF]}$$

$$[H_3O^+] = [F^-] = (0.079)(0.10M) = 0.0079M$$

The amount of HF remaining undissociated is

$$(0.10M - 0.0079M) = 0.092M$$

$$K_a = \frac{[7.9 \times 10^{-3}][7.9 \times 10^{-3}]}{[9.2 \times 10^{-2}]}$$

$$K_a = 6.8 \times 10^{-4} \text{ mole liter}^{-1}$$

12–3 Bases

A base reacts with water producing hydroxide ions (HO$^-$), according to Equation 12–11

$$\text{B:} + \text{H}_2\text{O} \rightleftarrows \text{BH}^+ + \text{HO}^- \tag{12–11}$$

The basicity constant, K_b, is defined by

$$K_b = \frac{[\text{BH}^+][\text{HO}^-]}{[\text{B}]} \tag{12–12}$$

Once again, the concentration of water is included in K_b values for bases in water solution. The meaning of the term "strong base" is a little loose. All alkali metal hydroxides (MOH) are called strong bases. In terms of the Brønsted–Lowry definitions, we would say that all of these compounds contain the same strong base, the hydroxide ion. The hydroxide ion is the strongest base that can exist in water solution, because any more basic species would be converted to its conjugate acid and hydroxide ion by Reaction 12–11. For example, indirect methods indicate that K_b for the amide ion (H$_2$N$^-$) in water solution should be about 10^8. Actually, when a compound such as sodium amide (NaNH$_2$) is dissolved in water, conversion to ammonia and hydroxide ion is so nearly complete that it is nearly impossible to determine experimentally the concentration of H$_2$N$^-$ ion remaining in solution at equilibrium.

Example 12–4. Calculate the equilibrium concentrations of H$_2$N$^-$, NH$_3$, and HO$^-$ in a 0.01F solution of NaNH$_2$.

Solution. When x is the concentration of H$_2$N$^-$ that reacts with H$_2$O, the equation for the reaction and the equilibrium concentration relationships are

$$\text{H}_2\text{N}^- \quad + \quad \text{H}_2\text{O} \quad \rightleftarrows \quad \text{NH}_3 \quad + \quad \text{HO}^-$$
$$0.01 - x \qquad\qquad\qquad\quad x \qquad\quad x$$

$$10^8 = K_b = \frac{x^2}{0.01 - x}$$

The problem is analogous to that of nitric acid in the preceding section. The value of K_b is so large that x must be very close to 0.01 mole liter^{-1}. Consequently, we approximate the solution

$$10^8 \cong \frac{(0.01)^2}{[\text{H}_2\text{N}^-]}$$

$$[\text{H}_2\text{N}^-] \cong \frac{10^{-4}}{10^8} = 10^{-12} \text{ mole liter}^{-1}$$

and

$$[NH_3] = [HO^-] \cong 10^{-2} \text{ mole liter}^{-1}$$

Example 12–5. Explain why the odor of hydrogen sulfide gas (H_2S) is present when a salt such as NaHS is dissolved in water. K_b for HS^- is 1.1×10^{-7}.

Solution. When NaHS dissolves, Na^+ and HS^- ions form. The hydrogen sulfide anion is a base, reacting with water according to the equation

$$HS^- + H_2O \rightleftarrows H_2S + HO^-$$

Thus, H_2S will form, and some of the gas will escape from the solution. To calculate the equilibrium concentrations, without taking account of the escape of hydrogen sulfide into the atmosphere, we will assume an arbitrary, but reasonable, concentration of $0.5F$ for NaHS. Then, if x is the concentration of HS^- that reacts with water, we can write

$$1.1 \times 10^{-7} = \frac{x^2}{0.5 - x}$$

Since 0.5 is significantly larger than 1.1×10^{-7}, we approximate

$$0.5 - x \cong 0.5$$
$$x^2 \cong 0.5 \, (1.1 \times 10^{-7})$$
$$x \cong 2.3 \times 10^{-4} \text{ mole liter}^{-1} = [H_2S]$$

The solubility of H_2S in water is about $0.1M$ at 1 atm H_2S pressure. A solubility of $2 \times 10^{-4}M$ corresponds to a partial pressure of 2×10^{-3} atm, or 2000 parts of H_2S in a million parts of air. The H_2S vapor can be detected by the human nose at levels of a few parts per million in air, and it is recognized easily; H_2S is responsible for the odor of rotten eggs. We caution

Table 12–3.   Weak Bases

Base	Formula	$K_b(moles\ liter^{-1})$
Ammonia	NH_3	1.8×10^{-5}
Methylamine	CH_3NH_2	3.5×10^{-4}
Phenylamine (aniline)	$C_6H_5NH_2$	4.6×10^{-10}
Acetate	$CH_3C\!\!\begin{array}{c}\diagup O \\ \diagdown O^-\end{array}$	5.4×10^{-10}
Fluoride	F^-	1.5×10^{-11}

against careless experimental verification of this calculation, since hydrogen sulfide is a deadly poison.

The hydrogen sulfide ion, referred to in the preceding example, is only one of many weak bases of importance in chemistry. A few other examples are listed in Table 12–3. The most important weak bases are ammonia and its organic derivatives, the amines, and the anions of organic acids. These materials are not only important in laboratory work and as items of commerce, but they also play leading roles in biological chemistry.

Table 12–3 contains one datum that should not be new to you. The acetate ion is the conjugate base of acetic acid, so K_b for acetate must be directly related to K_a for acetic acid. To see the relationship, write equations for both K_a and K_b

$$HOAc + H_2O \rightleftarrows AcO^- + H_3O^+$$

$$K_a = \frac{[AcO^-][H_3O^+]}{[HOAc]}$$

$$AcO^- + H_2O \rightleftarrows HOAc + HO^-$$

$$K_b = \frac{[HOAc][HO^-]}{[AcO^-]}$$

The two relationships cannot be independent, since the ion product of water is a constant (Section 10–2)

$$K_w = [H_3O^+][HO^-] = 1.0 \times 10^{-14} \text{ mole}^2 \text{ liter}^{-2}$$

Now, we replace $[HO^-]$ in the equation for K_b with the equivalent value in terms of $[H_3O^+]$

$$[HO^-] = \frac{K_w}{[H_3O^+]} \qquad K_b = \frac{[HOAc]K_w}{[AcO^-][H_3O^+]}$$

Comparison with the equation for K_a shows that

$$K_a \times K_b = \frac{[AcO^-][H_3O^+]}{[HOAc]} \times \frac{[HOAc][HO^-]}{[AcO^-]} = [H_3O^+][HO^-] = K_w$$

or

$$K_b = \frac{K_w}{K_a}$$

Similarly, the K_b values for other weak bases can be given as K_a values for the conjugate acids. That is, with the $K_a \times K_b = K_w$ relationship, we can calculate K_a if we know K_b, and if we know K_b, we can calculate K_a.

Example 12–6. Calculate the value of K_a for the methylammonium ion ($CH_3NH_3^+$).

Solution. From Table 12–3, we find K_b for $CH_3NH_2 = 3.5 \times 10^{-4}$. Therefore,

$$K_a = \frac{K_w}{K_b} = \frac{1.0 \times 10^{-14}}{3.5 \times 10^{-4}} = 2.9 \times 10^{-11}$$

This is an important specific example of a general principle mentioned in Section 10–2. The equilibrium constant for the sum of two chemical reactions is the product of the equilibrium constants for the two reactions. This arises fundamentally from the fact that free energies of reactions are additive, and free energies are related to the logarithms of equilibrium constants. In this case, we have

$$\text{HA} + \text{H}_2\text{O} \rightleftarrows \text{A}^- + \text{H}_3\text{O}^+ \qquad \Delta G_a^0 = -2.303\ RT \log K_a \tag{12–13}$$

$$2\text{H}_2\text{O} \rightleftarrows \text{HO}^- + \text{H}_3\text{O}^+ \qquad \Delta G_w^0 = -2.303\ RT \log K_w \tag{12–14}$$

Subtracting the water self-ionization reaction from the acid dissociation reaction, we get

$$\text{HA} + \text{HO}^- \rightleftarrows \text{A}^- + \text{H}_2\text{O}$$

$$\Delta G^0 = \Delta G_a^0 - \Delta G_w^0 = -2.303\ RT\ (\log K_a - \log K_w)$$

$$= -2.303\ RT \log \frac{K_a}{K_w}$$

The standard base dissociation reaction is just the reverse of this

$$\text{A}^- + \text{H}_2\text{O} \rightleftarrows \text{HA} + \text{HO}^-$$

$$\Delta G_b^0 = -2.303\ RT \log K_b = -\Delta G^0 = -2.303\ RT \log \frac{K_w}{K_a} \quad \text{or} \quad K_b = \frac{K_w}{K_a}$$

12–4 The pH Scale

Because the ion product of water is constant, the total acid–base balance in any water solution can be expressed by stating the hydronium ion concentration. It does not matter how many acid–base pairs there are in the medium; the ratios [Acid]/[Base] will be defined completely by the [H_3O^+] concentration, if the K_a (or K_b) values are known. To see this clearly, consider the ionization of any acid, as was done earlier

$$K_{\text{HA}} = \frac{[\text{H}_3\text{O}^+][\text{A}^-]}{[\text{HA}]}$$

$$\frac{[\text{HA}]}{[\text{A}^-]} = \frac{[\text{H}_3\text{O}^+]}{K_{\text{HA}}} \tag{12–15}$$

The rates of many reactions, both in chemical laboratories and in living tissues, are controlled by the acid–base level of the system. Therefore, the acidity is an important characteristic of samples from many sources and of interest for many reasons. The acidity level usually is expressed in pH units. The pH of a solution is the negative logarithm of the hydronium ion concentration

$$\text{pH} = -\log [\text{H}_3\text{O}^+] \tag{12–16}$$

Example 12–7. Calculate the pH of solutions that are (a) 0.001M in H_3O^+ and (b) 0.005M in HO^-.

Solution.

(a) $[H_3O^+] = 0.001$ mole liter$^{-1} = 1 \times 10^{-3}M$

$$pH = -\log\ [H_3O^+] = -\log(1 \times 10^{-3})$$

$-\log 1 = 0;\ -\log 10^{-3} = 3.0$

$$pH = 3$$

(b) $[HO^-] = 5 \times 10^{-3}$ mole liter^{-1}

$$[H_3O^+] = \frac{1 \times 10^{-14}}{5 \times 10^{-3}} = 0.2 \times 10^{-11}$$

$$pH = -\log\ [H_3O^+] = -\log(2 \times 10^{-12})$$

$-\log 2 = -0.3;\ -\log 10^{-12} = 12$

$$pH = 11.7$$

There are several reasons for using pH units. Most important is that the values are usually small positive numbers, since a negative pH would correspond to $[H_3O^+] > 1$, an uncommon situation in water solutions. The pH of almost every conceivable kind of system—rivers, people's stomachs, mill effluents, and even rainfall—is measured, often by people who have no familiarity with exponential notation. Consequently, the use of small positive numbers (i.e., the pH scale) simplifies tabulating data and keeping records. Furthermore, the literature of chemistry, biology, medicine, and agriculture is now so filled with pH values that the scale could not be abandoned, even if anyone wanted to do so. Another scale, pOH, is automatically defined by the pH scale

$$pOH = -\log\ [OH^-] \tag{12–17}$$

Since

$$K_w = [H_3O^+][HO^-] = 1.0 \times 10^{-14}$$
$$-\log K_w = -\log\ [H_3O^+] - \log\ [HO^-] = 14$$

and

$$pH + pOH = 14 \tag{12–18}$$

Figure 12–1 shows the common ranges of acidity of water solutions expressed both in concentration units and in terms of pOH and pH. *In a neutral water solution, pH = pOH = 7.*

Example 12–8. The pH of an aqueous solution of the strong acid HNO_3 is 2.47. Calculate $[H_3O^+]$ and $[HO^-]$ for this solution.

Solution. Since the pH is given, we can use the definition to calculate $[H_3O^+]$

$$pH = 2.47 = -\log\ [H_3O^+]$$
$$\log\ [H_3O^+] = -2.47 = -3.00 + 0.53$$
$$\text{antilog } 0.53 = 3.39;\ \text{antilog} -3.00 = 10^{-3}$$

Therefore,

$$[H_3O^+] = 3.39 \times 10^{-3} \text{ mole liter}^{-1}$$

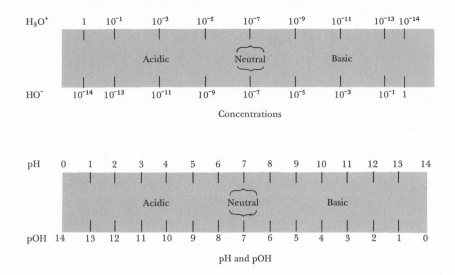

Figure 12–1. *Acid–base scales in water solution. Concentrations are in mole liter^{-1}.*

We can calculate $[HO^-]$ using either of the relationships

$$[HO^-] = \frac{1.0 \times 10^{-14}}{[H_3O^+]} = \frac{1.0 \times 10^{-14}}{3.39 \times 10^{-3}} = 2.9 \times 10^{-12} \text{ mole liter}^{-1} \text{ or,}$$

$$pOH = 14 - pH = 14 - 2.47 = 11.53$$
$$\log [HO^-] = -11.53 = -12 + 0.47$$
$$[HO^-] = (\text{antilog } 0.47) \times 10^{-12} = 2.9 \times 10^{-12} \text{ mole liter}^{-1}$$

12–5 Buffers

In the preceding section, we pointed out that for a conjugate acid–base pair the ratio [acid]/ [base] is fixed by the hydrogen ion concentration (pH) of the solution. Equation 12–15 also shows a way in which the acidity of a solution can be fixed within narrow limits. If an acid and its conjugate base are added to a solution in a known ratio, the pH of the medium will automatically assume the value calculated by Equation 12–15. In a practical system, the acid and base are added in concentrations higher than those of all other acids and bases in the solution. A mixture prepared in this way is called a *buffer*. If there are small amounts of other acids or bases present, they will react with the components of the buffer pair and change the [acid]/ [base] ratio. However, the use of the buffer mixture in a relatively high concentration will assure that the ratio will not change very much.

Suppose that we prepare a solution 0.1F in both sodium acetate and acetic acid. Since K_a for acetic acid is 1.85×10^{-5}, we know from Equation 12–15 that

$$[H_3O^+] = \frac{[HOAc]}{[AcO^-]} \times K_a = \frac{0.1}{0.1}(1.85 \times 10^{-5}) \tag{12–19}$$

$$[H_3O^+] = 1.85 \times 10^{-5} \text{ mole liter}^{-1}$$

or,

$$pH = 4.73$$

Now suppose that we add to this solution enough hydrochloric acid to make the HCl concentration 0.01F, if the system contained no buffer. If the addition had been made to pure water, $[H_3O^+]$ would have become 10^{-2} mole liter^{-1}; that is, the solution would have a pH of 2. In the buffered system, the HCl will neutralize an equivalent amount of acetate, producing acetic acid. We can calculate the change in the buffered solution as follows:

$$\text{new } [AcO^-] = 0.1 - 0.01 = 0.09 \text{ mole liter}^{-1}$$
$$\text{new } [HOAc] = 0.1 + 0.01 = 0.11 \text{ mole liter}^{-1}$$

$$[H_3O^+] = \frac{0.11}{0.09}(1.85 \times 10^{-5}) = 2.26 \times 10^{-5} \text{ mole liter}^{-1}$$

or,

$$pH = 4.65$$

In the unbuffered system, the addition of HCl produced a pH change from 7 (neutral water) to 2, or a change of 5 units. In the buffered system, the same addition of HCl produced a change from 4.73 to 4.65, or only 0.08 pH unit.

Because of the influence of acid–base balance on chemical phenomena, buffers are commonly used to fix the pH of solutions. Nearly all the chemical tests used in medical diagnosis are performed in buffered solutions, and biochemical research depends heavily on the use of buffers to obtain reproducible results. The fluids in the human body are buffered by the balance of organic acids and bases, and acute physiological malfunction may result if the pH of the various fluids varies much from the normal.

An interesting story can be built around the pH levels normally found in the different body fluids. For example, the interior of the stomach has a variable but rather high acidity. A pH of 1.4 is considered normal. By way of contrast, the pH of blood is maintained close to 7.4, slightly on the alkaline side of neutral. The high acidity in gastric juices (stomach fluids) is required to optimize the rates of hydrolysis (Section 12–6) of proteins to amino acids. The hydrolysis is catalyzed by the so-called tryptic enzymes, and it is known from laboratory studies with purified enzymes that the rates of the catalyzed reactions are sensitive to pH. The hydrolysis must occur because proteins, being very large molecules, cannot pass through the intestinal walls. However, the amino acids formed by hydrolysis can diffuse through the intestinal walls to be absorbed by the bloodstream.

Maintaining the pH of blood at a nearly constant level is important for many reasons. One reason is the role that blood plays in the transport of carbon dioxide. The blood carries oxygen to cells of the body by binding oxygen molecules to the iron atoms in hemoglobin, the red blood pigment. In the cells, oxygen burns nutrients, producing carbon dioxide, which must be taken by the blood back to the lungs where it is released. The iron in hemoglobin does not bind carbon dioxide, so transport must depend upon the solubility of carbon dioxide in blood. The solubility of gaseous carbon dioxide in acidic aqueous solutions is rather low, but it increases as the pH increases (i.e., as the solution becomes more basic). This is because CO_2 reacts with water to give a weakly acidic solution

$$CO_2 \ + \ 2H_2O \ \rightleftarrows \ HCO_3^- \ + \ H_3O^+ \tag{12-20}$$
$$\text{bicarbonate}$$
$$\text{ion}$$

If the equilibrium constant for Equation 12–20 is labeled K_{CO_2}, the ratio of CO_2 to HCO_3^- is given by the relationship

$$K_{CO_2} = \frac{[HCO_3^-][H_3O^+]}{[CO_2]}$$

$$\frac{[HCO_3^-]}{[CO_2]} = \frac{K_{CO_2}}{[H_3O^+]}$$

If the blood became too acid, the CO_2 would not be picked up from the tissues; if the blood were too basic, CO_2 would not be released properly in the lungs.

12–6   Hydrolysis

Hydrolysis is an antiquated term that is still in common chemical use. Literally, the word means "splitting by water." When it was observed originally that some compounds react with water to produce new species, it was assumed that water splits the compounds added to the water to give the new materials. For example, a solution of sodium acetate produces acetic acid and hydroxide ions. At one time, the equation for this process would have been written

$$AcONa \ + \ H_2O \ \rightleftarrows \ AcOH \ + \ NaOH \tag{12-21}$$
$$\text{sodium} \qquad\qquad \text{acetic} \quad\ \text{sodium}$$
$$\text{acetate} \qquad\qquad \text{acid} \quad\ \text{hydroxide}$$

The meaning of "hydrolysis" is clear in relation to Equation 12–21. However, we now realize that the following pair of equations are a better model for the solution chemistry of sodium acetate

$$AcONa \rightleftarrows AcO^- + Na^+ \tag{12-22}$$
$$AcO^- + H_2O \rightleftarrows AcOH + HO^- \tag{12-23}$$

The actual splitting into ions in Reaction 12–22 is essentially complete, and the acid–base Reaction 12–23 occurs only to a small extent. Now, the second reaction is called "hydrolysis," and we say that the acetate ion is slightly "hydrolyzed" in water to form acetic acid and hydroxide ion. Similarly, chemists say that ammonium ion (NH_4^+) is hydrolyzed by water to form ammonia and hydronium ion

$$H_4N^+Cl^- \xrightarrow{\ \sim 100\%\ } H_4N^+ + Cl^- \tag{12-24}$$
$$H_4N^+ + H_2O \rightleftarrows NH_3 + H_3O^+ \tag{12-25}$$

The term "hydrolyze" is used to include almost any reaction with water. Thus, Reaction 12–20 is sometimes referred to as the hydrolysis of carbon dioxide. Curiously, the dissociation of substances into ions in solution is not usually called hydrolysis, although it well fits the semantic origin of the word.

Example 12–9. What will be the pH of a solution if 20 ml of a $0.200F$ NH_3 solution is neutralized with an equal volume of $0.200F$ HCl? K_b for NH_3 is 1.8×10^{-5}.

Solution. The neutralization of NH_3 with HCl results in the formation of NH_4Cl. Since the combined volume of the acid and base is 40 ml, the concentration of the salt, NH_4Cl, is only half the concentration of the NH_3 and HCl originally present, or $0.100F$. NH_4Cl dissociates into NH_4^+ and Cl^- ions. The ammonium ion is an acid and reacts with water (hydrolyzes) in the following manner

$$NH_4^+ + H_2O \rightleftarrows H_3O^+ + NH_3$$

$$K_a = \frac{[H_3O^+][NH_3]}{[NH_4^+]} = \frac{K_w}{K_b}$$

$$K_a = \frac{1.0 \times 10^{-14}}{1.8 \times 10^{-5}} = 0.56 \times 10^{-9}$$

Let x be the amount of NH_4^+ that reacts; then

$$5.6 \times 10^{-10} = \frac{[x][x]}{[0.100 - x]} \text{ or } x^2 \cong 5.6 \times 10^{-11}$$

$$x^2 \cong 56 \times 10^{-12}$$

$$x \cong 7.5 \times 10^{-6} \cong [H_3O^+]$$

$$pH \cong -\log 7.5 \times 10^{-6} \cong 6 + (-0.88)$$

$$pH \cong 5.12$$

Although the NH_3 has been "neutralized" by the HCl, the solution is not neutral, because the pH is 5.12 and not 7.

12–7　Nonaqueous Solvents

Acid–base reactions occur in other media, as well as in water. Such systems are important, because a choice of different solvents allows us to prepare mixtures of much higher absolute acidity (greater protonating ability) than can be achieved in water. In aqueous solutions, the H_3O^+ ion is the most powerful protonating agent that can exist in appreciable concentration, and HO^- is the strongest base. In other solvents, the achievable acidity and basicity levels also are determined by the properties of the conjugate acid and base of the solvent.

In liquid ammonia (boiling point $-33°C$) the ions produced by self-ionization are the ammonium ion (H_4N^+) and the amide ion (H_2N^-)

$$2NH_3 \rightleftarrows H_4N^+ + H_2N^- \tag{12–26}$$

The ammonium ion is the strongest acid that can be produced in liquid ammonia. Since the ion is a weak acid compared with H_3O^+, solutions in liquid ammonia never can be strongly acidic in comparison with acidic aqueous solutions. However, the amide ion is a much more powerful base than HO^-, so solutions in liquid ammonia can be far more basic than any solution in water.

Acetic acid as a solvent shows the opposite effect. Self-ionization occurs according to the equation

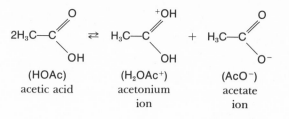

$$\underset{\substack{\text{(HOAc)}\\\text{acetic acid}}}{2H_3C-C} \quad\rightleftarrows\quad \underset{\substack{\text{(H}_2\text{OAc}^+)\\\text{acetonium}\\\text{ion}}}{H_3C-C} \quad+\quad \underset{\substack{\text{(AcO}^-)\\\text{acetate}\\\text{ion}}}{H_3C-C} \tag{12-27}$$

The acetonium ion is a much more potent proton donor than H_3O^+. Therefore, a solution of perchloric acid ($HClO_4$) in acetic acid will neutralize very weak bases that are essentially inert toward aqueous perchloric acid.

Superacidic and superbasic solutions are sometimes useful for performing reactions requiring catalysis by very powerful acids and bases. They are also useful in chemical analysis, since they can be used for the titration of acids and bases too weak to react with any water solutions.

12-8   Thermodynamics of Acid–Base Reactions

Following the notation of Chapter 10, we can express the standard free energy of acid–base reactions in terms of K_a

$$HA + H_2O \rightleftarrows H_3O^+ + A^-$$

$$K_a = \frac{[H_3O^+][A^-]}{[HA]}$$

$$\Delta G_a^0 = -2.303\ RT \log K_a \tag{12-28}$$

Notice that in the definition of the standard free energy, ΔG_a^0, we have considered the concentration of water a constant and included it in K_a, as before (Equation 12–7). Equation 12–28 and others related to it have considerable use. The equation leads to a definition of a quantity called pK_a, which is analogous to pH and pOH.

If

$$pK_a = -\log K_a$$

then,

$$\Delta G_a^0 = 2.303\ RT\ pK_a \tag{12-29}$$

This says that the pK_a is essentially the free energy for the dissociation of an acid. Since free energies of reactions are additive, pK values are also additive, and calculations of equilibrium constants are often simplified by use of pK values. Consider the ionization of phosphoric acid, a tribasic acid

$$\Delta G_1^0 \quad \underset{\substack{\text{phosphoric}\\\text{acid}}}{H_3PO_4} + H_2O \underset{}{\overset{K_1}{\rightleftharpoons}} H_3O^+ + \underset{\substack{\text{dihydrogen}\\\text{phosphate}}}{H_2PO_4^-} \tag{12-30}$$

$$\Delta G_2^0 \quad H_2PO_4^- + H_2O \overset{K_2}{\rightleftharpoons} H_3O^+ + \underset{\substack{\text{monohydrogen} \\ \text{phosphate}}}{HPO_4^{2-}} \tag{12–31}$$

$$\Delta G_3^0 \quad HPO_4^{2-} + H_2O \overset{K_3}{\rightleftharpoons} H_3O^+ + \underset{\text{phosphate}}{PO_4^{3-}} \tag{12–32}$$

We can formulate the equilibrium for the overall ionization process

$$\Delta G_{123}^0 \quad H_3PO_4 + 3H_2O \overset{K_{123}}{\rightleftharpoons} 3H_3O^+ + PO_4^{3-} \tag{12–33}$$

Since Reaction 12–33 is the sum of Reactions 12–30, 12–31, and 12–32,

$$\Delta G_{123}^0 = \Delta G_1^0 + \Delta G_2^0 + \Delta G_3^0 \tag{12–34}$$

Since the values of ΔG^0 are directly proportional to pK_a values by Equation 12–29, the pK_a's are also additive, and

$$pK_{123} = pK_1 + pK_2 + pK_3 \tag{12–35}$$

If we rewrite Equation 12–35 using the definition of pK_a, we obtain

$$-\log K_{123} = -\log K_1 - \log K_2 - \log K_3 \tag{12–36}$$

$$= -\log K_1 K_2 K_3 \tag{12–37}$$

and consequently,

$$K_{123} = K_1 K_2 K_3 \tag{12–38}$$

Equation 12–38 is a useful relationship that can be derived by direct examination of the various K_a values. The route that we have used to obtain the expression illustrates two points: (1) The free energy change in any process is the sum of the free energy changes for any component steps and (2) the addition of free energy terms is, in a sense, equivalent to multiplication of equilibrium constants.

Another common use of pK_a values is worth noting. They can be used with the pH scale to give relationships between a buffer ratio and the pH that it will establish

$$pK_a = -\log \frac{[H_3O^+][A^-]}{[HA]} \tag{12–39}$$

$$= -\log [H_3O^+] - \log \frac{[A^-]}{[HA]} \tag{12–40}$$

Since $-\log [H_3O^+]$ is pH, we can write

$$pK_a + \log \frac{[A^-]}{[HA]} = pH \tag{12–41}$$

Example 12–10. If the pK_a for phosphoric acid is 2.12, what ratio of $[H_3PO_4]$ to $[H_2PO_4^-]$ will be needed to buffer a solution at pH 2.52?

Solution.

$$\log \frac{[H_2PO_4{}^-]}{[H_3PO_4]} = pH - pK_a = 0.40$$

$$\frac{[H_2PO_4{}^-]}{[H_3PO_4]} = 2.5$$

12–9 Structure and Acidity

One of the most significant objectives of chemistry is to obtain useful models relating molecular structure to chemical reactivity. As an ambition, we would like to look at a structural formula and predict quantitatively all of the reactions of the material represented by the formula. This would include prediction of both free energy changes (equilibrium constants) and rates of all conceivable reactions. We quickly confess that we are now a long way from being able to accomplish this objective, which is another indication that much remains to be done in chemistry. However, we have made a start, and we can use a comparison of the pK_a values of various acids to illustrate a little about the methods that are used.

Proton acid–base reactions have one simplifying characteristic—their great speed. Reactions in which protons (H^+) are transferred from one electronegative element (O, N, S, P, and the halogens) to another are usually so fast that their rates can be measured only by special techniques. Reactivity in acid–base reactions such as those discussed in this chapter is really a matter of considering equilibrium relationships. This situation is unusual, since reaction rates (Chapter 13) are often a controlling factor in determining reactivities.

Consider the ionization of an acid (HA) in water solution

$$HA + H_2O \rightleftarrows H_3O^+ + A^- \tag{12–42}$$

All the following factors must be of importance in determining the reaction energetics:
1 The energy required to separate H—A into H^+ and A^-.
2 The energy released by placing the H^+ on H_2O.
3 The free energy of solvation of HA in water.
4 The free energy of solvation of H_3O^+ in water.
5 The free energy of solvation of A^- in water.

There are ways of estimating all of the Factors 1–5, but none is precise. In fact, the overall estimate probably will have an uncertainty of several pK_a units. An estimate that uncertain is of little use to us.

However, all is not lost. If we examine Factors 1–5 more closely, we note that Factors 2 and 4 will be the same for all acids in water solution. We can eliminate these two terms by *comparing* two acids, HA_1 and HA_2

$$K_1 = \frac{[H_3O^+][A_1{}^-]}{[HA_1]} \tag{12–43}$$

$$K_2 = \frac{[H_3O^+][A_2{}^-]}{[HA_2]} \tag{12–44}$$

$$\frac{K_1}{K_2} = \frac{[A_1{}^-]}{[HA_1]} \times \frac{[HA_2]}{[A_2{}^-]} \tag{12–45}$$

Equation 12–45 is the equilibrium constant for the reaction that directly compares the acidities of HA_1 and HA_2

$$HA_1 + A_2^- \rightleftarrows HA_2 + A_1^- \tag{12–46}$$

If A_1^- and A_2^- have similar structures, Factors 3 and 5 having to do with solvation of the acids and their conjugate bases, may come close to canceling, and we may be able to see a relationship between the values of pK_1 and pK_2 that relate, in a rational way, to differences in structure. This is the most common route now taken to the analysis of structure–reactivity relationships.

Table 12–4 shows the formulas of some acids in which the acidic proton is bound to oxygen; each acid can be thought of as having the structure Z—OH. The various Z groups obviously exert a profound influence on the acidity of the protons of the hydroxyl groups. The table shows that there is a simple correlation between acidity and the number of unprotonated oxygen atoms attached to the central atom. This kind of empirical relationship usually inspires some explanation in terms of molecular structure, and we can use this case as an example of the kind of qualitative rationalization that can be developed.

Consider the ionization of perchloric acid

$$HClO_4 + H_2O \rightleftarrows H_3O^+ + ClO_4^- \tag{12–47}$$

Compared with the other acids, there is an unusually large decrease in free energy associated with taking a proton off the perchlorate ion and putting it on a water molecule. We believe that this is explained by the structure of the perchlorate ion. Intuition and experiment both tell us that the ion must be symmetrical, with all of the oxygen atoms equivalent. The negative charge on the anion is not constrained to stay on one oxygen atom, but can be spread symmetrically among all four. This delocalization can be shown by the resonance notation

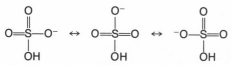

four equivalent resonance formulas for perchlorate ion

We could give an equivalent explanation in terms of molecular orbitals. Either explanation makes use of the 3d orbitals of chlorine as acceptors of electrons from oxygen. Examination of resonance formulas for HSO_4^- and $H_2PO_4^-$ reveals a decreasing opportunity for delocalization

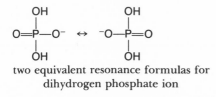

three equivalent resonance formulas for hydrogen sulfate ion

two equivalent resonance formulas for
dihydrogen phosphate ion

This explanation associates high acidity with electron delocalization that stabilizes the anions. The correlation is qualitative, but it allows some predictions to be made. For example, we can

Table 12–4.   Strengths of Some Neutral Oxyacids

Acid	Formula	Number of unprotonated oxygens	pK_a
Perchloric	$O=\overset{\displaystyle O}{\underset{\displaystyle O}{\|\|}}Cl-OH$	3	−10
Sulfuric	$O=\overset{\displaystyle O}{\underset{\displaystyle OH}{\|\|}}S-OH$	2	−3
Phosphoric	$O=\overset{\displaystyle OH}{\underset{\displaystyle OH}{\|}}P-OH$	1	2
Telluric	$\begin{array}{c} H \\ HO \quad O \quad OH \\ \diagdown \mid \diagup \\ Te \\ \diagup \mid \diagdown \\ HO \quad OH \quad OH \end{array}$	0	6

predict that $Te(OH)_6$ will have very low acidity, because the anion $Te(OH)_5O^-$ will be unable to delocalize its negative charge.

The preceding structure–reactivity correlation was possible because we chose to compare acids that all have the same net charge (zero). An extremely important factor affecting proton reactivity is the overall electrical charge on the acid. All other things being equal, increasing the negative charge on a species decreases its acidity. For example, HSO_4^- has a pK_a of 2, compared with −3 for H_2SO_4. We would have been in trouble if we had tried to rationalize the acidity of HSO_4^- in the same series with $HClO_4$, H_2SO_4, and H_3PO_4. If we ignore the negative charge on HSO_4^-, the acid is analogous to $HClO_4$, in that it has three unprotonated oxygens. Thus, the extra negative charge on HSO_4^- decreases its acidity by 12 pK_a units in comparison to $HClO_4$! The same effect is shown clearly by comparing the ammonium ion (NH_4^+), a weak acid in water, with ammonia (H_3N), which has no measurable acidity in aqueous solution.

Exercise. Predict the relative acidities of HSO_4^- and $H_2PO_4^-$. What is the basis for your prediction? Would HSO_4^- be a stronger, or weaker, acid than HPO_4^{2-}? What factor dominates here?

An effect related to changing the charge should occur if we introduce dipolar groups into a molecule. Consider the acidities of acetic and chloroacetic acids,

Table 12–5.   Strengths of Chloroacetic Acids

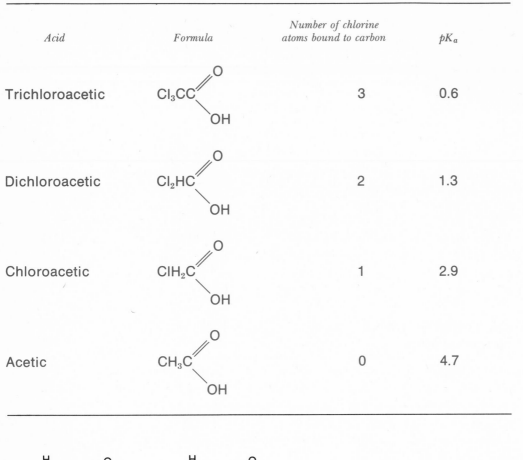

Acid	Formula	Number of chlorine atoms bound to carbon	pK_a
Trichloroacetic		3	0.6
Dichloroacetic		2	1.3
Chloroacetic		1	2.9
Acetic		0	4.7

acetic acid   chloroacetic acid

Chloroacetic acid has a dipolar carbon–chlorine bond. The small positive charge on the carbon ($Cl^{\delta-}$—$C^{\delta+}$) should help stabilize the negative charge on the oxygens of the conjugate base and make chloroacetic acid stronger than acetic acid. This small positive charge on the carbon, which is adjacent to the —C̈—OH group, will increase as more chlorines are substituted for hydrogens, thus allowing more negative charge to be "pulled," or delocalized, from the oxygens in the conjugate base. Thus, as Table 12–5 shows, acid strength increases as chlorines are added in the series of chloroacetic acids.

12–10   Summary

1   We have considered acid–base reactions of the type

$$acid_1 + base_2 \rightleftarrows base_1 + acid_2$$

in which the reaction is a proton transfer.

2   In water solution, the ionization constant of an acid is defined as

$$K_a = \frac{[H_3O^+][A^-]}{[HA]}$$

3   The basicity constant of a base is defined as

$$K_b = \frac{[BH^+][HO^-]}{[B]}$$

4   The values of K_a and K_b for a conjugate pair are related by the ion product of water

$$K_a = \frac{K_w}{K_b}$$

5   A buffer is a mixture of an acid and its conjugate base in a ratio suitable for controlling the pH of a solution at, or close to, a fixed value.

6      $pH = -\log[H_3O^+]$

7   Superacidic and superbasic solutions can be prepared in solvents other than water.

8   Thermodynamic formulation of ionization equilibria shows that ΔG_a^0 is proportional to pK_a.

9   It is possible to correlate molecular structure with proton reactivity for a series of closely related acids.

12–11   Postscript: pH and Yellow Leaves

In the eastern United States, especially in heavily forested areas, soils tend to be acidic. The pH is usually below 7, and frequently as low as 5. Some plants, such as rhododendron, thrive in strongly acid soils, but most plants suffer root damage and decline in soils with the pH as low as 5. To adjust the pH to the optimum value for plant growth, farmers and the backyard gardeners add limestone ($CaCO_3$) to the soil. The carbonate ion not only consumes acid, producing HCO_3^-, but also acts as a buffering agent, because HCO_3^- in the soil can release or absorb hydrogen ions to maintain a roughly constant pH.

In the semi-arid and arid areas of the West, the opposite condition persists. Because of limited rainfall, there is little vegetation and little organic material in the soil. Rainfall leaches alkaline materials from the soil and runs off to basins or plains where it evaporates, leaving a salty, alkaline residue. These basins or plains are typically farming areas (California's Imperial Valley, the Great Salt Lake basin in Utah, and the San Luis Valley in Colorado). In these soils, the pH can be as high as 8.5. Most vegetables and grains thrive in this environment, if they can get enough mineral nourishment, but in many areas this is a difficult requirement to meet.

Consider the mineral nutrient iron, for example. Iron has two common valences, two and three. Divalent iron usually is present in solution as the ion $Fe(H_2O)_6^{2+}$. Trivalent iron can exist as the ion $Fe(H_2O)_6^{3+}$. However, $Fe(H_2O)_6^{3+}$ is a relatively strong acid ($pK_a = 4$) and readily forms $Fe(H_2O)_5(OH)^{2+}$, $Fe(H_2O)_4(OH)_2^+$, and, finally, $Fe(H_2O)_3(OH)_3$, which is insoluble in water. In addition, both divalent and trivalent iron form insoluble neutral compounds with bases such as carbonate, and the divalent iron hydroxide, $Fe(OH)_2$, is also fairly insoluble.

Even this picture presents a simplified view of the many iron-containing compounds present in the soil that potentially can be used by plants. A diagram of the various common iron compounds is shown in Figure 12–2. The horizontal axis shows the pH of the soil. The vertical axis is essentially a measure of the amount of oxygen in the soil. Oxygen reacts with divalent iron in

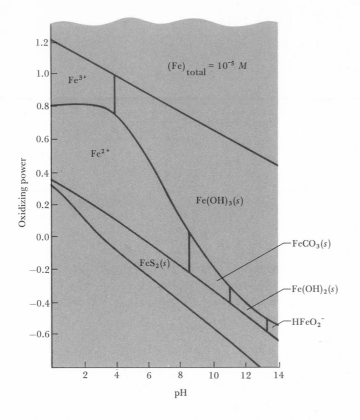

Figure 12–2.   Iron species in soils as a function of pH and oxidizing power.

its many forms to produce compounds of trivalent iron. The higher the oxygen concentration in the soil, the lower the concentration of divalent iron species. The diagram is easy to read. Any oxidizing power and any pH define a point. The only iron species that can exist at that point in any measurable concentration is the compound given for the field which contains the point. For example, at pH 2, with oxidizing power +0.9, only Fe^{3+} is present.

Very little is known about the mechanism of transport and biological use of metal nutrients such as iron. It probably is true that some compounds of iron are better foods than others. But one thing is certain: Insoluble substances in the soil stay in one spot and cannot diffuse through the soil solution in the way that water-soluble substances can. Therefore, insoluble iron in soils is effectively unavailable as plant food. From the diagram, it is clear that below about pH 7 there is a wide range of oxidizing power over which the soluble forms Fe^{2+} and Fe^{3+} are the main iron species present. But in alkaline soils above pH 8, only insoluble iron compounds exist under any oxidizing conditions (soils are never alkaline enough to produce $HFeO_2^-$).

Both important food crops, such as corn, and plants for pleasure, such as shade trees, suffer in such an environment. The general symptoms are yellow leaves, leaf splitting and dropping, and impaired growth. Growth and maturity are most severely affected in annual fast growth crops such as corn, alfalfa, and so on. After three weeks of growth, an iron-deficient corn plant may be less than half the size of a healthy well-nourished one.

There is a chemical solution to this problem that is equally applicable to huge farms and sick-looking front lawns. It is usually impractical to try to lower the pH of the soil. Impossibly large amounts of acid would have to be added (for example, 100 tons of HCl per acre). An alternative is to add some substance to the soil that will react with insoluble iron species to produce soluble iron species. An example of such a substance is sodium ethylenediaminetetraacetate (or Na_4L)

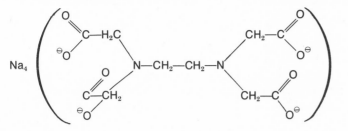

Reactions of the sort

$$Fe(OH)_3 + L^{4-} \rightleftarrows FeL^- + 3OH^-$$

have very large equilibrium constants. The negative ions FeL^- and FeL^{2-} are water soluble and good sources of iron for plants.

This illustrates the point that there is more to farming than mowing weeds, and more to acids and bases than routine laboratory operations. The chemistry of soils is a complex field full of unsolved theoretical and practical problems, and one that is largely ignored by professional academic chemists. However, it is an exciting field in which to work, for it combines the study of fundamental and difficult chemical problems with the practical application of chemistry to immediate problems of soil fertility.

New Terms

Buffer: A solution with the property that addition of relatively large amounts of an acid or a base will not change its pH appreciably.

Conjugate acid–base pairs: A pair of substances, one an acid, the other a base, related by the transfer of a proton. The acid has one more transferable proton than the base, and it becomes the base by transfer of a proton.

Hydrolysis: A term literally meaning "splitting by water," now used loosely to refer to reactions with water.

K_a (or K_{HA}): The equilibrium constant for the dissociation of an acid. The solvent base with which the acid reacts does not appear in the equilibrium expression.

K_b (or K_B): The equilibrium constant for the dissociation of a base. The solvent acid with which the base reacts does not appear in the equilibrium expression.

pH: A measure of acidity. $pH = -\log [H_3O^+]$.

pK: A measure of the free energy of a reaction. $pK = -\log K = \Delta G^0/2.303RT$.

Protonic acid (base): A Brønsted acid (base); see Chapter 11.

Strong acid (base): An acid (base) in water with $K_a (K_b) > 1$.

Weak acid (base): An acid (base) in water with $K_a (K_b) < 10^{-2}$.

Questions and Problems

1 Classify the following as Brønsted–Lowry acids, bases, or both, and write reactions with water to illustrate each: H_2O, S^{2-}, $H_2PO_4^-$, HCl, NH_3, NH_2^-, $(CH_3)_3NH^+$, HSO_3^-, and OAc^-.

2 Identify the partners in the conjugate pairs for the species in Problem 1.

3 Calculate the H_3O^+ concentration for a solution $1.5 \times 10^{-3}F$ in the strong acid $HClO_4$.

4 Calculate the H_3O^+ concentration for a solution $1.5 \times 10^{-3}F$ in the weak acid HCN; $K_a = 4.8 \times 10^{-10}$.

5 Calculate the percent dissociation of HCN in Problem 4.

6 Show that the dissociation reaction

$$H_3PO_4 + 3H_2O \rightleftarrows 3H_3O^+ + PO_4^{3-}$$

may be written in three steps. If K_1, K_2, and K_3 for these three steps are 7.1×10^{-3}, 6.3×10^{-8}, and 4.4×10^{-13}, respectively, what is the value of K, the equilibrium constant for the one-step process?

7 Hydrazine, NH_2—NH_2, reacts with water to produce the hydrazinium ion ($NH_2NH_3^+$) according to the equation

$$NH_2\!-\!NH_2 + H_2O \rightarrow NH_2\!-\!NH_3^+ + HO^-$$

K_b for hydrazine is 9.8×10^{-7}. If a solution is $0.25F$ in hydrazine, what will be the HO^- concentration? What will be the H_3O^+ concentration?

8 Calculate the HO^- concentration in the following cases:
(a) An aqueous solution $4.7 \times 10^{-3}M$ in H_3O^+ ion.
(b) Pure water.
(c) An aqueous solution prepared by dissolving 2.81 g $Ba(OH)_2$ in sufficient water to make one liter of solution.

9 Calculate the pH and pOH of solutions that are (a) $0.032M$ in H_3O^+ ion; (b) $7.5 \times 10^{-6}M$ in HO^- ion; and (c) $0.1F$ in HOAc, $K_a = 1.85 \times 10^{-5}$.

10 Calculate the H_3O^+ and HO^- ion concentrations in solutions that have (a) pH = 6.0; (b) pH = 2.4; and (c) pOH = 4.8.

11 A $0.500F$ solution of formic acid, HCO_2H, has H_3O^+ and $HC\overset{\displaystyle O}{\overset{\|}{-}}O^-$ concentrations of $8.9 \times 10^{-3}M$. Calculate pK_a for this weak acid.

12 Will the following sodium salts hydrolyze significantly in aqueous solution? Arrange the salts in order of increasing basicity of these solutions; K_a for the conjugate acid of each anion is given: $NaNO_3$ ($K_a \gg 1$); $NaCN$ ($K_a = 4.8 \times 10^{-10}$); $NaOAc$ ($K_a = 1.85 \times 10^{-5}$); NaF ($K_a = 6.9 \times 10^{-4}$); NaO_2CCH_2Cl ($K_a = 1.4 \times 10^{-3}$); and Na_2HPO_4 ($K_a = 6.2 \times 10^{-8}$).

13 Calculate the pH and pOH of a solution $0.15F$ in the salt NH_4NO_3; K_b for NH_3 is 1.8×10^{-5}.

14 Calculate the H_3O^+ concentration of a buffer solution $0.01F$ in HCN and $0.01F$ in NaCN; $K_a = 4.8 \times 10^{-10}$.

15 Suppose that to one liter of the solution in Problem 14 we add 10^{-4} mole of H_3O^+. What will be the final concentration of H_3O^+?

16 A buffer solution is prepared by adding enough sodium hydroxide to a solution which is $0.2F$ in HOAc to convert half of the HOAc into NaOAc. What will be the pH of this solution?

17 What solvent would be suitable to make HOAc act as a stronger acid than it does in water? What solvent would make NH_3 act as a stronger base than it does in water?

18 Using the data in Table 12–4, predict structures for the two acids H_3PO_3 (phosphorous acid) and H_3AsO_3 (arsenious acid), which have pK_a values of 2.0 and 9.2, respectively.

19 The dissociation constants for a series of substituted acetic acid derivatives, RCH_2CO_2H, where R represents any one of several groups varying from a simple H— to the molecular

unit C—OH, are given below. Can you offer a rationalization of the results?

Acid	K_a
$FCH_2C(=O)OH$	2.6×10^{-3}
$HOCH_2C(=O)OH$	1.5×10^{-4}
$BrCH_2C(=O)OH$	1.3×10^{-3}
$HCH_2C(=O)OH$	1.8×10^{-5}
$ICH_2C(=O)OH$	6.7×10^{-4}
$CH_3CH_2C(=O)OH$	1.3×10^{-5}
$ClCH_2C(=O)OH$	1.4×10^{-3}
$HO(=O)CCH_2C(=O)OH$	1.4×10^{-3}

RATES AND MECHANISMS OF CHEMICAL REACTIONS

13

Thermodynamics tells us what changes are possible in a chemical system. A reaction has the potential to proceed spontaneously only if it is accompanied by a loss in free energy. However, knowing the limits placed on chemical change by thermodynamics does not tell us everything about what will happen to a mixture of reactants. Some reactions that could release large amounts of free energy are so slow that they cannot be detected. This is the case for the oxygen–hydrogen mixture under most conditions (Section 10–6). Many important chemical changes in nature occur very slowly. Geological changes are much slower than the chemical changes occuring in living things. These changes in turn are much slower than many reactions performed in the laboratory. For some reactions, the approach to equilibrium is so rapid that the reaction appears to be instantaneous.

The study of the rates of chemical reactions is an important part of chemical dynamics, the study of chemical change. One aspect that makes this field so interesting is that understanding reaction rates offers the possibility of controlling them. This, in turn, holds the potential for controlling ourselves and our environment.

In this chapter, we will present some basic facts and theories about reaction rates. In reading this chapter, you should be careful to distinguish the facts from the theories. We are reasonably sure that the facts are true, because they come from careful experimentation. However, all the theories are unsatisfactory in some way, and they are used mainly to guide our thinking about what kinds of new questions need to be asked about chemical reactions. The field of chemical dynamics is a genuine frontier of science, in which there are far more questions than satisfactory answers.

13–1 Fast and Slow Reactions

The complexity of the problem of reaction rates is illustrated by the reactions with water of two apparently similar compounds, boron trifluoride (BF_3) and nitrogen trifluoride (NF_3). These hydrolysis reactions are superficially similar

$$BF_3 + 2H_2O \rightarrow HBO_2 + 3HF \tag{13-1}$$

boron trifluoride   water   boric acid   hydrogen fluoride

$$NF_3 + 2H_2O \rightarrow HNO_2 + 3HF \tag{13-2}$$

nitrogen   nitrous
trifluoride   acid

However, the detailed mechanisms and rates of these reactions are astonishingly different.

If boron trifluoride gas is bubbled into water, reaction seems to occur instantaneously with the evolution of heat. Not all of the boron trifluoride is converted immediately to boric acid; one of the first reaction products is the tetrafluoroborate ion, BF_4^-, which reacts with water rather slowly. The overall reaction can be broken into several steps

$$BF_3 + 2H_2O \rightarrow HBO_2 + 3HF$$
$$3HF + 3BF_3 \rightarrow 3BF_4^- + 3H^+$$

Adding these two equations, we obtain the equation for the first stage of the overall reaction

$$4BF_3 + 2H_2O \rightarrow HBO_2 + 3BF_4^- + 3H^+$$

The slow reaction of BF_4^- with water completes the overall reaction shown in Equation 13–1

$$H^+ + BF_4^- + 2H_2O \rightarrow HBO_2 + 4HF$$

In contrast to this reaction, when nitrogen trifluoride is bubbled through water, no appreciable change occurs. If the gas is heated to 100°C in contact with concentrated sulfuric acid in water, the hydrolysis of nitrogen trifluoride (Reaction 13–2) occurs very slowly.

We can rationalize the large difference in the reactivities of BF_3 and NF_3 by looking more closely at their electronic structures. Boron trifluoride is a Lewis acid (Section 11–2) and will readily form addition compounds with Lewis bases such as water. Thus, the first step in the reaction of BF_3 with water is the very rapid addition of H_2O to BF_3

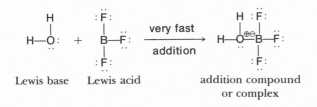

Lewis base   Lewis acid   addition compound
                                    or complex

The addition compound could then eliminate HF to give HOBF$_2$. For the next step, we could postulate the addition of another molecule of water, followed by the elimination of two molecules of HF to give the boric acid product

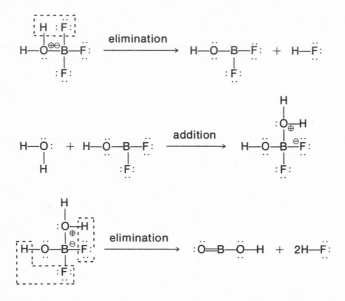

The sequence of addition and elimination steps for the hydrolysis of BF$_3$ constitutes a proposed mechanism, or pathway, for the overall substitution reaction. The addition compounds formed on the way to final products are called the *intermediates* in the reaction. Because of the Lewis acid nature of BF$_3$, it is easy to see how the initial addition step in the mechanism would be very fast. It is also possible to see how the BF$_4^-$ forms, because as HF is released it competes with H$_2$O for the BF$_3$ left: HF + BF$_3$ → H$^+$ + BF$_4^-$. (In water, HBO$_2$ reacts to form B(OH)$_3$. However, to simplify the preceding discussion, we ignored this reaction.)

By analyzing the BF$_3$ reaction in this way, we obtain some understanding of the difference in the behavior of BF$_3$ and NF$_3$. Nitrogen trifluoride is not a Lewis acid. It has an unshared electron pair, :NF$_3$, and it functions as a weak Lewis base. The initial addition of water to BF$_3$, which sets off the rapid substitution of B—O bonds for B—F bonds, is energetically very unfavorable for NF$_3$. Thus, a mechanism starting with an addition step is ruled out

The contrasting behavior of BF$_3$ and NF$_3$ is just one of thousands of examples of extreme variation in reaction behavior with compounds that at first glance appear similar to one another. We have seen the value of breaking the BF$_3$ hydrolysis reaction into the simple steps that might take place at the molecular level. A very plausible series of steps (or *mechanism*) was suggested that took advantage of the electronic nature of BF$_3$. Clearly, this mechanism was not reasonable for NF$_3$, so we successfully explained the enormous difference in the reaction rates of the two compounds. The lesson that we take from this example is that we need models for how reactions occur at the molecular level if we are to understand variations in chemical reactivity.

In the preceding example we have used simple intuition in setting up a model for the hydrolysis mechanism of BF_3. What experiments can we perform to check the consistency of a proposed mechanism, or to help us formulate a new one? One of the most important experiments involves determining the reaction *kinetics,* or how the rate of a reaction depends on the concentrations of the substances present. Another question that can be answered experimentally is how the reaction rate changes with variations in temperature and other conditions. The dependence of reaction rate on concentrations and temperature is kinetic information of great value in constructing and testing model mechanisms.

13-2 The Kinetics of Chemical Reactions

The rates of reactions in solution usually depend on the concentrations of reactants, sometimes on the concentrations of products, and often on the concentration of some other substance that does not appear in the overall equation for the reaction. A substance that affects the reaction rate but does not appear in the overall reaction equation is called a *catalyst* if it increases the reaction rate; it is called an *inhibitor* or *negative catalyst* if it decreases the rate.

The fact that reaction rates usually depend on the concentrations of the reactants should not seem surprising. For instance, when BF_3 is reacting with water, it is reasonable that the rate of the reaction will increase if in a given volume the number of BF_3 molecules present is increased. More surprising is the fact that this is not always so. For example, the rate of the reaction

$$\text{HO}^- \; + \; \underset{\substack{t\text{-butyl}\\ \text{chloride}}}{(\text{CH}_3)_3\text{CCl}} \; \xrightarrow[\text{C}_2\text{H}_5\text{OH}]{\text{H}_2\text{O}} \; \underset{\substack{t\text{-butyl}\\ \text{alcohol}}}{(\text{CH}_3)_3\text{COH}} \; + \; \text{Cl}^- \tag{13-3}$$

increases if the concentration of *t*-butyl chloride is increased, but it *does not change* when the concentration of hydroxide ion is changed. In contrast with this situation, the rate of the very similar reaction

$$\text{HO}^- \; + \; \underset{\substack{\text{ethyl}\\ \text{chloride}}}{\text{CH}_3\text{CH}_2\text{Cl}} \; \xrightarrow[\text{C}_2\text{H}_5\text{OH}]{\text{H}_2\text{O}} \; \underset{\substack{\text{ethyl}\\ \text{alcohol}}}{\text{CH}_3\text{CH}_2\text{OH}} \; + \; \text{Cl}^- \tag{13-4}$$

increases when either the hydroxide or ethyl chloride concentration is increased.

Reaction rates often depend on the concentration of catalysts. Catalysis by acids or bases is a common phenomenon, as mentioned in Chapter 12. An example is the bromination of acetone in the presence of the catalyst acetic acid

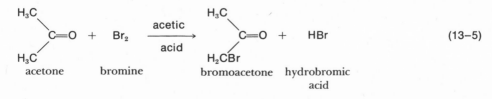

The reaction rate depends on the acetone and the acetic acid concentrations, but not on the bromine concentration, as long as bromine is present in reasonably high concentration.

Table 13–1. Some Reactions and Their Rate Laws

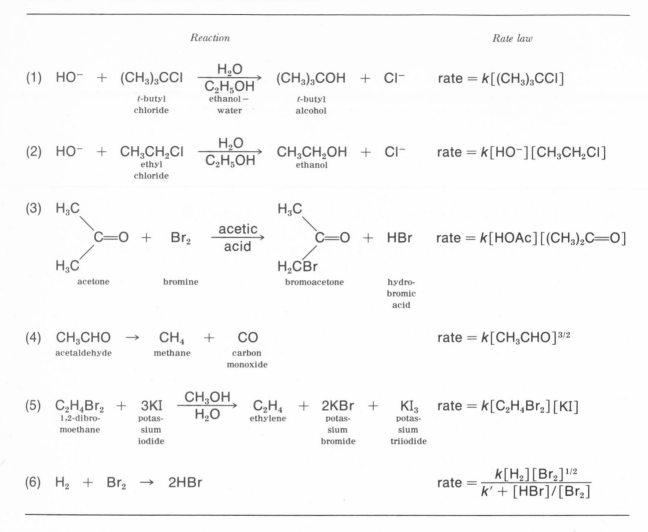

Reaction	Rate law
(1) $HO^- + (CH_3)_3CCl \xrightarrow[C_2H_5OH]{H_2O} (CH_3)_3COH + Cl^-$ *t*-butyl chloride / ethanol – water / *t*-butyl alcohol	rate $= k[(CH_3)_3CCl]$
(2) $HO^- + CH_3CH_2Cl \xrightarrow[C_2H_5OH]{H_2O} CH_3CH_2OH + Cl^-$ ethyl chloride / ethanol	rate $= k[HO^-][CH_3CH_2Cl]$
(3) acetone + Br₂ → bromoacetone + HBr	rate $= k[HOAc][(CH_3)_2C{=}O]$
(4) $CH_3CHO \rightarrow CH_4 + CO$ acetaldehyde / methane / carbon monoxide	rate $= k[CH_3CHO]^{3/2}$
(5) $C_2H_4Br_2 + 3KI \xrightarrow[H_2O]{CH_3OH} C_2H_4 + 2KBr + KI_3$ 1,2-dibromoethane / potassium iodide / ethylene / potassium bromide / potassium triiodide	rate $= k[C_2H_4Br_2][KI]$
(6) $H_2 + Br_2 \rightarrow 2HBr$	rate $= \dfrac{k[H_2][Br_2]^{1/2}}{k' + [HBr]/[Br_2]}$

If we know how the rate of a reaction depends on concentration, we can learn a great deal about how the reaction takes place. In the preceding example, the strange concentration dependency of the rate tells us immediately that the overall Reaction 13–5 must proceed by reaction steps that involve acetic acid. We will discuss this sort of reasoning in detail in the next section, but first let us be more precise about how reaction rates depend on concentration and introduce some useful terminology.

TERMINOLOGY

Extensive experience with thousands of different reactions has shown that the rate can be expressed in an equation called the *rate law,* which contains the concentrations of the substances involved. Some examples are given in Table 13–1. The proportionality constant between the rate of a reaction and the products of the concentrations of the substances present in the rate law is called the *rate constant,* and it is given the symbol k. The rate constant is a characteristic feature of a chemical reaction. Some reactions, such as (6) in Table 13–1, have complicated rate laws that require two or more rate constants.

The kinetic *order* of a reaction gives the overall concentration dependency of the reaction. In a first order reaction such as Reaction 1 in Table 13–1, the rate depends on one concentration. In a second order reaction (Reaction 2, Table 13–1), the rate depends on the product of two concentrations, or the square of one concentration. Further examples of rate laws are given in Table 13–2.

There are a few subtle points involved in classifying reactions according to order. For example, suppose that we were to carry out Reaction 2 in Table 13–1 in a buffered solution. Then, the hydroxide ion concentration would be constant, and the rate law would appear to be: rate $= k'$ $[CH_3CH_2Cl]$, which is the rate law for a first order reaction. In this case, $k[HO^-] = k'$. Only if the hydroxide ion concentration were varied would the proper rate law be found.

With reactions such as (4) and (6) in Table 13–1, the fact that a simple integer order rate law is not found shows that the actual reaction proceeds by two or more simple steps, having simple rates that combine to give the overall reaction. Examples such as (4) in Table 13–1 are referred to by their exponent, as in the preceding examples. Thus, Reaction 4 is a three halves order reaction. Examples such as (6) are referred to as complex order.

GAS PHASE REACTIONS

The rate laws for gas phase reactions have the same form as those for reactions in solution. The only difference is that in solutions concentrations usually are expressed in moles liter^{-1} (*M*), whereas gas phase concentrations usually are expressed in partial pressures (*p*).

Reaction rates usually are expressed in units of moles liter^{-1} sec^{-1} when concentrations are expressed in moles liter^{-1}, or in atm sec^{-1} when concentrations are expressed as partial pressures.

Table 13–2. Rate Laws Having Different Reaction Orders

Order	Sum of exponents of concentrations	Rate law
First order	1	rate $= k[A]$
	1	rate $= k[B]$
Second order	2	rate $= k[A]^2$
	2	rate $= k[A][B]$
	2	rate $= k[B]^2$
Third order	3	rate $= k[A][B][C]$
	3	rate $= k[A]^2[B]$
	3	rate $= k[A][C]^2$
	3	rate $= k[B]^3$

REACTIONS IN NONHOMOGENEOUS MIXTURES

Many extremely interesting and important reactions take place in mixtures containing more than one phase, such as a solid and a liquid, a liquid and a gas, two immiscible liquids, and so on. Among the important examples are the reactions used in the refining of petroleum, in which gases or liquids are passed over a catalyst, usually a metallic oxide, and the reactions that occur in batteries. These reactions are usually much more complex than homogeneous reactions, and for this reason we will not discuss them further.

13–3   Reaction Mechanisms

In the previous section, we stated that rate laws give information about mechanisms. A reaction mechanism is a description of the molecular structural changes that occur in the course of a reaction. There are two ways of describing these changes that must be distinguished carefully from each other:

1 Description of the *elementary chemical reactions* in the mechanism. An elementary chemical reaction is a one-step chemical change, such as the addition of H_2O to BF_3 in a previous example (Section 13–1).

2 Description of the details of *bond-breaking and bond-making* in the elementary reactions. We can best distinguish between these two descriptions of a mechanism by discussing some examples.

THE CONNECTION BETWEEN KINETICS AND MECHANISMS

In the next section, we will discuss some theories of reaction rates, but here let us use a few simple rules to show how to go from the rate law to a reaction mechanism. The first rule provides the direct connection between rates and mechanisms and shows us how to write the rate law. The rule states that, for an *elementary reaction,* the rate is the product of the concentrations of the reactants and a rate constant. Following are two examples:

$$I \quad + \quad H_2 \quad \rightarrow \quad HI \quad + \quad H \qquad \text{rate} = k[I][H_2]$$
iodine   hydrogen   hydrogen   hydrogen
atom   iodide   atom

$$H_3O^+ \quad + \quad HO^- \quad \rightarrow \quad 2H_2O \qquad\qquad \text{rate} = k[H_3O^+][HO^-]$$
hydronium   hydroxide   water
ion   ion

Notice that in these examples there are two reactants, not more. We believe that all elementary reactions are either *unimolecular* (involving only one reactant molecule) or *bimolecular* (involving two reactant molecules). Some further examples of these two types of reactions are given in Table 13–3.

If all *elementary* reactions are either unimolecular or bimolecular, then we know that any overall chemical reaction with more than two reactants *must* proceed by two or more elementary steps. One of these processes is often very slow in comparison with the other steps in the mechanism. This slow step, or *rate-determining* step, essentially determines how rapidly the reaction will proceed. For example, suppose the process

$$2A + B \rightarrow 2C$$

Table 13–3. Types of Elementary Reactions

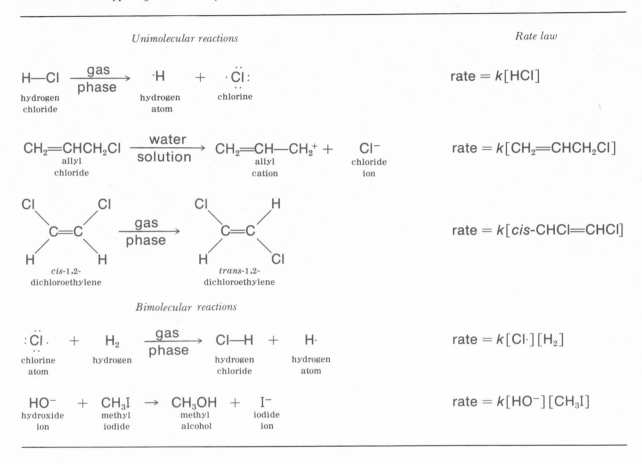

Unimolecular reactions	Rate law
$H-Cl \xrightarrow[\text{phase}]{\text{gas}} \cdot H + \cdot \ddot{Cl}:$ hydrogen chloride → hydrogen atom + chlorine	rate $= k[HCl]$
$CH_2{=}CHCH_2Cl \xrightarrow[\text{solution}]{\text{water}} CH_2{=}CH{-}CH_2^+ + Cl^-$ allyl chloride → allyl cation + chloride ion	rate $= k[CH_2{=}CHCH_2Cl]$
cis-1,2-dichloroethylene $\xrightarrow[\text{phase}]{\text{gas}}$ *trans*-1,2-dichloroethylene	rate $= k[\text{\textit{cis}-CHCl}{=}CHCl]$
Bimolecular reactions	
$:\ddot{Cl}\cdot + H_2 \xrightarrow[\text{phase}]{\text{gas}} Cl{-}H + H\cdot$ chlorine atom + hydrogen → hydrogen chloride + hydrogen atom	rate $= k[Cl\cdot][H_2]$
$HO^- + CH_3I \rightarrow CH_3OH + I^-$ hydroxide ion + methyl iodide → methyl alcohol + iodide ion	rate $= k[HO^-][CH_3I]$

follows the rate law

$$\text{rate} = k[A][B] \qquad\qquad (13\text{–}6)$$

The following mechanism composed of two elementary reactions is consistent with the rate law:

(1) $A + B \rightarrow C + D$ (slow)
(2) $D + A \rightarrow C$ (fast)

The first step is slow, but once the intermediate D is formed, it reacts immediately with A to give the product C. The rate law can be written from Step 1, using the rule which states that the rate of an elementary process is the product of the concentrations of the reactants and a rate constant

$$\text{rate} = k[A][B]$$

This equation is the same as the rate law given in Equation 13–6. An example of such a process is the reaction of fluorine with nitrogen dioxide

$$2NO_2 + F_2 \rightarrow 2NO_2F$$

which follows the rate law

$$\text{rate} = k[NO_2][F_2] \tag{13-7}$$

A proposed mechanism for this reaction is

(1) NO_2 + F_2 → NO_2F + F   (slow)
    (A)   (B)   (C)   (D)

(2) NO_2 + F → NO_2F (fast)
    (A)   (D)   (C)

The slow step leads to the rate law in Equation 13–7. In this example, we know that there must be at least two elementary reactions combining to give the overall reaction, because there are *three* reactant molecules ($2NO_2 + F_2$).

However, reactions that have only one or two reactants are not necessarily elementary. An example is the reaction of *t*-butyl chloride with hydroxide ion (Equation 13–3). The rate law for the overall reaction has been determined by experiment to be first order. Let us see if we can devise a reasonable set of elementary reactions that would combine to give the overall reaction observed, and that would be consistent with the observed rate law

$$\text{rate} = k[(CH_3)_3CCl] \tag{13-8}$$

Since the observed rate law depends only on the $(CH_3)_3CCl$ concentration, we must think of a *unimolecular* reaction of $(CH_3)_3CCl$ that will produce some intermediate substance that can react with HO^- to form the final product. One possibility is a slow dissociation reaction

$$(CH_3)_3CCl \xrightarrow{k} (CH_3)_3C^+ + Cl^- \tag{13-9}$$

followed by a rapid bimolecular reaction of the cation with hydroxide ion

$$(CH_3)_3C^+ + HO^- \xrightarrow{\text{fast}} (CH_3)_3COH \tag{13-10}$$

Adding these two reactions, we obtain the overall reaction

$$(CH_3)_3CCl + HO^- \rightarrow (CH_3)_3COH + Cl^- \tag{13-11}$$

The rate expression is correct, if Reaction 13–10 occurs so rapidly that the reverse of Reaction 13–9 can never occur. Note also that we cannot use kinetics to tell us how the intermediate *t*-butyl cation is consumed. It might react with water so that two fast steps, Equations 13–12 and 13–13, are required to arrive at the final product

$$(CH_3)_3C^+ + H_2O \rightarrow (CH_3)_3\overset{\oplus}{C}OH_2 \tag{13-12}$$

$$(CH_3)_3\overset{\oplus}{C}OH_2 + HO^- \rightarrow (CH_3)_3COH + H_2O \tag{13-13}$$

Measuring the kinetics of a reaction can only tell us about the mechanism of the slow or rate-determining step.

Example 13–1. Gaseous N_2O_5 decomposes according to the equation

$$2N_2O_5 \rightarrow 4NO_2 + O_2$$

The process is known to be first order with respect to $[N_2O_5]$; that is,

rate $= k[N_2O_5]$

Show that the following mechanism is consistent with the known rate law.

(1) $N_2O_5 \underset{k_{-1}}{\overset{k_1}{\rightleftarrows}} NO_2 + NO_3$ (fast)

(2) $NO_3 + NO_2 \xrightarrow{k_2} NO + NO_2 + O_2$ (slow)

(3) $NO_3 + NO \xrightarrow{k_3} 2NO_2$ (fast)

Solution. The second step in the mechanism is rate determining, and we should be able to write the rate law from it directly

rate $= k_2[NO_3][NO_2]$

Clearly, there is little similarity between this rate law and the experimental one. Notice, however, that Step 1 is an equilibrium process, and its equilibrium constant should have the form

$$K_{eq} = \frac{[NO_2][NO_3]}{[N_2O_5]}$$

We can see from this equilibrium expression that

$K_{eq}[N_2O_5] = [NO_3][NO_2]$

We can substitute into the rate law for $[NO_3][NO_2]$ and obtain

rate $= k_2K_{eq}[N_2O_5] = k[N_2O_5]$

where $k_2K_{eq} = k$, which is consistent with the experimental rate law.

13–4 Theory of Elementary Reaction Rates

The speed of an overall reaction depends mainly on the rates of the elementary reactions, not on their number. These rates vary, so being able to predict them is of considerable practical, as well as intellectual, value. Unfortunately, our present theories of reaction rates are too crude to allow us to make reliable quantitative predictions. However, the simple prediction that a certain reaction is too slow to be useful may save days or weeks of laboratory work. The use of a theory to predict reaction rates also should allow us to test and improve the theory by comparing predicted rates with measured rates for many reactions.

UNIMOLECULAR REACTIONS

We begin by describing a unimolecular reaction, the isomerization of *cis*-dichloroethylene. This example was discussed in detail in Chapter 10 from the equilibrium point of view.

Because the rate of interconversion of the dichloroethylene isomers is influenced strongly by temperature, and since increasing the temperature of a sample simply increases the energies

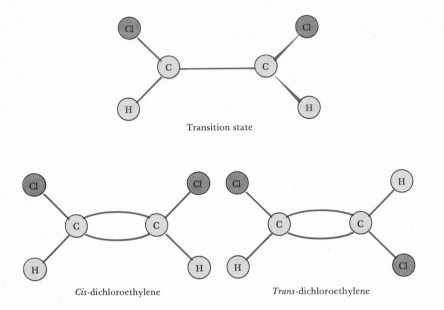

Transition state

Cis-dichloroethylene   *Trans*-dichloroethylene

Figure 13–1.   Cis- and trans-dichloroethylene and the high-energy transition state.

of the individual molecules, we infer that before a molecule can change from one form to the other, it must acquire a large amount of energy. By looking at models of the two forms (Figure 13–1), we can make a reasonable guess about the mechanism of the interconversion. A molecule of either *cis*- or *trans*-dichloroethylene twists around the carbon-carbon double bond into the higher energy *transition state* shown in Figure 13–1. Then the transition state can relax into either of the lower energy forms. We expect energy input would be required to produce the transition state because twisting will destroy the π bond (Section 7–12).

The energy of a molecule going through this conversion process is represented in Figure 13–2. A molecule of the cis compound, which is fairly stable (low energy), picks up energy as it changes to the transition state configuration, and then releases energy as it rolls downhill to the trans form. The abscissa of the graph, called the reaction coordinate, is a measure of the progress of the reaction. In this example, the reaction coordinate is the angle of twist about the central bond. In many cases, the reaction coordinate will be some fairly complex function of bond angles and distances.

For this reaction to take place, a molecule of *cis*-dichloroethylene must gain an amount of energy shown in Figure 13–2 as E_a, which is called the *activation energy* for the process. Notice that the activation energy is much larger than the energy difference between the two isomers, ΔE. This need not always be the case. There is, in fact, no regular and simple relationship between the activation energy (E_a) and the overall energy change for the reaction (ΔE).

The model helps to clarify the effect of temperature on the reaction. As the temperature of the gas is increased, the average energy of the molecules increases, and the fraction of molecules having energy greater than the average also increases. This effect was discussed in Chapter 2, and it is shown again in Figure 13–3.

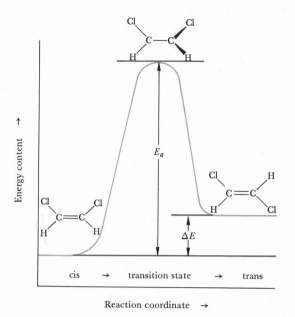

Figure 13–2. Energy profile for the conversion of cis- to trans-dichloroethylene.

Figure 13–3. Distribution of energy in gases at different temperatures.

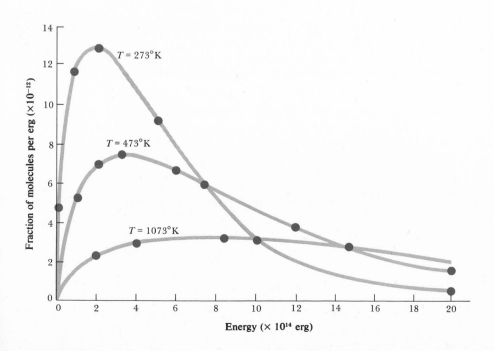

The activation energy for the conversion of *cis*- to *trans*-dichloroethylene is approximately 40 kcal mole^{-1}. When we increase the temperature to about 600°C, a substantial fraction of the molecules have energy higher than the activation energy, and the reaction proceeds at a rapid rate. However, this rate is not simply equal to the fraction of molecules having enough energy, for some of the energetic molecules will shed their excess energy and return to the starting state, rather than proceeding to form the reaction product.

We can, however, express the overall reaction rate as the product of the number of molecules having enough energy to react and the rate at which these energized molecules become products

$$\text{rate} \;=\; \begin{bmatrix} \text{rate for} \\ \text{molecules} \\ \text{with energy} \geq E_a \end{bmatrix} \times \begin{bmatrix} \text{number of} \\ \text{molecules} \\ \text{with energy} \geq E_a \end{bmatrix}$$

Alternatively, we could express the number of energized molecules as the product of the total number of molecules times the energized fraction

$$\text{rate} \;=\; \begin{bmatrix} \text{rate for} \\ \text{energized} \\ \text{molecules} \end{bmatrix} \times \begin{bmatrix} \text{fraction of} \\ \text{molecules} \\ \text{energized} \end{bmatrix} \times \begin{bmatrix} \text{total number} \\ \text{of} \\ \text{molecules} \end{bmatrix}$$

In concentration units, this becomes

$$\text{rate} \;=\; \begin{bmatrix} \text{rate for} \\ \text{energized} \\ \text{molecules} \end{bmatrix} \times \begin{bmatrix} \text{fraction of} \\ \text{molecules} \\ \text{energized} \end{bmatrix} \times \begin{bmatrix} \text{concentration} \\ \text{of} \\ \text{molecules} \end{bmatrix} \qquad (13\text{--}14)$$

We can calculate the fraction of the molecules in any sample that has energy greater than a given value. This fraction is an exponential expression

$$\begin{bmatrix} \text{fraction of molecules} \\ \text{having energy equal to} \\ \text{the activation energy} \end{bmatrix} \;=\; e^{-(E_a/RT)} \qquad (13\text{--}15)$$

Combination of Equation 13–14 and 13–15 gives Equation 13–16 for a unimolecular reaction

$$\text{rate} \;=\; \begin{bmatrix} \text{rate for} \\ \text{energized} \\ \text{molecules} \end{bmatrix} \times e^{-(E_a/RT)} \times [A] \qquad (13\text{--}16)$$

$[A]$ = concentration of reactant A

Experimental data for rates at various temperatures usually fit the predicted exponential dependence of temperature, inspiring some confidence in the general theory. However, we do not have any good theory for the prediction of the rate for energized molecules. Experiments indicate that this temperature-independent factor varies over wide limits for different reactions. The range of reported values is about 10^9–10^{16} sec^{-1}, but many results cluster around the value 10^{13} sec^{-1}.

The model gives conceptual meaning to the rate constant for a unimolecular reaction and predicts the form of its temperature dependence

rate $= k[A]$

$$k = pe^{-(E_a/RT)} \tag{13-17}$$

where

  $p =$ frequency factor

   $=$ rate for energized molecules

By measurement of rates at two or more temperatures, the value of E_a can be determined experimentally. The analysis is usually carried out using the logarithmic form of Equation 13-17

$$\ln k = \ln p - \frac{E_a}{RT}$$

or, using $\log_{10}$

$$2.303 \log k = 2.303 \log p - \frac{E_a}{RT} \tag{13-18}$$

Once E_a has been determined, the value of k can be calculated for any temperature.

Example 13-2. What is the activation energy, E_a, for a unimolecular reaction whose rate constant is ten times larger at 25°C than it is at 0°C? If the temperature is raised to 100°C, what will be the rate constant relative to that at 0°C?

Solution. It is convenient to eliminate the term $2.303 \log p$ in Equation 13-18 by subtracting the two equations that may be written for two different temperatures. Let k_1 be the rate constant at the lower temperature, T_1, and k_2 be the rate constant at T_2. Then we have

$$2.303 \log k_2 = 2.303 \log p - \frac{E_a}{RT_2}$$

$$2.303 \log k_1 = 2.303 \log p - \frac{E_a}{RT_1}$$

Subtracting the second equation from the first we obtain

$$2.303 \log k_2 - 2.303 \log k_1 = \left(-\frac{E_a}{RT_2}\right) - \left(-\frac{E_a}{RT_1}\right)$$

or

$$2.303 \log \frac{k_2}{k_1} = \frac{E_a(T_2 - T_1)}{RT_1 T_2}$$

We are now in a position to calculate E_a, since we know the ratio k_2/k_1 for two temperatures. T_2 and T_1 are absolute temperatures, E_a should be expressed in cal mole^{-1}, and the units of R must be cal mole^{-1} deg^{-1}. For this example,

  $T_2 = 273° + 25° = 298°K$
  $T_1 = 273° + \ \ 0° = 273°K$
  $R \ = 1.99$ cal mole^{-1} deg^{-1}

$$\frac{k_2}{k_1} = 10$$

Thus we have

$$2.303 \log 10 = \frac{E_a(25°)}{1.99 \text{ cal mole}^{-1} \text{deg}^{-1}\ (273°)(298°)}$$

$$E_a = \frac{2.303(1.99)(273)(298)}{25} \text{ cal mole}^{-1}$$

$$E_a = 14{,}900 \text{ cal mole}^{-1}, \text{ or } 14.9 \text{ kcal mole}^{-1}$$

Now we turn to the second part of the problem. At 100°C, or 373°K, we want to find the ratio k_2/k_1. Now k_2 and T_2 refer to 373°K. Since we have determined that $E_a = 14{,}900 \text{ cal mole}^{-1}$,

$$2.303 \log \frac{k_2}{k_1} = \frac{(14{,}900)(100)}{(1.99)(273)(373)}$$

$$\log \frac{k_2}{k_1} = 3.19$$

$$\frac{k_2}{k_1} = 1.55 \times 10^3$$

The rate constant at 100°C is 15510 times that at 0°C.

Example 13–3. The decomposition of an organic azo compound liberates N_2. An azo compound that has been studied carefully is 1,1'-azocyanocyclohexane

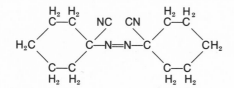

which we can abbreviate

R—N≡N—R

The molecule R—N≡N—R decomposes to R_2 and N_2, according to a first order rate law

$$\text{R—N≡N—R} \xrightarrow{k} \text{R—R} + \text{N}_2$$
$$\text{rate} = k[\text{R—N≡N—R}]$$

Using chlorobenzene (C_6H_5Cl) as a solvent, the activation energy for the reaction has been found to be 33 kcal mole^{-1} and the rate constant k is 8.3×10^{-6} sec^{-1} at 80°C. Suppose that we want to do an experiment in which R—N≡N—R will be decomposed considerably faster, with a k of 1.0×10^{-4} sec^{-1}. What temperature would we choose?

Solution. Starting with the equation

$$2.303 \log \frac{k_2}{k_1} = \frac{E_a(T_2 - T_1)}{RT_1T_2}$$

we want to calculate T_2

$k_2 = 1.0 \times 10^{-4} \text{ sec}^{-1}$

$k_1 = 8.3 \times 10^{-6} \text{ sec}^{-1}$

$T_1 = 273° + 80° = 353°\text{K}$

$E_a = 33 \text{ kcal mole}^{-1} = 33,000 \text{ cal mole}^{-1}$

Thus, we have

$$2.303 \log \frac{1.0 \times 10^{-4}}{8.3 \times 10^{-6}} = \frac{33,000(T_2 - 353°)}{1.99(353°)T_2}$$

$$\log 1.2 \times 10^1 = \frac{33,000(T_2 - 353°)}{4.58(353°)T_2}$$

$$T_2[(1.08)(4.58)(353°) - 33,000] = 33,000(-353°)$$

$$T_2 = 373°\text{K}$$

To obtain a k of $1.0 \times 10^{-4} \text{ sec}^{-1}$, we should run the reaction at 100°C (373°K).

BIMOLECULAR REACTIONS

In a bimolecular reaction, two molecules react with each other to form products. To react, they obviously must collide. We can calculate the collision rate of molecules in solution, or in the gas phase. These collision rates are much faster than any reaction rates we can measure. This means that for reactions with measurable rates, only a small fraction of collisions result in reaction. Another important feature of bimolecular reactions is that their rates increase with temperature in the same way as unimolecular reaction rates increase. These two facts suggest that most bimolecular reactions have activation energies just as do unimolecular reactions.

Consider Reaction 13–19, a typical bimolecular reaction occurring in solution. Since the hydroxide ion finally is attached to the carbon atom, the mechanism must involve an attack by the hydroxide oxygen on the carbon atom of the methyl group

$$\text{HO}^- + \text{CH}_3\text{I} \rightarrow \text{CH}_3\text{OH} + \text{I}^- \qquad \qquad (13\text{–}19)$$
$$\quad\quad\quad \text{methyl} \quad\quad \text{methyl}$$
$$\quad\quad\quad \text{iodide} \quad\quad \text{alcohol}$$

We can imagine a transition state that looks like

In this picture, the hydroxide and iodide are weakly attached to the carbon, or partially detached from it, depending on your point of view. This highly strained form looks like a methyl iodide molecule badly distorted by strong interaction with a hydroxide ion. Our model of the transition state predicts instability because the central carbon atom seems to be trying to accommodate *ten* valence electrons. The energy profile for this reaction is shown in Figure 13–4.

We can write an expression for the rate of a bimolecular reaction, using the concept of collisions between energetic molecules. For this reaction,

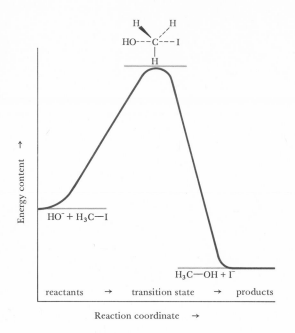

Figure 13–4.   Energy profile for the reaction of hydroxide ion with methyl iodide.

$A + B \rightarrow$ Products

the rate is

$$\text{rate} = \begin{bmatrix} \text{rate of collision} \\ \text{between A} \\ \text{and B} \end{bmatrix} \times \begin{bmatrix} \text{fraction of} \\ \text{collisions with} \\ \text{enough energy} \end{bmatrix} \times \begin{bmatrix} \text{fraction of energetic} \\ \text{collisions resulting} \\ \text{in reaction} \end{bmatrix}$$

The last factor in the rate expression is called the "orientation" factor. It expresses the fact that some energetic collisions do not result in reaction simply because the molecules are not oriented properly. This term corresponds to the p factor in the unimolecular rate expression (Equation 13–17). In the preceding example, collision of a hydroxide ion with the iodine atom of methyl iodide would not result in reaction. Only collisions involving direct contact of the oxygen atom with the carbon atom would be effective in causing reaction to occur. The collision rate is a number that depends on the concentrations of the molecules colliding

$$\text{collision rate} = \begin{bmatrix} \text{collision rate when} \\ [A] = [B] = 1M \end{bmatrix} \times [A] \times [B]$$

Thus, the rate expression can be written

$$\text{rate} = \begin{bmatrix} \text{collision rate} \\ \text{when} \\ [A] = [B] = 1M \end{bmatrix} \times \begin{bmatrix} \text{fraction of} \\ \text{collisions with} \\ \text{enough energy} \end{bmatrix} \times \begin{bmatrix} \text{fraction of energetic} \\ \text{collisions resulting} \\ \text{in reaction} \end{bmatrix} \times [A] \times [B]$$

or

$$\text{rate} = pe^{-(E_a'/RT)}[A][B] \qquad (13-20)$$

The bimolecular rate constant for the reaction

$$A + B \xrightarrow{\ k\ } \text{products}$$

in this model is

$$k = pe^{-(E_a/RT)}$$

The expression for k is similar to the one for unimolecular processes; the difference is in the interpretation and units of p. For bimolecular rate processes, p has units of liter mole^{-1} sec^{-1}, whereas for unimolecular reactions, p is given in sec^{-1}.

We have given some examples of values of E_a and p for chemical reactions. It is useful to state two handy rules of thumb from the theories that we have discussed to help with routine predictions:

1   A typical reaction will proceed at an appreciable rate at room temperature if E_a is 10 kcal mole^{-1} or less.

2   For such a reaction, the fraction of high-energy molecules increases with temperature in such a way that a 10° rise in temperature will roughly double the reaction rate.

Example 13–4. Verify the preceding Rule 1.
Solution. The general equation for the rate constant is

$$k = pe^{-(E_a/RT)}$$

where $p \cong 10^{13}$ sec^{-1}, $R = 1.99$ cal mole^{-1} deg^{-1}, and here $E_a = 10,000$ cal mole^{-1}, $T \cong 300°$K. Values of p and T are chosen to be typical values. Using these numbers,

$$k = 10^{13}e^{-[10,000/(1.99 \times 300)]}$$
$$k = 10^{13}10^{-[100/(2.303 \times 1.99 \times 3)]}$$
$$k = 10^{13} \times 10^{-7.3} \cong 10^6 \text{ sec}^{-1}$$

Suppose that we have solutions $1M$ in the reactants. Then the reaction rate is

$$\text{rate} = 10^6 \text{ moles liter}^{-1} \text{ sec}^{-1}$$

which is a very fast rate. If the frequency factor, p, were as low as 10^9, the reaction rate would still be 100 moles liter^{-1} sec^{-1}, which is fast enough to ensure that equilibrium will be reached in a short time.

13–5   Catalysis

There are chemical processes that have equilibrium states strongly favoring the products, but because they achieve equilibrium very slowly, they are of little practical synthetic value. In many cases, the time required to reach equilibrium can be reduced dramatically by the addition of a substance that is involved in the reaction, but which is not consumed. Such substances that affect the rate of a chemical reaction are called *catalysts*. The way in which a catalyst becomes involved in the mechanism of a reaction is in some cases well understood, but in many situations

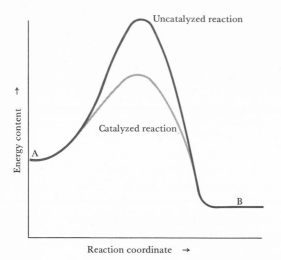

Figure 13–5. Energy profile of a catalyzed and uncatalyzed reaction.

the exact role of the catalyst is not well defined. We can make the following generalization concerning the action of a catalyst, however. A catalyst lowers the activation energy of the reaction process (see Figure 13–5), causing a much larger fraction of the reactant molecules to achieve the required energy, E_a. This causes the rate of the reaction to increase. Notice that the activation energy of the reverse process is lowered too. Since both forward and reverse rates are affected in the same way, the time needed to achieve the equilibrium state is reduced. There are substances that, when added to a reaction, slow it down. We can interpret this behavior as resulting from an increase in the activation energy of the process. Such substances are called *negative catalysts* or *inhibitors.*

   A catalyst may be a solid metal surface upon which the reaction takes place, a hydronium ion (H_3O^+) in aqueous solution, or any of a wide variety of other substances. As an example of the first, a finely powdered mixture of metallic oxides is used in breaking long-chain hydrocarbons into smaller chain ones

$$C_{18}H_{38} \xrightarrow[\text{heat}]{\substack{\text{metallic} \\ \text{oxide catalyst}}} C_8H_{18} \ + \ C_{10}H_{20}$$

This process is used in the petroleum industry to prepare lower boiling hydrocarbons for fuels. An example of the use of an acid catalyst, is the conversion of an alcohol to an alkyl bromide

$$\underset{\substack{\text{propyl} \\ \text{alcohol}}}{CH_3CH_2CH_2OH} \ + \ HBr \ \xrightarrow[\text{catalyst}]{H_2SO_4} \ \underset{\substack{\text{propyl} \\ \text{bromide}}}{CH_3CH_2CH_2Br} \ + \ H_2O \qquad (13\text{–}21)$$

The sulfuric acid is believed to function as a catalyst by increasing the acidity level, thereby increasing the extent to which the alcohol is converted to its conjugate acid

$$CH_3CH_2CH_2\!-\!\overset{\overset{\displaystyle H}{|}}{\underset{}{O}}\!\overset{\oplus}{-}\!H$$

The Br⁻ easily can displace the water molecule to produce $CH_3CH_2CH_2Br$ and H_2O.

In living systems, enzymes are the catalysts that keep chemical processes going at a proper rate, and allow us to operate effective chemical plants in our bodies.

AN ANALOGY

The schematic energy profiles in Figures 13–2 and 13–4 look much like the cross section of a mountain chain separating two valleys, and, in fact, that is a worthwhile analogy. Just as *cis*-dichloroethylene is more stable (and hence more favored) than *trans*-dichloroethylene, so the fertile Po Valley in northern Italy is more hospitable, and therefore more densely populated, than the high rocky valleys of southern Switzerland. Just as the rate of reaction from the cis to the trans form is hindered by a high energy barrier, so travel between the Swiss valleys and the Po Valley is hindered by the forbidding Alps. Reaction between cis and trans isomers takes place over the lowest possible barrier; most traffic across the Alps goes over the lowest route, the St. Gotthard Pass. Raising the temperature increases the reaction rate, and temperature formerly had a dramatic effect on travel between Switzerland and Italy because, until recently, traffic across the Alps came to a complete halt during the winter months. To complete the analogy, the new all-weather St. Gotthard Pass road is like a catalyst. It will not change the relative populations of the high mountain valleys and the low river valley, but it will greatly increase the rate of exchange between them.

13–6 The Correlation of Structure with Reactivity

In Chapter 12, we discussed the general objective of understanding the relationships between molecular structure and chemical reactivity. We also illustrated the fact that it is feasible to search for logical trends among groups of related compounds, rather than to attempt to predict the absolute values of individual acidity constants. The examples given were all equilibrium reactions, but the same approach is used in trying to understand patterns of behavior in reaction rates. Following are examples of two ways in which we go about establishing and using correlations:

1 If two similar reactions have the same mechanism, can we predict from the structures of the compounds involved which reaction will be faster?

Example. The reactions

$$CH_3I + HO^- \rightarrow CH_3OH + I^- \tag{13–22}$$

and

$$CH_3CH_2I + HO^- \rightarrow CH_3CH_2OH + I^- \tag{13–23}$$

are similar bimolecular reactions. The rate laws are

$$\text{rate}_1 = k_1[CH_3I][HO^-]$$

and

$$\text{rate}_2 = k_2[CH_3CH_2I][HO^-]$$

Which is greater, k_1 or k_2?

2   If two similar reactions have different mechanisms, can we explain from the structures of the compounds involved why one mechanism is favored in one case and another in the other case?

Example. The reactions

$$CH_3CH_2Cl + HO^- \rightarrow CH_3CH_2OH + Cl^-$$ (13–24)

and

$$(CH_3)_3CCl + HO^- \rightarrow (CH_3)_3COH + Cl^-$$ (13–25)

are similar, yet Reaction 13–24 is an elementary bimolecular reaction, whereas Reaction 13–25 occurs by a unimolecular dissociation reaction, followed by a bimolecular reaction. Why?

Let us try to answer these questions in turn.

DIFFERENT RATES FOR REACTIONS WITH THE SAME MECHANISM

To predict whether Reaction 13–22 or 13–23 is faster, let us look at the model for the transition states of the reactions. Remember, the higher the activation energy, the lower the value of k. The transition states look like this

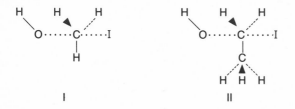

I   II

There are at least two effects that would make this type of transition state have very high energy. The first is *electronic;* there are too many electrons around the central carbon atom. A second effect is *steric;* that is, the effect has to do with the arrangement of the atoms in space. When a carbon atom is close to five other groups, instead of the usual maximum of four, the space around the carbon atom becomes crowded with atoms that are likely to bump into each other, or be forced into strained high-energy positions. Because Structure II has a CH_3 group in place of a hydrogen atom in Structure I, Structure II is more crowded, and therefore of higher energy. The steric factor suggests that Reaction 13–22 should be faster than Reaction 13–23, or k_1 greater than k_2. A simple model correlates qualitatively with the experimental facts.

DIFFERENT MECHANISMS FOR SIMILAR REACTIONS

We have discussed Reaction 13–25, which is the same as Reaction 13–3, and have shown that the kinetics suggest a two-step mechanism, with ionization as the first step. Let us see whether we can rationalize the fact that ethyl chloride reacts by the bimolecular mechanism, whereas *t*-butyl chloride chooses the two-step path.

First, consider the models for the bimolecular transition states, shown in Figure 13–6. We already have attributed the difference between the rates of reaction of methyl iodide and ethyl iodide to steric hindrance by the methyl group to the approach of the hydroxide ion to the reac-

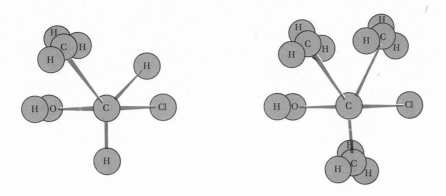

Figure 13–6. Transition states for bimolecular reactions of ethyl chloride and t-butyl chloride with hydroxide ion.

tive carbon atom. If real, the effect should become far greater with *t*-butyl chloride, since the incoming hydroxide ion must fight its way through a forest of hydrogen atoms carried by the three methyl groups. If *t*-butyl chloride did react by the bimolecular mechanism, it would react much slower than ethyl chloride.

The energy profile for the bimolecular mechanism should look like Figure 13–7. Since the reaction occurs in two steps, we have put in two transition states (energy maxima). However, the second hump must be lower than the first, if the first reaction is rate determining. Note that we also make the *t*-butyl cation a high-energy intermediate, since we know that the rate with which it passes over the second barrier must be very fast.

A reasonable model for the first transition state would be one in which the carbon–chlorine bond is stretched, and the *t*-butyl group has started to assume the geometry of the *t*-butyl cation. The signs δ+ and δ− indicate that partial electrical charges are being concentrated on carbon and chlorine

The *t*-butyl cation should be planar, in order to minimize repulsion among the pairs of electrons in the three C—C bonds (Section 7–1). In going from *t*-butyl chloride to the *t*-butyl cation, the C—C—C bond angles will increase from 109° to 120°. If the methyl groups repel each other because of their size, this steric repulsion should decrease on going to the transition state. In ethyl chloride, there is only one methyl group and two hydrogen atoms. Steric strain in the tetrahedral molecule will surely be smaller, so there will be less strain (and perhaps none) to be relieved by ionization to a planar cation. We could, therefore, account for the fact that *t*-butyl chloride reacts more rapidly than ethyl chloride in the two-step ionization mechanism. This is all we need to rationalize the results. If ethyl chloride is expected to react more rapidly by one mechanism, and *t*-butyl chloride more rapidly by the other, we can predict that they *may* choose different reaction paths.

That the rationalization is enough to account for the facts does not mean that we have considered all of the factors involved. There could be many others. Actually, other comparisons of

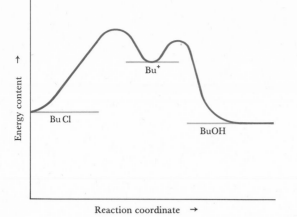

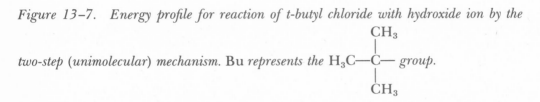

Figure 13–7. Energy profile for reaction of t-butyl chloride with hydroxide ion by the two-step (unimolecular) mechanism. Bu *represents the* H_3C—$\underset{\underset{CH_3}{|}}{\overset{\overset{CH_3}{|}}{C}}$— *group.*

reactivity, and theory as well, indicate that there is a powerful electronic effect that favors ionization of *t*-butyl chloride. Without giving any concrete evidence, we will simply point out that replacement of the hydrogen atoms of the methyl cation (H_3C^+) by carbon groups always has a powerful stabilizing influence. We believe that the effect occurs because electrons in σ bonding orbitals within the attached groups become slightly delocalized and help relieve the electron deficiency at the positive carbon atom. In the *t*-butyl cation,

electrons from the nine C—H bonds of the methyl groups "leak" into the vacant *p* orbital of the central carbon atom. The effect is called "hyperconjugation."

13–7 Rates of Reactions and Chemical Equilibrium

In Chapter 10, we explained dynamic chemical equilibrium and discussed how the position of equilibrium is described by the equilibrium constant or the free energy change for the reaction. One of the examples we used was the isomerization of *cis*-dichloroethylene. In this chapter, we have discussed the same reaction as an example of a unimolecular elementary reaction.

The dynamic chemical equilibrium between the cis and trans forms is established by the occurrence of the unimolecular elementary reactions

$$\text{cis} \xrightarrow{k_1} \text{trans} \tag{13–26}$$
$$\text{rate}_1 = k_1[\text{cis}]$$

and the reverse reaction

$$trans \xrightarrow{k_2} cis \qquad\qquad\qquad (13-27)$$
$$rate_2 = k_2[trans]$$

The system is at equilibrium when the amounts of cis and trans isomers do not change. Since Reactions 13–26 and 13–27 both are occurring continuously at equilibrium, the *rates* of these reactions must be the same; that is,

$$rate_1 = rate_2$$
$$k_1[cis]_{eq} = k_2[trans]_{eq}$$

or

$$\frac{[trans]_{eq}}{[cis]_{eq}} = \frac{k_1}{k_2}$$

This means that at equilibrium the ratio of concentrations of the cis and trans isomers is equal to the ratio of the rate constants.

But we also know that the reaction

$$cis \underset{k_2}{\overset{k_1}{\rightleftharpoons}} trans$$

has an equilibrium constant, K_{eq}, with

$$\frac{[trans]_{eq}}{[cis]_{eq}} = K_{eq}$$

at equilibrium. Therefore,

$$\frac{k_1}{k_2} = K_{eq}$$

Because of the relationship between rates and equilibrium, we could, if we wished, discuss all equilibrium problems in terms of rates. To do so turns out to be unprofitable in most cases. Rate laws are frequently far more complex than equilibrium constants, although the added complexities will disappear when we take the ratios of the forward and reverse rate laws. To talk about equilibria in terms of component rates is to disregard the simplification that comes from knowing that the standard free energy change for a reaction is independent of the path used to carry out the change. Furthermore, we frequently know equilibrium relationships in systems for which we have no knowledge of the rates and mechanisms involved in establishing equilibrium. An example of this is the acid–base reaction (Chapter 12), in which the rates of proton transfers are too fast to measure by common methods.

13–8 Summary

1   Rates of reactions usually depend on the concentrations of the substances present. The nature of this dependency must be determined experimentally, and it is usually expressed mathematically in the rate law of the type

$$rate = k[A]^m[B]^n$$

2   Reactions may be described as proceeding through a series of single steps called elementary reactions. These steps constitute the mechanism of the reaction.

3   The rate of an elementary reaction is the product of the concentrations of the reactants and a rate constant. Elementary reactions are either unimolecular or bimolecular (i.e., they involve either one or two molecules).

4   Going from reactants to products is viewed as proceeding through a high-energy transition state. The increased energy required to achieve this short-lived transition state is called the activation energy, which is given the symbol E_a. Molecules in the transition state may convert into products or revert to reactants.

5   The energetics of a reaction may be shown schematically in terms of a diagram called an energy profile.

6   A catalyst increases the rate of a chemical reaction by lowering the activation energy of the reaction.

7   Correlations between the structures of molecules and their chemical reactivity can be constructed by comparing experimental data and using reasonable models for transition state structures.

8   The equilibrium constant, K_{eq}, for a reaction involving elementary forward and reverse processes, is the ratio of the forward and reverse rate constants, k_1 and k_2

$$K_{eq} = \frac{k_1}{k_2}$$

13–9   Postscript: A Dynamic Chemical Problem

We have speculated about the use of chemical reactions in the design of man-made machines (see the Postscript to Chapter 11). A living organism is a chemical machine of unbelievable complexity. Chemical change is involved in everything that we do; living, seeing, thinking, and procreating are largely chemical functions. Small changes in body chemistry can have an enormous range of effects. For example, the introduction of a small amount of carbon monoxide into the air we breathe can be fatal, because carbon monoxide binds more strongly than oxygen to hemoglobin (Section 12–5).

The fields of nutrition and pharmaceutical medicine (see Postscript to Chapter 14) are basically concerned with providing fuel for the machine and tinkering with the rates of some of the chemical reactions in the body. In many cases, we are quite unclear as to the details of the effects, but in others we have a glimmer of understanding. For example, we believe that riboflavin, or vitamin B_2, in the diet is turned into a compound known as flavin adenine dinucleotide (FAD). FAD plays a critical role in the oxidation of sugars. Supply of the vitamin at a proper level controls the rate of the process in which energy is released by the combustion of foodstuffs in the cells of the body.

We have no clear idea of the number of distinct chemical reactions that occur in a living organism, and perhaps we will never know them all. Even if we did know all of the reactions, and the factors that control their rates, we would still find the task of describing a living cell in terms of its chemistry a monumental task. The reactions do not operate independently, but form a vast network of coupled changes. The products of one reaction are reactants in others, some reactions make the enzymes that catalyze other processes, enzymes digest one another, and so on. Building a model for the biochemical systems analysis (see Postscript to Chapter 5) may be one of the most complex problems that man will ever undertake.

Assume for the moment that the job of analysis will be done, or at least enough of it, to allow man to make decisions about what he will control in the chemistry of life. What controls will we exercise? Many people think first of the aging process, which is almost surely some kind of cumulative chemical change. If we could, would we really want to halt aging in all people, or just in ourselves, or not at all? We already are concerned about the level of world population, and know that people must die if there is to be room for new people. There is something rather grim about the thought of a world in which chemical control has eliminated death by all natural causes; such a world would know death caused only by accident, starvation, or violence.

Chemical biologists now are vigorously initiating study of brain chemistry, and they have already uncovered much of the story of chemical control of genetic characteristics. There is little doubt that the time will come when chemistry will be able to influence the thinking and emotional potential of new-born infants. If this occurs, what qualities will be asked for, and who will decide? Imagine, for example, the possible consequences of developing a race with far greater intellectual analytical prowess than we now have. That decision might be judged to be desirable by standards created using our present intellects, but would it conform to the standards of the new people, with their improved intellectual apparatus?

Such imponderable philosophical questions may seem remote and esoteric. Yet we see the high probability that they eventually will be encountered in some form. Man now has a biochemical apparatus that leads him to be curious, and that curiosity leads to ever increasing knowledge of himself and his environment. We are programed to learn, so learn we will. The desire to tinker and try things may be almost as rigidly fixed in us. Consequently, we can be virtually certain that experiments in chemical modification of the quality of life will continue. Some people are horrified by the prospect of continued biochemical manipulation of man. Their dilemma arises because the only reasonable way that we can see to stop the process is to find a pill that will change the nature of curious man. Would *you* manipulate a man to make *him* nonmanipulative?

New Terms

Activation energy: The amount of energy that the reactants in a chemical process must gain to reach the transition state.

Bimolecular reaction: An elementary reaction that involves two reactant molecules.

Catalyst: A substance that, when added to a reaction mixture, affects the rate of the reaction; it is normally not consumed in the process. Substances increasing the reaction rate are called catalysts; substances decreasing the reaction rate are usually called negative catalysts or inhibitors.

Electronic effects: Effects on reaction rates or mechanisms due to the electronic structures of the reacting substances.

Elementary reaction: A concerted one-step reaction, or one of the steps in the mechanism of a complex reaction. The rate of an elementary reaction equals the product of the concentrations of the reactants and a rate constant.

Energy profile or reaction-coordinate diagram: A graph indicating how the energy of a reacting system changes during the course of the reaction.

Frequency factor: The rate of reaction of activated molecules in a unimolecular reaction, or of collided molecules in a bimolecular reaction.

Kinetics: The way in which the rate of a reaction depends on the concentration of the substances present.

Rate constant: A temperature-dependent constant, *k,* which is characteristic of each reaction. The proportionality constant linking reaction rate to concentrations.

Rate-determining step: The slowest elementary reaction in the mechanism of an overall reaction.

Rate law: The mathematical relationship that describes how the rate depends on the concentrations of the substances present.

Reaction coordinate: The measure of the progress of a reaction from reactants to products; usually a complicated function of bond angles and bond lengths. The reaction coordinate is the abscissa in an energy profile diagram.

Reaction mechanism: The series of elementary reactions through which a reaction proceeds from starting materials to products; the detailed description of the bond-making and bond-breaking processes that occur in the reaction.

Reaction order: The sum of the exponents to which the reactant concentrations are raised in the rate law.

Steric effects: Effects on rates or mechanisms due to the physical arrangements of atoms in space.

Transition state: A short-lived state through which reactants pass on their way to products. The activation energy is the energy required to achieve this transition state.

Unimolecular reaction: An elementary process that involves only one reactant molecule.

Questions and Problems

1 The rate laws are given for the following reactions. State the reaction order with respect to each component in the reaction, and give the overall order of the reaction.

    (a) $2NO + O_2 \rightarrow 2NO_2$
        rate $= k[NO]^2[O_2]$

    (b) $NO_2 + CO \rightarrow CO_2 + NO$
        rate $= k[NO_2][CO]$

    (c) $(CH_3)_3CCl + HO^- \rightarrow (CH_3)_3COH + Cl^-$
        rate $= k[(CH_3)_3CCl]$

    (d) $O_2NNH_2 \rightarrow N_2O + H_2O$

$$\text{rate} = k\,\frac{[O_2NNH_2]}{[H^+]}$$

2 Explain what is meant by the following terms: (a) bimolecular process, (b) elementary process, (c) rate-determining step, (d) catalyst, and (e) rate constant.

3 Nitric oxide and hydrogen react to form nitrogen and water, according to the equation

$$2NO + 2H_2 \rightarrow N_2 + 2H_2O$$

The rate of the reaction has been observed to quadruple when the concentration of the NO is doubled, and to increase by ninefold when NO is tripled in concentration. The reaction rate only doubles, however, when the H_2 concentration is doubled. Write the rate law that describes this reaction. What is the overall order of this reaction?

4 What effect on the rate of a chemical reaction does an increase in temperature have? Explain this change in terms of the energetics of the system.

5 Explain why it is not accurate to equate the rate of a reaction with the total number of colli-sions between the reacting molecules in the system. How does the structure of a reacting molecule affect the rate of its chemical reactions?

6 Consider the reaction of the general type

$$2A + B \rightarrow C$$

for which the following experimental data are available:

Experiment	$[A]$ (moles liter^{-1})	$[B]$ (moles liter^{-1})	Rate (moles liter^{-1} sec^{-1})
1	0.1	0.1	0.75
2	0.1	0.2	1.50
3	0.1	0.3	2.25
4	0.2	0.2	3.0
5	0.4	0.2	6.0
6	0.6	0.2	9.0

Write a rate law for this reaction that is consistent with the data. Does the following mecha-nism agree with the rate law?

$$A + B \rightarrow A\!-\!B \quad \text{(slow step)}$$
$$A\!-\!B + A \rightarrow C \quad \text{(fast step)}$$

7 Draw an energy profile diagram for the exothermic bimolecular reaction $A + A \rightleftarrows B$. What is the order of the reverse reaction? Indicate the activation energies for the forward and re-verse reactions.

8 For the reaction $N_2O_5 \rightarrow 2NO_2 + \frac{1}{2}O_2$, the following rate constants were measured at 273°K and 328°K, respectively; $k_1 = 7.87 \times 10^{-7}$ sec^{-1} and $k_2 = 1.50 \times 10^{-3}$ sec^{-1}. Calculate the value of the activation energy for this reaction.

9 Explain how a catalyst alters the time required for a reaction to reach equilibrium. Explain why a catalyst does not shift the position of the equilibrium state.

10 In examining the rates of reaction of a series of related molecules, what effect would you expect an increase in steric crowding to have, if the rate-determining step is unimolecular? What effect would steric crowding have on a bimolecular step?

11 Consider the hypothetical reaction

$$A + 2B + C \rightarrow D$$

which follows the rate law

$$\text{rate} = k[A][B]^2$$

Show that the following mechanism is consistent with the rate law:

$$A + B \rightleftarrows A\!-\!B \quad \text{(fast)}$$
$$A\!-\!B + B \rightarrow A\!-\!B\!-\!B \quad \text{(slow)}$$
$$A\!-\!B\!-\!B + C \rightarrow D \quad \text{(fast)}$$

STRUCTURES AND REACTIONS OF COMPOUNDS OF CARBON AND SILICON

The number of known compounds containing carbon outnumber all other known compounds by a very large margin. The principal constituents of all living tissue are compounds in which carbon is combined with other elements, notably hydrogen, oxygen, nitrogen, phosphorus, and sulfur. The chemistry of carbon compounds is so extensive that it is designated as a separate field of chemistry, known as *organic chemistry;* carbon-containing compounds are called organic compounds. The name "organic" is derived from the universal presence of carbon compounds in living organisms, and it was once thought that organic compounds contained some mystical, vital force associated with life. Now we know that there is no special chemical difference between living and nonliving materials. In fact, many important materials in our lives are organic compounds not found in living things. Examples are plastics, synthetic fibers, and drugs.

Because of the great importance of carbon compounds, we have chosen to discuss them in a special chapter, and to illustrate periodic changes in the elements by contrasting the chemistry of carbon with that of silicon.

14-1 Classes of Compounds

As was noted in Chapter 6, the strength of carbon–carbon bonds allows the existence of essentially an infinite number of organic compounds containing carbon atoms linked together in chains and rings. Even an introductory discussion of the reactions of carbon compounds requires some subdivision of these compounds into structural classes. We will discuss only a few of the more important classes and a few of their reactions.

HYDROCARBONS

Compounds containing only carbon and hydrogen are called hydrocarbons. Hydrocarbons are the principal constituents of petroleum, so the chemistry of hydrocarbons is basic to the technology of refining natural petroleum to produce high-grade gasoline and other products.

Hydrocarbons are divided into several subclasses. One subclass is the *alkanes,* which are "saturated" hydrocarbons that contain the maximum number of hydrogen atoms that can be held by the carbon atoms in the molecule. Examples of alkanes are

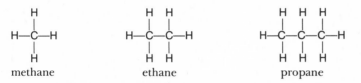
methane   ethane   propane

Another subclass is the *alkenes.* These are hydrocarbons that contain carbon-to-carbon double bonds, and they are called "unsaturated," because they do not contain the maximum number of hydrogen atoms per carbon atom. Two examples are

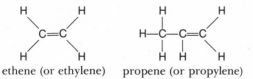
ethene (or ethylene)   propene (or propylene)

Alkynes form another subclass of unsaturated hydrocarbons. These compounds contain carbon-to-carbon triple bonds. Two examples are

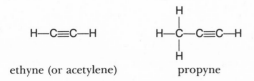
ethyne (or acetylene)   propyne

In addition to compounds in which carbon atoms are linked in chains (straight or branched), there are also carbon compounds in which carbon atoms are bonded to form rings. A subclass of saturated hydrocarbons that contains ring compounds is the *cycloalkanes.* These are cyclic hydrocarbons that contain no carbon-to-carbon double or triple bonds. An example is

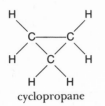

cyclopropane

An important subclass of hydrocarbons is the unsaturated, *aromatic* hydrocarbons. These are cyclic hydrocarbons in which alternating single and double bonds structures can be formulated (Section 7–14). Benzene is the "parent" compound of the aromatic hydrocarbons

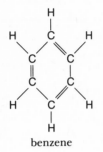

benzene

HYDROCARBON DERIVATIVES

There are several classes of organic compounds that can be considered to be derivatives of hydrocarbons. We will give a few of the most common classes.

Alcohols belong to a class of compounds containing simple —OH groups attached to a carbon skeleton. You probably are familiar with some of these compounds. A few examples are

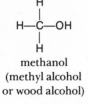

methanol
(methyl alcohol
or wood alcohol)

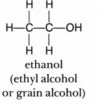

ethanol
(ethyl alcohol
or grain alcohol)

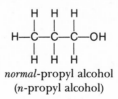

normal-propyl alcohol
(*n*-propyl alcohol)

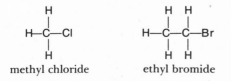

isopropyl alcohol
(*i*-propyl alcohol)

Halocarbons are those compounds that contain carbon, hydrogen, and halogen atoms. We have used examples of this class of compounds in our previous discussions. Two simple examples are

H—C—Cl

methyl chloride

H—C—C—Br

ethyl bromide

An example of an aromatic halocarbon is

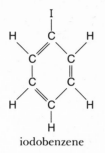

iodobenzene

Many pesticides, such as DDT, fall in this class of compounds.

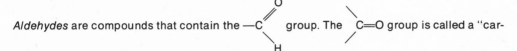

Aldehydes are compounds that contain the —C group. The C=O group is called a "car-

bonyl" group. Two examples of aldehydes are

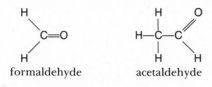

formaldehyde   acetaldehyde

Ketones also contain carbonyl groups, but they differ from aldehydes in that the carbonyl group is attached to two carbon atoms. Acetone is a familiar example of a ketone

acetone

Carboxylic acids are compounds that contain the —C—OH group. Two examples are

acetic acid

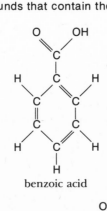

benzoic acid

Esters are compounds having the structure R—C—O—R', where R and R' are carbon groups. An ester can be thought of as being derived by reaction of an alcohol with a carboxylic acid, with the elimination of a molecule of water. Many esters are, in fact, prepared in this manner. An

equation illustrating this reaction is

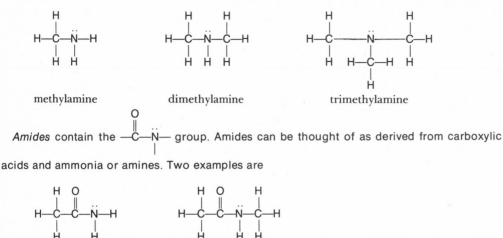

$$\text{acetic acid} \quad + \quad \text{methyl alcohol} \quad \rightarrow \quad \text{methyl acetate} \quad + \quad H_2O \tag{14-1}$$

Amines contain nitrogen bound only to carbon or hydrogen atoms. Some examples are

methylamine   dimethylamine   trimethylamine

Amides contain the —C—N— group. Amides can be thought of as derived from carboxylic

acids and ammonia or amines. Two examples are

acetamide   *N*-methylacetamide

There are many other classes of compounds, but those we have shown are the most important. Many compounds contain more than one set of the characteristic groups. The groups are called "functional groups" *because they are responsible for many of the chemical reactions of organic compounds.* A molecule containing two or more functional groups is called "polyfunctional." Hydroxyacetic acid (or glycolic acid), is an example of a polyfunctional (bifunctional) compound

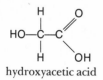

hydroxyacetic acid

14-2 Nomenclature

There are several systems for naming organic compounds. Some of the systems are sufficiently systematic to allow an unambiguous assignment of a name to any one of the several million known compounds. In this brief introduction, we will not attempt to discuss the systems of naming compounds, but we hope that students will appreciate that such systems exist and are necessary so that chemists can communicate. Some examples of the systematic relationships can be seen in the previous examples. Note that the names of hydrocarbons and alcohols having the same number of carbon atoms have the same roots in their names

CH_4	*meth*ane	CH_3OH	*methy*l alcohol
CH_3CH_3	*eth*ane	CH_3CH_2OH	*ethy*l alcohol

Notice also that the names of acids and aldehydes have common roots, and that the names of esters come from the names of alcohols and carboxylic acids from which they are derived

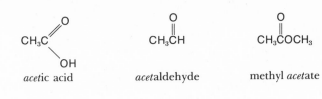

acetic acid   acetaldehyde   methyl acetate

14–3 Reactions of Saturated Hydrocarbons

The fully hydrogenated, or saturated, hydrocarbons are not very reactive under ordinary conditions. They can be burned in air, although the reaction must be initiated by a flame or an electric spark. The spark method is used to initiate combustion in the cylinders of an automobile engine. The combustion products are water and carbon dioxide

$$2C_2H_6 \;+\; 7O_2 \;\rightarrow\; 4CO_2 \;+\; 6H_2O \qquad\qquad (14\text{–}2)$$
ethane

Unless conditions are exactly right, traces of incompletely burned carbon compounds will be produced along with carbon dioxide. These products of incomplete combustion, including carbon monoxide and unsaturated hydrocarbons, are released into the atmosphere in automobile exhausts. When sufficient concentrations of such products accumulate in the atmosphere, they may undergo further slow oxidative reactions to form the noxious mixture of atmospheric pollutants known as smog. The oxidative changes leading to smog are initiated photochemically by sunlight. The smog problem is especially serious in the Los Angeles basin for three reasons: (1) A large number of automobiles are driven in the basin every day; (2) the basin is surrounded on three sides by mountains so that wind does not blow the polluted atmosphere away; and (3) there are many hours of bright sunshine during much of the year.

At high temperatures and in the presence of certain solid catalysts, hydrocarbons undergo less drastic reactions, which are important in petroleum technology.

14–4 Reactions of Unsaturated Hydrocarbons

The characteristic reactions of unsaturated hydrocarbons fall into two groups. Alkenes and alkynes undergo a large number of addition reactions, in which double bonds are converted to single bonds, and triple bonds are converted to double or single bonds. Most of the reagents that react readily with unsaturated hydrocarbons are *electrophilic* or electron-seeking. Examples of electrophilic compounds are the halogens and protonic acids

$$H_2C{=}CH_2 \;+\; Br_2 \;\rightarrow\; BrCH_2CH_2Br \qquad\qquad (14\text{–}3)$$
1, 2-dibromoethane

$$H_2C{=}CH_2 \;+\; HCl \;\rightarrow\; CH_3CH_2Cl \qquad\qquad (14\text{–}4)$$
ethyl chloride
(chloroethane)

In the presence of acids, the elements of water can be added to carbon–carbon double bonds. The reaction takes place in steps, with the intermediate formation of carbon cations, known as *carbonium ions.* Addition to unsymmetrical alkenes is usually highly selective, and takes a course indicating that the more highly branched of the possible carbonium ions is formed preferentially (Section 13–6)

$$\underset{\text{(14–5)}}{H_3C-\overset{\overset{\displaystyle H}{|}}{C}=CH_2 \;+\; H_2O \xrightarrow[\substack{\text{(or other}\\ \text{acid catalyst)}}]{H_2SO_4} H_3C-\overset{\overset{\displaystyle H}{|}}{\underset{\underset{\displaystyle\text{isopropyl alcohol}}{OH}}{C}}-CH_3}$$

The mechanism for this reaction is

$$H_3C-\overset{\overset{\displaystyle H}{|}}{C}=CH_2 \;+\; H^+ \;\rightarrow\; H_3C-\underset{\underset{\displaystyle\text{a carbonium ion}}{\oplus}}{\overset{\overset{\displaystyle H}{|}}{C}}-CH_3 \qquad (14\text{–}6)$$

$$H_3C-\underset{\oplus}{\overset{\overset{\displaystyle H}{|}}{C}}-CH_3 \;+\; H_2O \;\rightarrow\; H_3C-\underset{\overset{\oplus}{O}H_2}{\overset{\overset{\displaystyle H}{|}}{C}}-CH_3 \qquad (14\text{–}7)$$

$$H_3C-\underset{\overset{\oplus}{O}H_2}{\overset{\overset{\displaystyle H}{|}}{C}}-CH_3 \;\rightarrow\; H_3C-\underset{OH}{\overset{\overset{\displaystyle H}{|}}{C}}-CH_3 \;+\; H^+ \qquad (14\text{–}8)$$

The mechanism for this reaction does *not* involve the formation of the less branched carbonium ion as shown by the Equation 14–9

$$H_3C-\overset{\overset{\displaystyle H}{|}}{C}=CH_2 \;+\; H^+ \;\;\times\!\!\!\!\longrightarrow\;\; H_3C-\overset{\overset{\displaystyle H}{|}}{\underset{\underset{\displaystyle H}{|}}{C}}-\overset{\overset{\displaystyle H}{|}}{\underset{\underset{\displaystyle H}{|}}{C}}\!{}^\oplus \qquad (14\text{–}9)$$

Aromatic hydrocarbons are less reactive than alkenes and alkynes. However, under strongly acidic conditions, many substitution reactions occur. An important example is nitration, a reaction with nitric acid, usually with concentrated sulfuric acid to increase the rate of the reaction

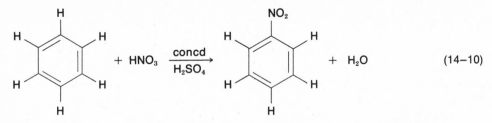

$$+ \; HNO_3 \xrightarrow[H_2SO_4]{\text{concd}} \cdots + H_2O \qquad (14\text{–}10)$$

Applied to toluene, the nitration reaction gives trinitrotoluene (TNT), an explosive

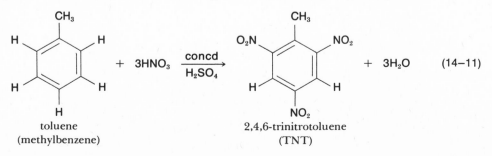

$$+ \; 3HNO_3 \xrightarrow[H_2SO_4]{\text{concd}} \cdots + 3H_2O \qquad (14\text{–}11)$$

toluene
(methylbenzene)

2,4,6-trinitrotoluene
(TNT)

The mechanism of the aromatic substitution reaction is more similar to addition reactions of alkenes than would appear at first glance. In the presence of powerful acids, nitric acid acts as a base and ionizes to produce a reactive electrophilic reagent, the nitronium ion (NO_2^+). This ion then adds to the aromatic hydrocarbon, forming a carbonium ion. Instead of reacting with some *nucleophilic* (nucleus-seeking; Lewis base) species to give an adduct, the carbonium ion loses a proton to regenerate the stable aromatic system. A mechanism for such a reaction is

$$HO\!-\!NO_2 \xrightarrow{H^+} H_2O^+\!-\!NO_2 \rightarrow H_2O + NO_2^+ \qquad\qquad (14\text{--}12)$$
$$\text{nitronium ion}$$

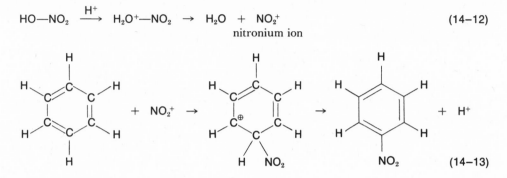

$$(14\text{--}13)$$

14–5 Ligand Substitution Reactions

We might expect organic compounds bearing nucleophilic ligands to undergo substitution reactions. Such reactions do occur, and they are quite useful in synthetic work. However, the rates are much slower than the rates of nucleophilic substitution involving BF_3 and other Lewis acids of the Group III elements (Section 11–2). The difference arises because carbon compounds have filled valence electronic shells. A nucleophilic reagent cannot be added to carbon without either putting ten valence electrons around carbon, or causing the simultaneous departure of one of the original ligands with a pair of electrons. An example of ligand substitution is the reaction of halocarbons with hydroxides to form alcohols

$$\begin{array}{cccc} CH_3CH_2Br & +\ HO^- & \xrightarrow{\hspace{3cm}} & CH_3CH_2OH & +\ Br^- \\ \text{ethyl} & & \text{dioxane-water} & \text{ethyl} \\ \text{bromide} & & \text{solvent at } 50°C & \text{alcohol} \end{array} \qquad (14\text{--}14)$$

The reaction must be carried out in a mixed solvent because ethyl bromide, like most organic compounds, is not soluble in water. Dioxane is an organic liquid that is entirely miscible with water, because the molecules can hydrogen bond with water (Chapter 9)

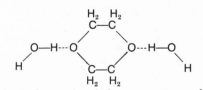

dioxane hydrogen bonded to two water molecules

The reactivity of organic halogen compounds in nucleophilic substitution reactions varies widely. For example, carbon tetrachloride cannot be hydrolyzed, despite the fact that the reaction has a negative standard free energy change

$$CCl_4\ +\ 2H_2O \ \not\!\!\longrightarrow\ CO_2\ +\ 4HCl \qquad\qquad (14\text{--}15)$$

The resistance of carbon tetrachloride to ligand substitution illustrates the importance of rates of reactions. It is another of the many reactions that can occur, but in fact does not occur in a reasonable length of time. The chemical balance of living things depends critically on reaction rates. The chemical changes in metabolism must proceed at the "right" rate, or the organism will die either by suffocation or by burning itself up! Because now there is no satisfactory reaction rate theory, much of the work devoted to predicting rates involves drawing correlations between structure and reactivity for large numbers of compounds. Examples of the sort of arguments used were given in Section 13–6. The study of the relationships between chemical structure and reaction rates is one of the most active and significant branches of organic chemistry.

14–6   Addition and Substitution at Carbonyl Groups

The carbonyl groups of aldehydes, ketones, carboxylic acids, and esters all are moderately reactive toward nucleophilic reagents. The general course of the reactions can be predicted

from consideration of the electronic structure of $\diagdown$C=O groups. Oxygen atoms have a greater

affinity than carbon atoms for electrons, so the group is polarized in the following sense

$$\xrightarrow{\;\delta+\quad\delta-\;}$$
$$\diagup^{\diagdown}\!\!C\!=\!O$$

If a nucleophilic reagent (Nu:⁻) is added at the carbon atom, a pair of electrons can be transferred entirely to the oxygen atom, thereby avoiding overloading the valence shell of the carbon atom

$$\text{Nu:}^- + \ \diagup^{\diagdown}\!\!C\!=\!\ddot{O}\!: \ \rightarrow \ \text{Nu}\!-\!\overset{|}{\underset{|}{C}}\!-\!\ddot{O}\!:^- \qquad (14\text{--}16)$$

If an electrophilic species, often a proton, is added to the oxygen atom an adduct is formed

$$\text{Nu:}^- + \ \diagup^{\diagdown}\!\!C\!=\!\ddot{O}\!: \ \rightarrow \ \text{Nu}\!-\!\overset{|}{\underset{|}{C}}\!-\!\ddot{O}\!:^- \ \xrightarrow{\;H^+\;} \ \text{Nu}\!-\!\overset{|}{\underset{|}{C}}\!-\!\ddot{O}H \qquad (14\text{--}17)$$

$$\text{adduct}$$

If the intermediate loses one of the original ligands (L), a substitution reaction occurs

$$\text{Nu:}^- + \ \overset{R}{\underset{L}{\diagup^{\diagdown}\!\!C}}\!=\!\ddot{O}\!: \ \rightarrow \ \text{Nu}\!-\!\overset{R}{\underset{L}{\overset{|}{C}}}\!-\!\ddot{O}\!:^- \ \rightarrow \ \overset{R}{\underset{Nu}{\diagup^{\diagdown}\!\!C}}\!=\!\ddot{O}\!: \ + \ L^- \qquad (14\text{--}18)$$

Such reactions usually are not very rapid, and often they are catalyzed by acids or bases. Addition of an acid may produce the conjugate acid of a carbonyl compound, generating a new species much more reactive toward nucleophiles

$$\diagup^{\diagdown}\!\!C\!=\!O \ + \ H^+ \ \rightleftarrows \ \diagup^{\diagdown}\!\!C\!=\!OH^+ \qquad (14\text{--}19)$$

$$\text{conjugate}$$
$$\text{acid}$$

$$Nu:^- + \quad \diagdown C{=}OH^+ \quad \rightarrow \quad Nu{-}\underset{\displaystyle |}{\overset{\displaystyle |}{C}}{-}OH \qquad (14\text{-}20)$$

However, the addition of a base may produce a new, reactive nucleophilic species

$$NuH + HO^- \rightleftharpoons Nu:^- + H_2O \qquad (14\text{-}21)$$

An interesting kind of substitution occurs with aldehydes and ketones. The reaction is illustrated by reactions with hydroxylamine (NH_2OH), which reacts with aldehydes and ketones to form oximes. Oximes are usually crystalline compounds that are solid at room temperature. Examples of these reactions are

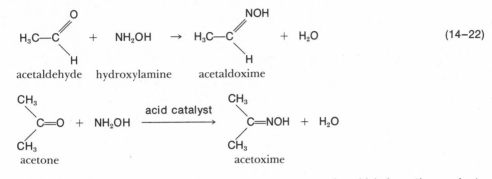

These reactions involve formation of unstable addition compounds, which form the products by decomposition with the elimination of molecules of water

$$(CH_3)_2CO + NH_2OH \rightarrow (CH_3)_2C\diagup^{OH}_{\diagdown NHOH} \rightarrow (CH_3)_2C{=}NOH + H_2O \qquad (14\text{-}23)$$

unstable
adduct

The role of acids and bases is shown by a study of the reactions of alcohols and carboxylic acids to produce esters, and the reverse reactions, in which esters are converted to acids and alcohols

$$\underset{\text{acetic}\atop\text{acid}}{CH_3\overset{O}{\overset{||}{C}}OH} + \underset{\text{methyl}\atop\text{alcohol}}{CH_3OH} \rightleftharpoons \underset{\text{methyl}\atop\text{acetate}}{CH_3\overset{O}{\overset{||}{C}}OCH_3} + H_2O \qquad (14\text{-}24)$$

The preceding reaction is very slow if only acetic acid and methanol are present. However, the addition of a small amount of a strong acid, such as sulfuric acid, causes the reaction to proceed at a moderate rate until equilibrium is established. The mechanism involves fast proton transfer steps, as well as addition and elimination

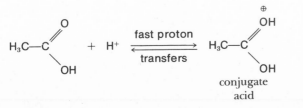

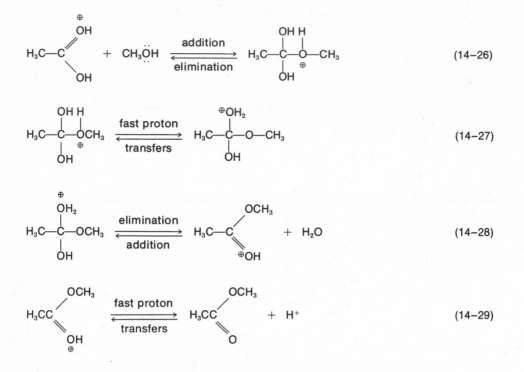

This five-step mechanism, operating in reverse direction, is also the mechanism for the acid-catalyzed hydrolysis of esters. The equilibrium constant for Reaction 14–24 can be formulated in the usual way

$$K_1 = \frac{[CH_3CO_2CH_3][H_2O]}{[CH_3CO_2H][CH_3OH]} \tag{14–30}$$

The values of K_1 are not far from one for many common esters. Consequently, the ratio of the concentrations of organic compounds can be controlled by controlling the water concentration

$$\frac{[CH_3CO_2CH_3]}{[CH_3CO_2H][CH_3OH]} = \frac{K_1}{[H_2O]} \tag{14–31}$$

A mixture of alcohol and acid can be converted mainly to the ester, if a way is found to remove water from the reaction mixture so that $[H_2O]$ is kept small. However, if an ester is heated in the presence of an acid catalyst with a large excess of water, there can be high conversion to acid and alcohol in the ester hydrolysis.

Esters are also hydrolyzed by treatment with water solutions of strong bases. The hydroxide ions play two roles. They are responsible for the primary attack on the carbonyl groups, and they also neutralize the carboxylic acid formed by cleavage, thus driving the reaction to high conversion. The net equation for this reaction is

$$CH_3CO_2CH_3 + HO^- \underset{}{\overset{K_2}{\rightleftharpoons}} CH_3CO_2^- + CH_3OH \tag{14–32}$$

and the mechanism for the reaction can be written in three steps

STRUCTURES AND REACTIONS OF COMPOUNDS OF CARBON AND SILICON

357

$$CH_3C\overset{O}{\underset{OCH_3}{\diagdown}} + HO^- \underset{\text{elimination}}{\overset{\text{addition}}{\rightleftarrows}} H_3C-\overset{O^-}{\underset{OCH_3}{\overset{|}{C}}}-OH \tag{14-33}$$

$$H_3C-\overset{O^-}{\underset{OCH_3}{\overset{|}{C}}}-OH \underset{\text{addition}}{\overset{\text{elimination}}{\rightleftarrows}} CH_3C\overset{O}{\underset{OH}{\diagdown}} + CH_3O^- \tag{14-34}$$

$$CH_3O^- + CH_3CO_2H \underset{\text{transfers}}{\overset{\text{fast proton}}{\rightleftarrows}} CH_3OH + CH_3CO_2^- \tag{14-35}$$

The equilibrium relationship for this reaction is

$$K_2 = \frac{[CH_3CO_2^-][CH_3OH]}{[CH_3CO_2CH_3][HO^-]}$$

The value of K_2 can be compared with K_1, using the ionization constant of acetic acid and the ion product of water

$$CH_3CO_2H \overset{K_a}{\rightleftarrows} CH_3CO_2^- + H^+ \tag{14-36}$$

$$K_a = \frac{[CH_3CO_2^-][H^+]}{[CH_3CO_2H]} = 1.8 \times 10^{-5} \tag{14-37}$$

$$H_2O \rightleftarrows H^+ + HO^- \tag{14-38}$$

$$K_w = [H^+][HO^-] = 1 \times 10^{-14} \tag{14-39}$$

$$K_2 = \frac{K_a[H_2O]}{K_1 K_w} \cong \frac{10^{11}}{K_1}; \quad [H_2O] \cong 56 \text{ moles liter}^{-1} \tag{14-40}$$

This relationship shows that the equilibrium conditions in ester hydrolysis can be made enormously more favorable to cleavage in basic solution than in acidic solution. Advantage is taken of this fact in soap making. Soaps usually are sodium salts of carboxylic acids that occur in nature. The salts are obtained by the hydrolysis of fats, using aqueous sodium hydroxide solutions. Fats are esters in which the alcohol is a trihydroxy compound known as glycerol

$$\begin{matrix} H_2C-OC(CH_2)_{16}CH_3 \\ HC-OC(CH_2)_{16}CH_3 \\ H_2C-OC(CH_2)_{16}CH_3 \end{matrix} + 3NaOH \rightarrow \begin{matrix} CH_2OH \\ CHOH \\ CH_2OH \end{matrix} + 3CH_3(CH_2)_{16}CO_2Na \tag{14-41}$$

glycerol tristearate, a fat   glycerol   sodium stearate, a soap

14-7 Amides

Amides are an especially important group of carbonyl compounds in both chemical technology and in biological chemistry. Nylon and a number of other synthetic fibers and films are amides, as are proteins. In these examples, the substances are made of very large molecules, in which amide groups are the principal building blocks.

Nylon is a polyamide made by the reaction of a *dibasic* acid with a diamine

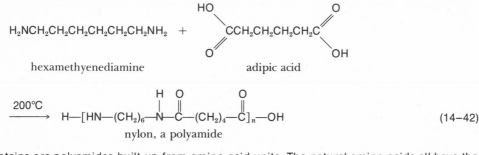

$H_2NCH_2CH_2CH_2CH_2CH_2CH_2NH_2$ +

hexamethyenediamine   adipic acid

$$\xrightarrow{200°C} \quad H—[HN—(CH_2)_6—\overset{H}{N}—\overset{O}{\overset{\|}{C}}—(CH_2)_4—\overset{O}{\overset{\|}{C}}]_n—OH \tag{14-42}$$

nylon, a polyamide

Proteins are polyamides built up from amino acid units. The natural amino acids all have the general structure

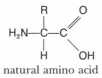

natural amino acid

The 20 natural amino acids have 20 different R groups. In the simplest, glycine, R is hydrogen, and in the others, R can be a variety of hydrocarbon groups, some of which contain other atoms, such as oxygen, nitrogen, and sulfur.

We can think of proteins being formed by a series of nucleophilic substitution reactions of the type

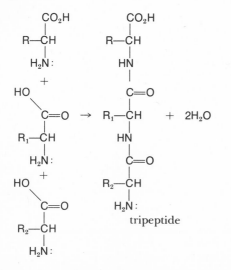

$$\tag{14-43}$$

tripeptide

The product of the preceding reaction is called a *tripeptide,* because it contains three amino acid groups. Polypeptide is a general term for a molecule that is built from a number of amino acid fragments linked together through peptide bonds

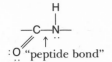

The peptide bond is relatively strong because it has partial double bond character, as shown by combining resonance formulas I and II

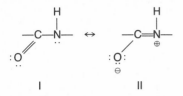

A protein is a very large polypeptide, containing hundreds and sometimes even thousands of amino acid units.

14-8 Silicon Compounds

Comparison of the chemistry of carbon and silicon shows relationships typical of those generally found in chemical families. As we develop this brief section, it may appear that the differences are very large; you may well wonder if we have over advertised the value of grouping elements into families. This is a matter of focus. If the comparison were made with neighboring elements in the periodic table, such as boron, aluminum, nitrogen, and phosphorus, we would see that silicon and carbon, although different, are much more similar to each other than to their nearest neighbors. Both elements form many compounds having the formula MX_4, both form stable hydrides, halides, and oxides, and both become key links in large, ramified structures, having potentially infinite variation. However, carbon-based giant molecules are built up on networks of carbon–carbon bonds, whereas in macromolecules containing silicon, oxygen bridges are always interspersed between silicon atoms. Although silicon–silicon bonds are fairly strong (70 kcal mole^{-1}), there are not a large number of compounds in which the structure is found. Instead, the chemistry of silicon is dominated by the great strength (88 kcal mole^{-1}) of Si—O bonds.

There are no known compounds containing silicon–silicon double bonds. This turns out to be a very general phenomenon, since there are only a few compounds containing double bonds between atoms of elements other than those in the middle of the second row of the periodic table (i.e., carbon, nitrogen, and oxygen).

The rates of ligand substitution reactions of silicon compounds are much faster than those of carbon. This undoubtedly reflects the fact that SiX_4 compounds do not really have closed shells, since $3d$ orbitals can be used in bonding to nucleophilic ligands. There are, in fact, a number of known compounds in which five or six ligands are bound to silicon. A notable example is the anion, SiF_6^{2-}.

The vast array of silicon–oxygen compounds consists mostly of solid materials of very large molecular size. A few low molecular weight compounds containing Si—O bonds are known. An example is tetramethyl silicate, $Si(OCH_3)_4$.

Sand is mostly silica, or silicon dioxide. Unlike carbon dioxide, which is a gaseous, monomeric substance at room temperature, silica is a very high melting solid (1710°C) due to its infinite three-dimensional structure

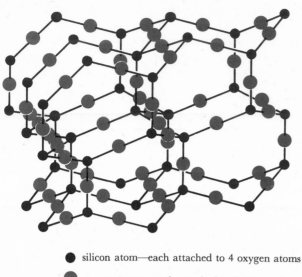

● silicon atom—each attached to 4 oxygen atoms

● oxygen atom—each attached to 2 silicon atoms

Each silicon atom in the infinite structure has a tetrahedron of oxygen atoms around it, and each oxygen is bound to two silicon atoms.

Many silicate rocks are known that have a variety of structures, in which atoms of some metallic element are substituted for some of the silicon atoms. Asbestos, $CaMg_3Si_4O_{12}$, for example, is made up of long fibers, in which double-stranded silicate chains are held together by calcium and magnesium cations.

Simple silicon compounds, such as silicon tetrachloride, are hydrolyzed rapidly on contact with water. The product is a gelatinous material called silicic acid. If the hydrolysis were performed in very dilute solution, simple $Si(OH)_4$ would probably be obtained

$$SiCl_4 \; + \; 4H_2O \; \xrightarrow[\text{solution}]{\text{very dilute}} \; Si(OH)_4 \; + \; HCl \qquad\qquad (14-44)$$

Under usual conditions, a mixture containing many Si—O—Si bridges is produced

$$SiCl_4 \; + \; H_2O \; \xrightarrow[\text{conditions}]{\text{ordinary}} \; (HO)_3Si\!-\!O\!-\!\underset{\underset{\displaystyle OH}{|}}{\overset{\overset{\displaystyle OH}{|}}{Si}}\!-\!O\!-\!Si(OH)_3 \; + \; HCl \qquad (14-45)$$

+ many other silicic acids

Equation 14–45 cannot be written in balanced form, since different amounts of water are consumed in the formation of the different products.

There is a large number of known compounds in which carbon-containing groups are bound to silicon. The following series of methylchlorosilanes are examples:

$(CH_3)_4Si$ $(CH_3)_3SiCl$
tetramethylsilane   trimethylchlorosilane

$(CH_3)_2SiCl_2$ $(CH_3)SiCl_3$
dimethyldichlorosilane   methyltrichlorosilane

These compounds were first regarded as chemical curiosities and were prepared mostly for the purpose of comparing the properties of silicon compounds with those of carbon. However, some now have become very important commercially. Hydrolysis of dialkyldichlorosilanes leads to linear, polymeric chains

$$R_2SiCl_2 \;+\; H_2O \;\rightarrow\; \overset{\displaystyle R}{\underset{\displaystyle R}{-O-\underset{|}{\overset{|}{Si}}-O-}}\overset{\displaystyle R}{\underset{\displaystyle R}{\underset{|}{\overset{|}{Si}}-O-}}\overset{\displaystyle R}{\underset{\displaystyle R}{\underset{|}{\overset{|}{Si}}-O-}} \;+\; HCl \qquad (14\text{-}46)$$

These polymers are known as *silicones.* Various structural modifications produce useful flexible films and tubing, lubricants, electrical insulators, and even bouncing putty. These materials are highly resistant to chemical action, are water repellant, and are stable at rather high temperatures. Their preparation depends upon the fact that Si—C bonds are resistant to nucleophilic substitution, whereas Si—Cl bonds are very reactive.

14-9   Summary

1   Carbon has a valence of four, and may satisfy this valence by bonding to four other atoms, or to fewer than four atoms by forming double and triple bonds.
2   A double bond is made up of one σ and one π bond, whereas the triple bond has one σ and two π bonds.
3   Much of the chemistry of organic molecules occurs at positions where atoms other than carbon are located. These atoms, or groups of atoms, are the reactive sites and are called functional groups.
4   A useful way to categorize organic reactions is under the headings addition, elimination, substitution, and isomerization.
5   Giant molecules are built up on networks of carbon–carbon bonds, whereas in macromolecules containing silicon, oxygen bridges are always interspersed between silicon atoms.

14-10   Postscript: Chemistry and the Drug Scene

Much is said and written about the growing drug culture in the Western World, and we see many of its consequences. In the halcyon days of only a few years ago, new medicinal drugs to relieve man's suffering were developed in great numbers and received enthusiastically by the public. Chemists were then pleased to receive credit for their ingenious contribution to the synthesis of new materials for medicinal evaluation. Now that a cloud hovers over the scene, we become less enthusiastic about claiming responsibility. Yet, chemistry is so inextricably a part of the action that we in the field must feel special concern, if not direct responsibility.

The desire to find materials that will alter the chemistry of the human body is as old as history, and it has spread through all cultures. The search usually has taken the form of extraction of vegetable products to obtain substances of great physiological activity. A witch doctor who boils herbs to obtain a brew to cure or kill his patients is practicing one of the simplest forms of chemistry in medicine, and we probably should not disparage his effort.

During the latter half of the last century, a new approach to drug therapy developed. Isolation of pure compounds from natural products revealed that many substances have profound physiological effect, even when administered in very small doses. This led to a new kind of goal, synthetic production of materials having profound influence on people. Three lines of work were initiated and remain important in chemical science. First, great effort was made to determine the exact chemical structure of natural products. Second, efforts to produce materials identical to natural substances by laboratory synthesis were launched, even before chemists had a very clear idea of the structures they were trying to duplicate. Finally, there began a search for synthetic materials not found in nature, but which have effects similar to those of potent natural materials. For example, it has long been known that morphine is the principal active ingredient in opium, the extract of poppy seeds used since time immemorial to relieve physical pain and to provide escape for those who find the emotional pain of life unbearable. It has also been known for a long time that minor chemical modification of morphine changes its action. On the one hand it can be converted easily to codeine, which has far less addictive threat while retaining some of the analgesic (pain killing) properties. On the other hand, morphine is just as easily turned into heroin, the most feared of the powerfully addictive drugs. For nearly a hundred years, chemists have sought to synthesize materials having the properties desired in heroin, but lacking those that are feared. The same drive has been directed toward the production of drugs that will combat disease of many kinds. Judged by simple criteria, the effort has been remarkably successful, and pharmaceutical chemistry has totally changed the practice of medicine. Another outgrowth of drug research has been the development of the field of synthetic organic chemistry as one of the most creative and sophisticated fields of science.

Some consequences of pharmaceutical chemistry are now causing serious second thoughts. Is the quality of a man's life really improved by the availability of a pill to put him to sleep, another to wake him up, and still a third to arrest his midday depression? This behavior is not entirely dependent on chemical research, since alcohol and coffee (also chemicals) can be used to maintain the same drab life pattern. However, we are becoming increasingly aware that addiction can be psychological as well as physiological, and that continued use of supposedly therapeutic drugs may simply perpetuate an emotional condition fatal to the spirit, if not to the body.

Chemistry is very much a part of the current scene in illegal drugs. Lysergic acid diethylamide (LSD) was discovered by two Swiss chemists, A. Stoll and A. Hofman, as a part of a program to prepare derivatives of lysergic acid, a natural product known for years to be a powerful agent for causing termination of unwanted pregnancies. The fact that conversion of lysergic acid to LSD is easy, and that the acid is readily available, made experimentation with this powerfully hallucinogenic drug very easy, until legal restrictions were imposed because of reports of very bad trips, lasting psychological effects, and possible damage to human chromosomes. Chemical analyses are frequently accepted as the most powerful evidence for identification of illegal drugs in court rooms. We do not, incidentally, feel highly elated by this particular scientific practice. Use of chemical analysis in a saliva test to detect drugging of race horses does make sense, but legal reliance on chemical testing to identify marijuana, when a kilo of the material is available to be seen and smelled, seems rather ritualistic.

Recent debate concerning marijuana has restimulated chemical study of the material. The most potent active ingredient, tetrahydrocannabinol (THC), has been made synthetically, but not by a route that is easily repeated in one's garage. Careful study has also been made of the abundance of THC and closely related compounds in the flowers and leaves of various hemp plants. Such studies have shown, for example, that the "low quality" of the plants growing wild in the midwestern United States is due more to their parentage than to soil conditions. Curiously, the study now seems to have as its most logical application the testing of the quality of marijuana, should it become legalized.

One cannot resist moralizing on a subject that is discussed with as much intensity as drugs are in our society. However, we have little to add to what has been said many times. Anyone who has seen the deterioration of a "speed freak" cannot have unlimited pride in the accomplishments of chemical synthesis. We can wish that the conversion of morphine to heroin were chemically difficult, but a chemist cannot repeal natural law and make a chemical reaction stop occurring. There is no doubt that many drugs in common use induce personality changes, at least temporarily. But whether this is undesirable depends a good deal on the value placed on the old personality. Like many other people, we are troubled by the inconsistency of our laws and customs that levy much more severe penalties on those who possess marijuana than on those who obtain apparently more dangerous synthetic drugs through illicit channels. Although chemistry is deeply involved in the drug issue both currently and historically, chemists find themselves caught in the same pattern of conflicting evidence and emotion as any thinking people when they consider the really important questions of the day. Statement of the various personal convictions of the authors on the issues would do little to resolve these conflicts.

New Terms

Alcohol: A compound of the type ROH.
Aldehyde: A compound of the type

Alkane: A saturated hydrocarbon.
Alkene: A compound containing carbon–carbon double bonds.
Alkyne: A compound containing carbon–carbon triple bonds.
Amide: A compound of the type

The hydrogen atoms on the nitrogen atom may be replaced by R groups.
Amine: A compound of the type RNH_2, in which nitrogen is bound only to carbon or hydrogen.
The hydrogen atoms on the nitrogen atom may be replaced by R groups.
Carbonium ion: A positive carbon ion; for example, $H_3C^{\oplus}$.

Carbonyl: The functional group $\diagdown C{=}O$.

Carboxylic acid: A compound of the type

R—C with double bond O and OH.

Cycloalkanes: Alkanes containing rings of carbon atoms.
Dibasic acid: An acid with two ionizable protons.
Electrophilic: Electron-seeking; electrophiles are Lewis acids.

Ester: A compound of the type

Functional group: A reactive part of an organic compound that determines its chemical behavior.

Halocarbon: A hydrocarbon that in addition to carbon and hydrogen contains one or more
halogen atoms.

Hydrocarbon: A compound containing only hydrogen and carbon.

Ketone: A compound of the type

R
 \
  C=O.
 /
R′

Nitration: The replacement of hydrogen or some group with —NO_2.

Nucleophilic: Nucleus-seeking; nucleophiles are Lewis bases.

Organic chemistry: The chemistry of carbon-containing compounds.

Photochemical: Having to do with the transfer of light energy to and from compounds and the
effects of the energy transfer on reactions.

Polyfunctional: Having more than one functional group.

R—:   Symbol used to denote a carbon-containing fragment of a molecule.

Saturated: Having the maximum number of bonded atoms. Carbon compounds having no
double or triple bonds are said to be saturated.

Unsaturated: Having multiple bonds.

Questions and Problems

1 Classify the following compounds as alkanes, alkenes, or alkynes:

(a) $CH_3CH_2CH=CH_2$

(b) $CH_3C≡CH$

(c) $CH_3(CH_2)_6CH_3$

(d)
```
        CH₃
        |
  H₃C—C—H
        |
        CH₃
```

(e) C_6H_{14}

(f)
```
   CH₃       CH₃
     \       /
      C=C
     /       \
    H         H
```

(g)
```
          CH₃
          |
   H₃C—C—C≡CH
          |
          CH₃
```

(h) C_8H_{16}

2 How many different ways can you arrange the atoms in the compound having the formula C_5H_{12}? What term is used to describe each different compound having the same molecular formula?

3 Give an example of a cyclic hydrocarbon.

4 Classify the following compounds according to the kind of functional group present:

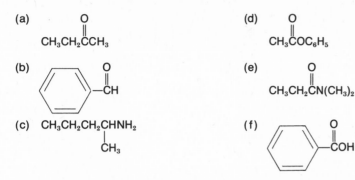

(a)
$$CH_3CH_2\overset{\overset{\displaystyle O}{\|}}{C}CH_3$$

(d)
$$CH_3\overset{\overset{\displaystyle O}{\|}}{C}OC_6H_5$$

(b)

(e)
$$CH_3CH_2\overset{\overset{\displaystyle O}{\|}}{C}N(CH_3)_2$$

(c) $CH_3CH_2CH_2\underset{\underset{\displaystyle CH_3}{|}}{C}HNH_2$

(f)

5 Predict what product would result from the reaction of 1-butene ($CH_3CH_2CH{=}CH_2$) with water in the presence of H_2SO_4

$$CH_3CH_2CH{=}CH_2 \ + \ H_2O \ \xrightarrow{\ H_2SO_4\ }$$

6 Give an example of a carbonium ion.

7 What is meant by the terms nucleophilic and electrophilic? Show how the polarizations of the electrons in the carbonyl group $\left(\ \overset{}{\underset{}{C}}{=}O\ \right)$ make the carbon susceptible to attack by a nucleophilic reagent.

8 Write a series of stepwise reactions showing four glycine amino acid molecules (Section 14–7) undergoing self-condensation to form the tetrapeptide molecule.

9 What is one of the most important differences between the way carbon atoms bond together to form large macromolecules and the way silicon atoms do?

10 Compare the reactivities of CCl_4 and $SiCl_4$ in water.

APPENDIX 1
SELECTED EQUATIONS EXPRESSING
IMPORTANT PHYSICAL AND
CHEMICAL RELATIONSHIPS

Density

$$d = \frac{W}{V}$$

d = density
W = weight
V = volume

standard units: d in g liter^{-1}, W in g, V in liters
d in g ml^{-1}, W in g, V in ml

Mass–energy relationship

$$E = mc^2$$

E = energy
m = mass
c = speed of light

standard units: E in ergs, m in g, $c = 3.00 \times 10^{10}$ cm sec^{-1}

Combined gas law (or ideal gas law)

$$PV = nRT$$

P = pressure
V = volume
n = number of moles
R = gas constant
T = absolute temperature

standard units: P in atm, V in liters, $R = 0.0821$ liter atm
mole^{-1} °K^{-1}, T in °K

Average kinetic energy

$$\overline{KE} = N(\tfrac{1}{2}m\overline{v^2})$$
$$= \tfrac{3}{2}RT$$

$\overline{KE}$ = average kinetic energy per mole
N = Avogadro's number
m = mass of one molecule

$$v = \text{velocity}$$
$$R = \text{gas constant}$$
$$T = \text{absolute temperature}$$

standard units: $\overline{KE}$ in ergs mole^{-1}, $N = 6.022 \times 10^{23}$ molecules mole^{-1}, m in g, v in cm sec^{-1}
$\overline{KE}$ in ergs mole^{-1}, $R = 8.31 \times 10^7$ ergs mole^{-1} °K^{-1}, T in °K
$\overline{KE}$ in cal mole^{-1}, $R = 1.99$ cal mole^{-1} °K^{-1}, T in °K

Molecular velocity and temperature

$$v_{\text{rms}} = \sqrt{\frac{3RT}{\text{mol wt}}}$$

$v_{\text{rms}} = $ root mean square velocity
$= \sqrt{\overline{v^2}}$
$R = $ gas constant
$T = $ absolute temperature
mol wt $=$ molecular weight

standard units: v_{rms} in cm sec^{-1}, $R = 8.31 \times 10^7$ ergs mole^{-1} °K^{-1}, T in °K, mol wt in g mole^{-1}

Kinetic energy

$$KE = \tfrac{1}{2}mv^2$$

$KE = $ kinetic energy of a particle
$m = $ mass of the particle
$v = $ velocity of the particle

standard units: KE in ergs, m in g, v in cm sec^{-1}

Wavelength and frequency of electromagnetic radiation (light)

$$c = \lambda\nu = 3.00 \times 10^{10} \text{ cm sec}^{-1}$$

c = velocity of light
λ = wavelength of light
ν = frequency of light

standard units: $c = 3.00 \times 10^{10}$ cm sec^{-1}, λ in cm, ν in sec^{-1}

Energy

$$E = h\nu$$

E = energy of one quantum
h = Planck's constant
ν = frequency

standard units: E in ergs, $h = 6.63 \times 10^{-27}$ erg sec, ν in sec^{-1}

Rydberg equation

$$\frac{1}{\lambda} = R\left(\frac{1}{n_1^2} - \frac{1}{n_2^2}\right)$$

λ = wavelength of emitted or absorbed light
R = Rydberg's constant
$n_1 = 1, 2, 3, 4, \ldots$
$n_2 = 2, 3, 4, 5, \ldots$

standard units: λ in cm, $R = 109{,}678$ cm^{-1}

Centrifugal force

$$F_c = \frac{mv^2}{r}$$

F_c = centrifugal force of an orbiting particle
m = mass of particle
v = velocity of particle
r = radius of orbit

standard units: F_c in dynes, m in g, v in cm sec^{-1}, r in cm

Coulomb's law

$$F_e = \frac{(q_1e)(q_2e)}{r^2}$$

F_e = electrostatic force between two charged particles
q_1e = electrostatic charge on particle 1
q_2e = electrostatic charge on particle 2
r = distance between particles 1 and 2

standard units: F_e in dynes, q_1e in esu, q_2e in esu, r in cm

$$PE = \frac{(q_1e)(q_2e)}{r}$$

PE = potential energy of two charged particles

standard units: PE in ergs, q_1e in esu, q_2e in esu, r in cm

Heisenberg's uncertainty principle

$$(\Delta x)(\Delta mv) > \frac{h}{4\pi}$$

Δx = uncertainty in x component of the position of a particle

Δmv = uncertainty in momentum of the particle

h = Planck's constant

$\pi = 3.14$

standard units: Δx in cm, Δmv in g cm sec^{-1}, $h = 6.63 \times 10^{-27}$ erg sec

Energies of the principal quantum levels of atomic hydrogen

$$E = -\frac{13.6}{n^2}$$

E = energy of quantum level

n = principal quantum number (1, 2, 3, 4 . . .)

standard units: E in eV

Dipole moment

$$\mu = (qe)R$$

μ = dipole moment

qe = separated electrostatic charge

R = distance of separation

standard units: μ in esu cm, qe in esu, R in cm, 1D $= 10^{18}$ esu cm

Equilibrium constant for a reaction

$$K_{eq} = \frac{[C]^c[D]^d}{[A]^a[B]^b}$$

K_{eq} = equilibrium constant for the reaction $aA + bB \rightleftarrows cC + dD$

[A] = concentration of A

a = number of moles of A in a balanced equation

standard units: K_{eq} in (moles liter^{-1})$^{c+d-a-b}$
[A], [B], [C], [D] in moles liter^{-1}
K_{eq} in (atm)$^{c+d-a-b}$
[A], [B], [C], [D] in atm

Change in free energy, heat, and entropy

$$\Delta G = \Delta H - T\Delta S$$

$\Delta G =$ change in free energy
$\Delta H =$ change in heat content
$\Delta S =$ change in entropy
$T =$ absolute temperature

standard units: ΔG in cal mole^{-1}, ΔH in cal mole^{-1}, T in °K, ΔS in cal mole^{-1} °K^{-1}

Standard free energy change and the equilibrium constant

$$\Delta G^0 = -2.303RT \log K_{eq}$$

$\Delta G^0 =$ change in free energy for a reaction in which one mole of reactant in its standard state is converted to product in its standard state
$R =$ gas constant
$T =$ absolute temperature
$K_{eq} =$ equilibrium constant for the reaction

standard units: ΔG^0 in cal mole^{-1}, $R = 1.99$ cal mole^{-1} °K^{-1}, T in °K

Rate law for a reaction

$$rate = k[A]^m[B]^n$$

rate = rate of a reaction involving the molecules A and B.

k = rate constant for the reaction

$[A]^m$ = concentration of A to the mth power

$[B]^n$ = concentration of B to the nth power

standard units: rate in moles liter^{-1} sec^{-1},
k in (liter mole^{-1})$^{m+n-1}$ sec^{-1}

Rate constant and temperature

$$k = pe^{-(E_a/RT)}$$

k = rate constant

p = frequency factor or the rate for molecules with $E \geqslant E_a$

E_a = activation energy

R = gas constant

T = absolute temperature

standard units for unimolecular processes:

p in sec^{-1}, E_a in cal mole^{-1},
R = 1.99 cal mole^{-1} °K^{-1}, T in °K

standard units for bimolecular processes:

p in liter mole^{-1} sec^{-1}, E_a in cal mole^{-1},
R = 1.99 cal mole^{-1} °K^{-1}, T in °K

APPENDIX 2
USEFUL PHYSICAL CONSTANTS AND
CONVERSION FACTORS

Physical constants[1]

Avogadro's number	$N = 6.022 \times 10^{23}$ mole^{-1}
Electronic charge	$e = 4.8032 \times 10^{-10}$ esu
Faraday	$\mathscr{F} = 96{,}487$ coulombs mole^{-1}
Gas constant	$R = 8.3143 \times 10^7$ ergs °K^{-1} mole^{-1}
	$R = 0.082056$ liter atm °K^{-1} mole^{-1}
	$R = 1.98718$ cal °K^{-1} mole^{-1}
Planck's constant	$h = 6.6262 \times 10^{-27}$ erg sec
Rydberg's constant	$R = 109{,}678$ cm^{-1}
Velocity of light	$c = 2.9979 \times 10^{10}$ cm sec^{-1}

Conversion factors[1]

Energy

1 electron volt (eV) = 1.6022×10^{-12} erg

1 erg = 6.2420×10^{11} eV = 2.3901×10^{-11} kcal = 1 g cm^2 sec^{-2}

1 kcal = 4.1840×10^{10} ergs = 2.612×10^{22} eV

1 volt coulomb = 1 joule = 10^7 ergs = 0.23901 cal

1 eV molecule^{-1} = 23.056 kcal mole^{-1} = 8065 cm^{-1}

100 kcal mole^{-1} = 34,982 cm^{-1}

1 kcal = 1000 cal

2.303 RT = 1.364 kcal mole^{-1} at 298°K

Length

1 cm = 10^8Å

Volume

1 liter = 1000 ml

---

[1] Values taken from *Handbook of Chemistry and Physics,* 51st ed (Chemical Rubber Publishing Co., Cleveland, Ohio, 1970–71).

APPENDIX 3
MATHEMATICAL NECESSITIES
AND DEVICES[1]

Exponential numbers

Not only does writing a number in exponential form enable us to express significant-figure information with minimal confusion, it avoids many zeros for either small or large numbers.

We use exponential numbers to express quantities as multiple powers of 10. Thus, an exponential number consists of two parts: a coefficient (chosen to be between 1 and 10) and a power of 10. For example, the Avogadro number is written 6.022×10^{23}; 6.022 is the coefficient and 10^{23} is the power of 10.

A positive exponent, n, indicates that the coefficient should be multiplied by 10 n times; that is, the decimal should be moved n places to the right of its position in the coefficient. A negative exponent, $-m$, indicates that the coefficient should be divided by 10 m times; that is, the decimal should be moved m places to the left. For example,

$$0.0000000192 = 1.92 \times 10^{-8}$$
$$10 \text{ thousand} = 1 \times 10^{4}$$
$$96500 = 9.65 \times 10^{4}$$

To add or subtract exponential numbers, we must be certain that the powers of 10 are the same. Otherwise, the operation would be like adding different things: $2x + 2y = ?$, whereas $2x + 2x = 4x$. This requirement may force us to rewrite the exponential number. This revision is easy if we remember that each time the power of 10 becomes more positive by one unit it is the same as multiplying the number by 10 or moving the decimal in the coefficient one place to the right. Similarly, if the power of 10 is made more negative, the decimal in the coefficient must be moved to the left. For example,

$$6.022 \times 10^{23} + 7.8 \times 10^{21} = ?$$

Rewrite both numbers so they have the same power of 10; for example, 21. To write 6.022×10^{23} as a multiple of 10^{21} (the exponent has been decreased by two powers of 10) requires that

---

[1] Adapted, with the kind permission of Dr. Wilbert Hutton, from his book *A Study Guide to Chemical Principles.*

the coefficient be increased by two powers of 10. Therefore, its decimal point should be moved two places to the right

$$6.022 \times 10^{23} = 602.2 \times 10^{21}$$

Now the numbers can be added

$$\begin{array}{r} 602.2 \times 10^{21} \\ + \quad 7.8 \times 10^{21} \\ \hline 610.0 \times 10^{21} \text{ or } 6.100 \times 10^{23} \end{array}$$

Do the following exercises to test your understanding:

(1) Add 2.46×10^{-9} cm to 2.46×10^{-8} cm.
(2) Subtract 1.625×10^{-1} cm from 2.234×10^2 cm.
(3) Add 4.0074×10^3 ml to 6.23×10^2 ml.
(4) Subtract 1.725×10^{-1} g from 2.1623×10^1 g.

Answers: (1) 2.71×10^{-8} cm;   (2) 2.232×10^2 cm;
         (3) 4.630×10^3 ml;   (4) 2.1450×10 g.

In *multiplication,* you need only multiply the coefficients together and then multiply the powers of 10 together (add their exponents) to obtain the coefficient and power of 10 for the answer. For example,

$$6.02 \times 10^{23} \times 1.76 \times 10^{-2} = ?$$

The product of the coefficients to the proper number of significant figures is $6.02 \times 1.76 = 10.6$. The product of the powers of 10 (exponentials) are $10^{23} \times 10^{-2} = 10^{23+(-2)} = 10^{21}$. The answer to the multiplication is 10.6×10^{21} or, writing the coefficient in the preferred way as a number between 1 and 10, 1.06×10^{22}.

In *division,* the coefficients are divided separately and then exponentials are divided. Recall that in division of exponential numbers, the exponent of the divisor (the denominator) is subtracted from the exponent of the dividend (the numerator). For example,

$$\frac{6.022 \times 10^{23}}{5.976 \times 10^{27}} = ?$$

Dividing 6.022 by 5.976 gives 1.008 to the correct number of significant figures. Dividing the exponentials gives $10^{23}/10^{27} = 10^{(23-27)} = 10^{-4}$. The answer is therefore 1.008×10^{-4}.

The same general procedure is followed when raising an exponential to a power. The coefficients are done first, the exponentials next, and the results of these computations combined for the answer. Thus,

$$(6 \times 10^3)^3 = 6^3 \times (10^3)^3 = 2 \times 10^{11}$$
$$(5.1 \times 10^{-2})^2 = (5.1)^2 \times (10^{-2})^2 = 26 \times 10^{-4} = 2.6 \times 10^{-3}$$

To avoid fractional exponents when extracting a root, we must adjust the power of 10 so it becomes a whole-number factor of 2 if a square root is to be extracted, a whole-number factor of 3 if a cube root is to be extracted, and so forth. Therefore, to extract the cube root of Avogadro's number, $(6.02 \times 10^{23})^{1/3}$, you must first rewrite the number so the power of 10 is a whole-number multiple of three. Since $3 \times 7 = 21$ and $3 \times 8 = 24$, either 10^{21} or 10^{24} are suitable exponentials. Let us rewrite the number as a coefficient times 10^{21} by moving the decimal in the coefficient two places to the right and decreasing the exponent of 10 by two units: $(602 \times 10^{21})^{1/3}$. The cube root of 602 is 8.45; the cube root of 10^{21} is 10^7. The answer is 8.45×10^7.

As a self-test of your understanding, do the following exercises:

(1) $\dfrac{5.23 \times 10^{27}}{9.76 \times 10^3} = ?$

(2) $\dfrac{3.42 \times 10^{-29}}{6.704 \times 10^5} = ?$

(3) $\dfrac{(2.46 \times 10^3)(1.7 \times 10^{-5})}{3.25 \times 10^4} = ?$

(4) $(5.2 \times 10^{-3})^3 = ?$

(5) $(7.5 \times 10^{-5})^{1/2} = ?$

Answers: (1) 5.36×10^{23}; (2) 5.10×10^{-35}; (3) 1.3×10^{-6};
(4) 1.4×10^{-7}; (5) 8.7×10^{-3}.

Logarithms

There are many advantages in expressing numbers in scientific notation; that is, as $a \times 10^x$, where a, the coefficient, is a number between 1 and 10 and x is the exponent, or power, of 10. If x is zero, the entire exponential (10^0) equals 1 (any number raised to the zero power is unity) and is not written; for example, $2.54 \times 10^0 = 2.54 \times 1 = 2.54$. Also, if the coefficient a or the exponent x is unity, it may be understood and not written as such; that is, 1×10^4 is written 10^4 and 4.2×10^1 is written 4.2×10.

Now if $10^0 = 1$ and $10^1 = 10$, it follows that any number between 1 and 10 can be expressed as a power of 10 with a coefficient of unity by using fractional exponents. Moreover, if negative powers are included, any number can be given in this way. This is the basis for expressing numbers as logarithms. The *logarithm* of a number is the exponent that must be put on 10 to give the number. For example, the number 2 can be expressed as $10^{0.301}$, in which case 0.301 is the logarithm (or "log") of 2. In Table A3–1 the powers of 10 needed to express the numbers 1 through 10 are listed. In other words, a table of the logarithms of the numbers 1 through 10 is given.

Table A3–1

$1 = 10^0$, log 1 = 0	$6 = 10^{0.778}$, log 6 = 0.778
$2 = 10^{0.301}$, log 2 = 0.301	$7 = 10^{0.845}$, log 7 = 0.845
$3 = 10^{0.477}$, log 3 = 0.477	$8 = 10^{0.903}$, log 8 = 0.903
$4 = 10^{0.602}$, log 4 = 0.602	$9 = 10^{0.954}$, log 9 = 0.954
$5 = 10^{0.699}$, log 5 = 0.699	$10 = 10^1$, log 10 = 1.000

The exponent to which 10 must be raised to give the number is referred to as a logarithm to the *base* 10. These logarithms are the most convenient for general use since, for calculations, one can follow the simple rule that 1 is added to the logarithm of a number every time the number is multiplied by 10.

However, numbers other than 10 can be used for the base. Logarithms to the base 2.718 . . . , so-called "Naperian base" or "natural base" logarithms, are used frequently in science. This base, 2.718 . . . , is designated by the letter e. In terms of e, the number 2 can be expressed as $e^{0.693}$ (i.e., e to the 0.693 power), and we would say that the logarithm of 2 to the base e is 0.693. This last statement is written $\log_e 2 = 0.693$, or ln 2 = 0.693, where the abbreviation "ln" refers to the natural or Naperian base logarithm. Logarithms to the base 10 are generally indicated by the word "log," and no base is specified (e.g., log 2 = 0.301). A simple relationship between base 10 and base e logs allows us to convert a logarithm from one base to the other. The relationship is $\log_e x = 2.303 \log x$.

Logarithms are useful because they make the multiplication of numbers a simple addition operation; division, a simple subtraction operation; and the computation of powers and roots, a simple multiplication or division process. Thus, much of the tedious digit-writing of arithmetic can be discarded, and operations can be performed faster and with less chance of error. The reason for this simplification is that when two powers of 10 are multiplied, their exponents are added, and when two powers of 10 are divided, their exponents are subtracted. These exponents are the logarithms. So, if we have tables that give the logarithms of numbers, we can take advantage of these simpler operations.

The logarithms of numbers expressed as powers of 10 with coefficients of unity are easily determined. They are simply the exponent. For example,

$$1 \times 10^5 = 10^0 \times 10^5 = 10^{(0+5)} = 10^5 \qquad \log 10^5 = 5$$

Similarly,

$$\log(1 \times 10^6) = \log 10^6 = 6$$
$$\log(1 \times 10^{-7}) = \log 10^{-7} = -7$$

Now let us obtain the logarithms of some numbers that have been multiplied together. Recalling that multiplication of powers of 10 is carried out by adding exponents, we can see immediately that the log of a product 10^x times 10^y is simply $x + y$.

Example. What is the log of $(1 \times 10^7)(1 \times 10^5)$?

Solution.

$$(1 \times 10^7)(1 \times 10^5) = 1 \times 10^{12}$$
$$\log[(1 \times 10^7)(1 \times 10^5)] = \log(1 \times 10^{12}) = 12$$

or, if we omit the coefficients,

$$10^7 \times 10^5 = 10^{12}$$
$$\log(10^7 \times 10^5) = \log 10^{12} = 12$$

An alternate solution is to add the logarithms,

$$\log(10^7 \times 10^5) = \log 10^7 + \log 10^5$$
$$= \qquad 7 + \qquad 5$$
$$= 12$$

Example. What is the log of $(1 \times 10^9)(1 \times 10^{-7})$?

Solution.

$$\log(10^9 \times 10^{-7}) = \log 10^9 + \log 10^{-7}$$
$$= \quad 9 + \quad -7$$
$$= 2$$

Similarly, the logarithm of a quotient $10^x/10^y$ is $x - y$.

Example. What is the log of $(1 \times 10^9)/(1 \times 10^3)$?

Solution.

$$\log\left(\frac{1 \times 10^9}{1 \times 10^3}\right) = \log 10^6 = 6$$

An alternate solution is to subtract the logarithms,

$$\log\left(\frac{1 \times 10^9}{1 \times 10^3}\right) = \log(1 \times 10^9) - \log(1 \times 10^3)$$
$$= \quad 9 - 3$$
$$= 6$$

Example. What is the log of $(1000 \times 10)/0.0001$?

Solution.

$$\log\left(\frac{10^3 \times 10}{10^{-4}}\right) = \log 10^3 + \log 10 - \log 10^{-4}$$
$$= \quad 3 + \quad 1 - \quad (-4)$$
$$= 8$$

In the preceding examples we have used only numbers that have coefficients of unity and whole-number powers of 10. In general, digits other than 1 will occur in the coefficients; in such cases we must look for these in a log table. Log tables always list the kind of information in Table A3–1, namely, the number and the exponent to which 10 must be taken to obtain the number—the number's logarithm.

Your next step should be to look up numbers expressed by coefficients larger than 1 and powers of 10 greater than zero. A few examples should illustrate how this is done.

Example. What is the log of 2×10^3?

Solution.

$$\log(2 \times 10^3) = \log 2 + \log 10^3$$
$$= 0.301 + \quad 3$$
$$= 3.301$$

The log 2 is found in the table; the log 10^3 is merely the exponent.

Example. Now try a number with a negative exponent. Find the log of 7×10^{-4}.

Solution.

$$\log(7 \times 10^{-4}) = \log 7 + \log 10^{-4}$$
$$= 0.845 + \quad -4$$
$$= -3.155$$

No special operation is necessary to obtain logarithms for numbers with negative exponents. Now you should be ready to put all of these concepts into practice. To help you to begin, here is an example worked out in detail.

Example. What is the logarithm of

$$\frac{(2 \times 10^{-3})(6 \times 10^{23})}{5 \times 10^4}$$

Solution.

$$\log\left[\frac{(2 \times 10^{-3})(6 \times 10^{23})}{5 \times 10^4}\right] = \log 2 + \log 10^{-3}$$
$$+ \log 6 + \log 10^{23}$$
$$- (\log 5 + \log 10^4)$$

The imposing string of logarithms on the right of the equation may be a bit bewildering. Remember that the order in which a sequence of multiplications and divisions is carried out does not make any difference in the answer. You would obtain the same answer to the multiplication and division sequence if you divide the first factor in the numerator by the denominator and then multiply the result by the second factor in the numerator as you would if you multiply both factors in the numerator and then divide their product by the denominator. The same is true if you do the multiplication and division with logarithms. In this problem, the two factors in the numerator have been multiplied first, and then their product is divided by the denominator by using logarithms. Now let us find, from a four-place table of logarithms or a slide rule, the logarithms for the right side of the equation

$$= 0.301 + (-3) + 0.778$$
$$+ 23 - 0.699 - 4$$
$$= 0.380 + 16 = 16.380$$

Try these exercises to test your understanding:

(1) What is the log of

$$\frac{4 \times 10^8}{2 \times 10^6}$$

(2) What is the log of

$$\frac{(6 \times 10^{-6})(3 \times 10^{14})}{(2 \times 10^3)(1 \times 10^6)}$$

Answers: (1) 2.301; (2) −0.046

[*Note:* You may find the solutions to problems such as this easier if you work out the co-efficients on a slide rule, since we have attempted wherever possible to simplify the arithmetic.]

You will also need to be able to reverse this procedure, so that, given the logarithm, you can express the number as an integral power of 10. To illustrate how to do this, we shall examine first some simple examples and then some more complicated ones.

Example. What is the number whose log is 3? [*Note:* The term "antilog" indicates "the number whose log is."]

Solution.

$$\log x = 3$$
$$x = 10^3$$
$$x = 1 \times 10^3 = 1000$$

Example. What is the number whose log is 3.301?

Solution.

$$\log x = 3.301$$
$$x = 10^{3.301}$$
$$= 10^3 \times 10^{0.301}$$
$$= 10^3 \times 2 = 2 \times 10^3$$

The method here is to take the logarithm 3.301, extract the whole number (3), and use it to derive the power of 10 (10^3), and then use the decimal part (0.301) to derive the coefficient (2). This operation requires looking in a table of logarithms for the logarithm 0.301 and then noting what number corresponds to it.

Exercise. What are the numbers whose logarithms are 6.903, 4.699, and 3.845?

Answers: 8×10^6; 5×10^4; 7×10^3.

Finding the antilog of a negative logarithm is complicated because the table or slide rule contains positive numbers only.

Example. Find the number whose logarithm is -4.222.

Solution.

$$\log x = -4.222$$
$$x = 10^{-4.222}$$

Note that -4.222 can be written as $-5 + 0.778$. Make this substitution and proceed as usual.

$$x = 10^{-4.222} = 10^{-5+0.778}$$
$$= 10^{-5} \times 10^{0.778}$$

In effect, we have transferred all of the minus signs to the whole-number exponent of 10 and have made the decimal exponent a positive number, the antilog of which can be found in the table. The antilog of 0.778 is 6. Therefore, $x = 10^{-5} \times 10^{0.778} = 6 \times 10^{-5}$.

Extracting roots and raising numbers to powers

We can use logarithms to raise numbers to powers or to extract roots of numbers. For instance, the number a^x means that the number a is multiplied by itself x times. To do this with logarithms, we would write the log of a x times and add; the sum would be the log a^x. That is, $\log a^x = x \log a$.

The xth root of a number a is expressed $a^{1/x}$. Through reasoning similar to that used for raising numbers to powers, we find that $\log a^{1/x} = (1/x) \log a$.

Example. What is the cube root of 0.008?

Solution.

$$
\begin{aligned}
\log[(8 \times 10^{-3})^{1/3}] &= \tfrac{1}{3}(\log 8 + \log 10^{-3}) \\
&= \tfrac{1}{3}[0.903 + (-3)] \\
&= \tfrac{1}{3}(-2.097) \\
&= -0.699
\end{aligned}
$$

Now we need to find the antilog of -0.699; this will be the cube root of 0.008

$$10^{-0.699} = 10^{0.301-1} = 10^{0.301} \times 10^{-1}$$

We find the number equal to $10^{0.301}$; it is 2. Therefore,

$$(0.008)^{1/3} = 2 \times 10^{-1}$$

The quadratic formula

Equations of the second order containing a single unknown that is raised to the second power have two roots. These equations can always be written as

$$ax^2 + bx + c = 0 \tag{1}$$

where a, b, and c are numbers and x is the unknown. If $a = 0$, the equation is first order. It is also possible for b or c, or both, to be zero.

Once the equation is in the form of Equation 1, and a is not zero, it can be solved by substitution into the quadratic formula

$$x = \frac{-b \pm \sqrt{b^2 - 4ac}}{2a} \tag{2}$$

In doing chemistry problems, only one of the two roots is an answer to the problem; the other is unrealistic. You can generally decide which of the two roots apply by considering the problem's physical circumstances. For instance, a common situation is when you use a quadratic equation to find the amount of a substance remaining after a chemical reaction. The solution of the equation will give two roots, each of which is a possible value for the amount of the remaining substance. If one of the roots gives a value larger than the amount of substance present when the reaction started, or if it gives a negative or imaginary number, it is obvious that this value does not apply. Hence, the other root must be the correct answer.

Example. Consider the solution of the following equation in which x represents the amount (in moles per liter) of a substance B formed after a chemical reaction has reached equilibrium. Assume that we start with 1 mole liter^{-1} of a reacting substance and that 1 mole of it produces 1 mole of B and 1 mole of C. The equilibrium constant for the reaction is 54.

$$\frac{(x)(x)}{1-x} = 54$$

Solution. Rearrange the equation to the form of Equation 1 by multiplying both sides by $1-x$

$$x^2 = 54(1-x)$$

Multiply the two quantities on the right, collect terms, and equate to zero

$$x^2 + 54x - 54 = 0$$

Now substitute the corresponding quantities in the quadratic equation: $a = 1$, $b = 54$, and $c = -54$

$$x = \frac{-54 \pm [54^2 - 4(1)(-54)]^{1/2}}{2(1)} = \frac{-54 \pm (2916 + 216)^{1/2}}{2}$$

$$x = \frac{-54 \pm (3132)^{1/2}}{2} = \frac{-54 \pm 55.9}{2}$$

$$x = \frac{-54 + 55.9}{2} = \frac{1.9}{2} = 0.95$$

and

$$x = \frac{-54 - 55.9}{2} = \frac{-109.9}{2} = -55.0$$

Logic indicates which of these two solutions is correct for the problem: Since the reaction began with 1 mole of the substance, the amount left after equilibrium is reached must be less than 1 mole. The first root, $x = 0.95$, satisfies this requirement. The second root, -55.0, cannot possibly be correct because the concept of a negative amount of material has no physical reality. Thus, the correct answer is $x = 0.95$ mole liter^{-1}.

Quadratic equations and equations of higher order are often not necessary for solving problems in beginning chemistry. The reason is that frequently the conditions of the problem are such that one can foresee approximations that reduce the complexity of the calculations without introducing a significant error in the answer.

Example. Let us look at a problem similar to the preceding one. Here, however, let us propose that the reaction reaches equilibrium when very little of the 1.0 mole liter^{-1} of starting material reacts. Thus, the amount of new substance formed, x, will be small, and the ratio of terms involving x will be smaller. The expression

$$\frac{(x)(x)}{1.0 - x} = 1.2 \times 10^{-6} \tag{3}$$

fits these conditions and can illustrate our point.

Solution. As before, to solve this expression by the quadratic formula, rearrange the equation so it is in the form

$$ax^2 + bx + c = 0$$

To do this, multiply both sides of Equation 3 by $1.0 - x$.

$$x^2 = (1.2 \times 10^{-6})(1.0 - x)$$

Now multiply the two quantities on the right, collect terms, and equate to zero.

$$x^2 + (1.2 \times 10^{-6})x - 1.2 \times 10^{-6} = 0$$

Substitute the corresponding quantities in the quadratic equation: $a = 1.0$, $b = 1.2 \times 10^{-6}$, and $c = -1.2 \times 10^{-6}$.

$$x = \frac{-b \pm \sqrt{b^2 - 4ac}}{2a}$$

$$= \frac{-1.2 \times 10^{-6} \pm [(1.2 \times 10^{-6})^2 - 4(1)(-1.2 \times 10^{-6})]^{1/2}}{2(1)}$$

$$= \frac{-1.2 \times 10^{-6} \pm (1.44 \times 10^{-12} + 4.8 \times 10^{-6})^{1/2}}{2}$$

$$= \frac{-1.2 \times 10^{-6} \pm (4.8 \times 10^{-6})^{1/2}}{2}$$

$$= \frac{-1.2 \times 10^{-6} \pm 2.2 \times 10^{-3}}{2}$$

$$= 1.1 \times 10^{-3} \text{ and } -1.1 \times 10^{-3}$$

Since the concentration cannot be a negative number, the correct answer is $x = 1.1 \times 10^{-3}$ mole liter^{-1}. This is a small number; that is, only a small amount of product is formed by the time equilibrium is attained.

Consider the denominator of the original expression $(1.0 - x)$. Since we have calculated the value for x, let us now evaluate its denominator

$$1.0 - x = 1.0 - 1.1 \times 10^{-3} = 1.0 - 0.0011 \cong 1.0$$

Observe that if we follow the rules for subtracting significant figures, the result of subtracting the value of x from 1.0 is the same as if we had completely forgotten about the x. Its value is too small to be significant compared with 1.0.

What is the point of all this? Suppose that you know that the value for x was negligible compared with 1.0; thus, you omitted it from the $(1.0 - x)$ expression before you attempted to solve the problem. Here is the simpler equation you would have obtained

$$\frac{x^2}{1.0} = 1.2 \times 10^{-6}$$

$$x^2 = 1.2 \times 10^{-6}$$

$$x = (1.2 \times 10^{-6})^{1/2} = 1.1 \times 10^{-3}$$

And you would have calculated the same answer as you did from your laborious calculations with the quadratic formula.

The point should now be obvious. If you had a hint at the outset that the value of x might be negligible compared with 1.0, you could have omitted it from the denominator and saved yourself effort. Generally, enough information in the problem indicates the possibility that such an omission is justified. When the opportunity presents itself, take advantage of it. Simplify the equation and solve for the answer. To ensure that your assumption was justified, place your value in the term or terms from which it was omitted. If it indeed yields the value for the term, unaltered within the precision governed by the rules for determining significant figures, you can feel confident that your answer is adequate. If this does not succeed, then the alternative is to do the problem by the more tedious, yet more exact, method.

APPENDIX 4
LOGARITHM TABLE

	0	1	2	3	4	5	6	7	8	9
10	0000	0043	0086	0128	0170	0212	0253	0294	0334	0374
11	0414	0453	0492	0531	0569	0607	0645	0682	0719	0755
12	0792	0828	0864	0899	0934	0969	1004	1038	1072	1106
13	1139	1173	1206	1239	1271	1303	1335	1367	1399	1430
14	1461	1492	1523	1553	1584	1614	1644	1673	1703	1732
15	1761	1790	1818	1847	1875	1903	1931	1959	1987	2014
16	2041	2068	2095	2122	2148	2175	2201	2227	2253	2279
17	2304	2330	2355	2380	2405	2430	2455	2480	2504	2529
18	2553	2577	2601	2625	2648	2672	2695	2718	2742	2765
19	2788	2810	2833	2856	2878	2900	2923	2945	2967	2989
20	3010	3032	3054	3075	3096	3118	3139	3160	3181	3201
21	3222	3243	3263	3284	3304	3324	3345	3365	3385	3404
22	3424	3444	3464	3483	3502	3522	3541	3560	3579	3598
23	3617	3636	3655	3674	3692	3711	3729	3747	3766	3784
24	3802	3820	3838	3856	3874	3892	3909	3927	3945	3962
25	3979	3997	4014	4031	4048	4065	4082	4099	4116	4133
26	4150	4166	4183	4200	4216	4232	4249	4265	4281	4298
27	4314	4330	4346	4362	4378	4393	4409	4425	4440	4456
28	4472	4487	4502	4518	4533	4548	4564	4579	4594	4609
29	4624	4639	4654	4669	4683	4698	4713	4728	4742	4757
30	4771	4786	4800	4814	4829	4843	4857	4871	4886	4900
31	4914	4928	4942	4955	4969	4983	4997	5011	5024	5038
32	5051	5065	5079	5092	5105	5119	5132	5145	5159	5172
33	5185	5198	5211	5224	5237	5250	5263	5276	5289	5302
34	5315	5328	5340	5353	5366	5378	5391	5403	5416	5428
35	5441	5453	5465	5478	5490	5502	5514	5527	5539	5551
36	5563	5575	5587	5599	5611	5623	5635	5647	5658	5670
37	5682	5694	5705	5717	5729	5740	5752	5763	5775	5786
38	5798	5809	5821	5832	5843	5855	5866	5877	5888	5899
39	5911	5922	5933	5944	5955	5966	5977	5988	5999	6010
40	6021	6031	6042	6053	6064	6075	6085	6096	6107	6117
41	6128	6138	6149	6160	6170	6180	6191	6201	6212	6222
42	6232	6243	6253	6263	6274	6284	6294	6304	6314	6325
43	6335	6345	6355	6365	6375	6385	6395	6405	6415	6425
44	6435	6444	6454	6464	6474	6484	6493	6503	6513	6522
45	6532	6542	6551	6561	6571	6580	6590	6599	6609	6618
46	6628	6637	6646	6656	6665	6675	6684	6693	6702	6712
47	6721	6730	6739	6749	6758	6767	6776	6785	6794	6803
48	6812	6821	6830	6839	6848	6857	6866	6875	6884	6893
49	6902	6911	6920	6928	6937	6946	6955	6964	6972	6981
50	6990	6998	7007	7016	7024	7033	7042	7050	7059	7067
51	7076	7084	7093	7101	7110	7118	7126	7135	7143	7152
52	7160	7168	7177	7185	7193	7202	7210	7218	7226	7235
53	7243	7251	7259	7267	7275	7284	7292	7300	7308	7316
54	7324	7332	7340	7348	7356	7364	7372	7380	7388	7396

APPENDIX 4

LOGARITHM TABLE (Continued)

	0	1	2	3	4	5	6	7	8	9
55	7404	7412	7419	7427	7435	7443	7451	7459	7466	7474
56	7482	7490	7497	7505	7513	7520	7528	7536	7543	7551
57	7559	7566	7574	7582	7589	7597	7604	7612	7619	7627
58	7634	7642	7649	7657	7664	7672	7679	7686	7694	7701
59	7709	7716	7723	7731	7738	7745	7752	7760	7767	7774
60	7782	7789	7796	7803	7810	7818	7825	7832	7839	7846
61	7853	7860	7868	7875	7882	7889	7896	7903	7910	7917
62	7924	7931	7938	7945	7952	7959	7966	7973	7980	7987
63	7993	8000	8007	8014	8021	8028	8035	8041	8048	8055
64	8062	8069	8075	8082	8089	8096	8102	8109	8116	8122
65	8129	8136	8142	8149	8156	8162	8169	8176	8182	8189
66	8195	8202	8209	8215	8222	8228	8235	8241	8248	8254
67	8261	8267	8274	8280	8287	8293	8299	8306	8312	8319
68	8325	8331	8338	8344	8351	8357	8363	8370	8376	8382
69	8388	8395	8401	8407	8414	8420	8426	8432	8439	8445
70	8451	8457	8463	8470	8476	8482	8488	8494	8500	8506
71	8513	8519	8525	8531	8537	8543	8549	8555	8561	8567
72	8573	8579	8585	8591	8597	8603	8609	8615	8621	8627
73	8633	8639	8645	8651	8657	8663	8669	8675	8681	8686
74	8692	8698	8704	8710	8716	8722	8727	8733	8739	8745
75	8751	8756	8762	8768	8774	8779	8785	8791	8797	8802
76	8808	8814	8820	8825	8831	8837	8842	8848	8854	8859
77	8865	8871	8876	8882	8887	8893	8899	8904	8910	8915
78	8921	8927	8932	8938	8943	8949	8954	8960	8965	8971
79	8976	8982	8987	8993	8998	9004	9009	9015	9020	9025
80	9031	9036	9042	9047	9053	9058	9063	9069	9074	9079
81	9085	9090	9096	9101	9106	9112	9117	9122	9128	9133
82	9138	9143	9149	9154	9159	9165	9170	9175	9180	9186
83	9191	9196	9201	9206	9212	9217	9222	9227	9232	9238
84	9243	9248	9253	9258	9263	9269	9274	9279	9284	9289
85	9294	9299	9304	9309	9315	9320	9325	9330	9335	9340
86	9345	9350	9355	9360	9365	9370	9375	9380	9385	9390
87	9395	9400	9405	9410	9415	9420	9425	9430	9435	9440
88	9445	9450	9455	9460	9465	9469	9474	9479	9484	9489
89	9494	9499	9504	9509	9513	9518	9523	9528	9533	9538
90	9542	9547	9552	9557	9562	9566	9571	9576	9581	9586
91	9590	9595	9600	9605	9609	9614	9619	9624	9628	9633
92	9638	9643	9647	9652	9657	9661	9666	9671	9675	9680
93	9685	9689	9694	9699	9703	9708	9713	9717	9722	9727
94	9731	9736	9741	9745	9750	9754	9759	9763	9768	9773

APPENDIX 4
LOGARITHM TABLE (Continued)

	0	1	2	3	4	5	6	7	8	9
95	9777	9782	9786	9791	9795	9800	9805	9809	9814	9818
96	9823	9827	9832	9836	9841	9845	9850	9854	9859	9863
97	9868	9872	9877	9881	9886	9890	9894	9899	9903	9908
98	9912	9917	9921	9926	9930	9934	9939	9943	9948	9952
99	9956	9961	9965	9969	9974	9978	9983	9987	9991	9996

APPENDIX 5
GLOSSARY OF
NEW TERMS

The number in italic type at the end of each definition refers to the chapter in which the new term is discussed.

A

Absolute zero: The temperature at which all motion stops; the lowest possible temperature. Absolute zero is 0°K or −273°C. *2*

Acid–base reaction: A special type of substitution reaction in which protons are transferred. *11*

Actinides: The group of elements beginning with thorium and ending with lawrencium. *3*

Activation energy: The amount of energy the reactants in a chemical process must gain to reach the transition state. *13*

Addition reaction: A reaction in which simpler molecules combine to form a more complex molecule. *11*

Alcohol: A compound of the type ROH. *14*

Aldehyde: A compound of the type

   14

Alkali metals: The elements that constitute Group I in the periodic table. *3*

Alkaline earth metals: The elements that constitute Group II in the periodic table. *3*

Alkane: A saturated hydrocarbon. *14*

Alkene: A compound containing carbon—carbon double bonds. *14*

Alkyne: A compound containing carbon—carbon triple bonds. *14*

Allotrope: One of two or more different forms of the same element. For example, two allotropes of carbon are diamond and graphite. *8*

Alpha (α) ray: Emission from a radioactive material. The ray is composed of alpha particles, which are helium nuclei. Each alpha particle is made up of two protons and two neutrons, and has a charge of +2. *4*

Amide: A compound of the type $R-C\overset{\displaystyle O}{\underset{\displaystyle NH_2}{}}$

The hydrogens on nitrogen may be replaced by R groups. *14*

Amine: A compound of the type RNH_2, in which nitrogen is bound only to carbon or hydrogen. *14*

Ampere: A measure of electric current, or the rate at which charge is flowing. One ampere is equal to a flow rate of one coulomb per second. *4*

Anode: The positively charged electrode in a cathode ray tube or an electrolysis cell; the electrode to which the negatively charged ions (anions) migrate. *4*

Antibonding orbital: Molecular orbital of higher energy than the atomic orbitals that are used to construct it. The electron density between the bonded atoms in an antibonding orbital is small, and is zero at one point along the bond axis. Electrons placed in antibonding orbitals decrease molecular stability. *7*

Aromatic compound: A compound having special stability, which can be explained in terms of delocalized π electrons. *7*

Arrhenius acid: A proton (H^+) donor. *11*

Arrhenius base: A hydroxide (HO^-) donor. *11*

Atom: The smallest chemically indivisible particle of an element. *1*

Atomic number: The number according to which the elements are arranged in the periodic table. It is equal to the number of protons in the nucleus of an element and the number of electrons in the neutral atom. *4*

Atomic weight: The weight of an element relative to the carbon isotope ^{12}C, which is assigned the exact value 12. The atomic weight also is approximately equal to the combined weights of the protons and neutrons in the nucleus of the element. *4*

Avogadro's hypothesis: Equal volumes of gases contain equal numbers of molecules, at the same temperature and pressure. *2*

Avogadro's number (N): The number of molecules in one mole, 6.022×10^{23}. *1*

B

Beta (β) ray: Emission from a radioactive substance; an electron. *4*

Bimolecular reaction: An elementary reaction that involves two reactant molecules. *13*

Binary compound: A compound composed of two different elements, such as NaCl, Fe_2O_3, and BF_3. *3*

Boiling curve: The temperature–pressure curve, which gives the boiling point of a liquid as a function of pressure. *8*

Boiling point: The temperature at which the vapor pressure of the liquid is the same as the atmospheric pressure. The normal boiling point is the temperature at which the vapor pressure of the liquid is 1 atm (760 torr). *8*

Bond axis: The line joining two atoms in a molecule. *7*

Bond energy: The energy required to separate two atoms held together by a chemical bond. It also may be thought of as the energy released when two atoms come together to form a chemical bond. *6*

Bond length: The distance between two bonded nuclei in the stable configuration of a molecule. *6*

Boyle's law: At constant temperature, the volume of a gas sample varies inversely with the pressure, or $PV = A$. *2*

Brønsted–Lowry acid: A proton donor. *11*

Brønsted–Lowry base: A proton acceptor. *11*

Buffer: A solution with the property that addition of relatively large amounts of an acid or a base will not change its pH significantly. *12*

C

Calorie: The energy required to heat one gram of water from 14.5°C to 15.5°C. *4*

Canal rays: The stream of positively charged particles that moves toward the cathode inside a cathode ray tube. *4*

Carbonium ion: A positive carbon ion; for example, $H_3C^{\oplus}$. *14*

Carbonyl: The functional group $\diagdown$C$=$O. *14*

Carboxylic acid: A compound of the type

14

Catalyst: A substance that, when added to a reaction mixture, affects the rate of the reaction; it is normally not consumed in the process. Substances increasing the rate are called catalysts; substances decreasing the rate are usually called negative catalysts or inhibitors. *13*

Cathode: The negatively charged electrode in a cathode ray tube or an electrolysis cell; the electrode to which positively charged ions (cations) migrate. *4*

Cathode ray: The stream of negatively

charged particles that moves toward the anode inside a cathode ray tube; the stream is composed of electrons. *4*

Charles' law: At constant pressure, the volume of a gas sample is directly proportional to its temperature, or $V/T = B$. *2*

Chemical family: A group of elements that have very similar chemical properties and predictable trends in their physical properties. Another name for any of the groups shown in the periodic table. *3*

Combined gas law: The law can be expressed by the equation $PV = nRT$. *2*

Compound: A pure substance composed of elements in fixed proportions. *1*

Concentration: The amount of solute dissolved in a given weight or volume of solution. *9*

Condensed phase: A phase in which molecules are closely packed together; a liquid or solid, for example. *2*

Conjugate acid–base pair: A pair of substances, one an acid, the other a base, related by the transfer of a proton. The acid has one more transferable proton than the base, and becomes the base by transfer of a proton. *12*

Conjugated molecule: Molecule having alternating single and double bonds. *7*

Conservation of mass: The principle that the total mass of each element present does not change in a chemical reaction. *1*

Constant chemical composition: The property that characterizes a pure substance. A substance has constant chemical composition if the relative amounts by weight of all the elements present do not change. *1*

Coulomb: A basic unit of electrical charge, equivalent to the charge of 6.24×10^{18} electrons. *4*

Covalent bond: The bond that results when two atoms are bound together through the sharing of a pair of electrons. *6*

Covalent solid: A solid consisting of atoms bound together through a network of covalent bonds. Graphite and diamond are covalent solids. *8*

Critical temperature: The temperature above which a substance cannot exist as a liquid no matter how much pressure is applied to it. *8*

Current: The flow of electricity or electrons expressed in terms of the ampere. *4*

Cycloalkanes: Alkanes containing rings of carbon atoms. *14*

D

De Broglie wave theory: Particles have wave properties associated with them. The wave properties influence significantly the behavior of atomic particles, but become insignificant in the case of macroscopic particles. *5*

Debye unit: The unit used to express dipole moments; one debye (D) is equal to 1×10^{-10} Å esu or 1×10^{-18} cm esu. *6*

Delocalized molecular orbital: Molecular orbital extending over more than two atoms in a molecule. *7*

Density: The weight of an object divided by its volume, or the weight per unit volume. *1*

Diamagnetic: Slightly repelled by a magnetic field. Diamagnetism is characteristic of compounds with no unpaired electrons. *7*

Diatomic: Having two atoms, in reference to molecules. *3*

Dibasic acid: An acid with two ionizable protons. *14*

Dipole–dipole attraction: The attraction that results from the interaction of molecules that have permanent dipole moments. *8*

Dissociation constant: Equilibrium constant for a dissociation reaction. *10*

Distillation: The process of boiling a liquid and allowing the vapor to be removed from the vessel and condensed to the liquid phase in another vessel. It is a method of separation of the more volatile component of a mixture from the less volatile components. *8*

Double bond: The covalent bond that results when two pairs of electrons are shared between two atoms. *6*

Ductility: The ability to be drawn into thin wires. *3*

Dynamic chemical equilibrium: That state of a reacting chemical system in which there is no net change in the number of moles of products or reactants. *10*

E

Electric dipole: A pair of electric charges of equal magnitude but opposite sign separated by a small distance. *6*

Electrolysis: The decomposition of a substance by passing an electrical current through it. *4*

Electromagnetic radiation: Light and all other similar forms of radiation including x rays, γ rays, ultraviolet, visible, infrared, radio waves, and microwaves. *4*

Electron: A basic particle of atomic structure bearing a negative electrical charge and weighing 1/1835 as much as a proton; a β particle. *4*

Electron-deficient molecule: A molecule that does not contain enough electrons to form electron-pair bonds between its constituent atoms and still obey the octet rule. *6*

Electron microscope: An instrument for examining very small objects, using electron scattering. *7*

Electron volt: The energy acquired by a particle with the charge of one electron when it moves through an electrical potential difference of one volt. *4*

Electronegativity: A measure of the relative ability of an atom to attract electrons to itself in a bond. *6*

Electronic effects: Effects on rates or mechanisms due to the electronic structures of the substances reacting. *13*

Electrophilic: Electron-seeking; electrophiles are Lewis acids. *14*

Element: The basic chemical unit of matter. All chemical substances are made from atoms of the elements. *1*

Elemental analysis: The process of breaking down a chemical substance into its elements and finding their relative proportions. *1*

Elementary reaction: A one-step reaction, or one of the steps in the mechanism of a complex reaction. The rate of an elementary reaction equals the product of the concentrations of the reactants and a rate constant. *13*

Elimination reaction: A reaction in which a molecule breaks apart into two or more simpler molecules; the opposite of an addition reaction. *11*

End point (equivalence point): That point in a titration when equivalent amounts of reactants (e.g., HO^- and H^+) have been added. *9*

Endothermic: Consuming heat energy. An endothermic process has a positive value of ΔH. *8*

Energy profile or reaction coordinate diagram: A graph indicating how the energy of a reacting system changes during the course of the reaction. *13*

Entropy: A measure of the probability or disorder of a state for any chemical system. The higher the disorder (or probability) of a state the higher is the entropy of that state. The entropy, *S,* of a pure crystal is zero at 0°K, and the entropy change in a reaction is $\Delta S = S_{products} - S_{reactants}$. *10*

Equilibrium constant: An experimentally determined constant that is equal to the product of the concentrations of the products divided by the product of the concentrations of the reactants, with each concentration raised to the power of the coefficient of that substance in the balanced equation. Specifically, we have

$$aA + bB \rightleftarrows cC + dD$$

$$K_{eq} = \frac{[C]^c[D]^d}{[A]^a[B]^b}. \quad 10$$

Equilibrium vapor pressure: The pressure exerted by the vapor above a pure liquid when the two phases are in equilibrium with each other. The value depends on the temperature of the system, but at any given temperature it is independent of the amount of liquid present. *8*

Erg: An energy unit equivalent to g cm² sec⁻². One erg is the kinetic energy of a mass of 1 g moving at the velocity 1 cm sec⁻¹. *4*

Ester: A compound of the type

. *14*

Exchange reaction: A reaction in which a molecule gives up an atom or group of atoms in exchange for another. *11*

Excited state: A state of higher energy than the ground state. An electronic excited state occurs when an electron is promoted from a lower to a higher energy level in an atom or molecule. *7*

Exothermic: Releasing heat energy. An exothermic process has a negative value of ΔH. *8*

Extraction: The process of drawing a substance from one phase into another. *10*

F

Faraday: The amount of charge, 96,500 coulombs, required to electrolyze one gram equivalent weight of a substance; the total charge of one mole of electrons. *4*

Formal charge: The charge assigned to an atom in a molecule as determined by subtracting 1 for each shared pair of electrons and 1 for each unshared electron from the number of valence electrons of that atom. *6*

Formality: The number of formula weights of solute dissolved in one liter of solution. *9*

Free energy: A criterion for determining the point of equilibrium in a chemical system.

For a chemical reaction, $\Delta G = \Delta H - T\Delta S = G_{products} - G_{reactants}$. The free energies of the most stable forms of the elements are all zero under standard conditions. *10*

Freezing curve: The curve that gives the freezing point of a liquid as a function of pressure. *8*

Frequency: The number of complete waves that pass a point per unit time (usually per second). *4*

Frequency factor: The rate constant for the reaction of activated molecules in a unimolecular reaction, or of collided molecules in a bimolecular reaction. *13*

Functional group: A reactive part of an organic compound that determines its chemical behavior. *14*

G

Gamma (γ) ray: High-energy electromagnetic radiation emitted from radioactive materials. *4*

Gas constant (R): The proportionality constant in the combined gas law, $PV = nRT$. One way to express R is

$$R = 8.21 \times 10^{-2} \frac{\text{liter atm}}{^\circ K^{-1} \text{ mole}^{-1}} \quad 2$$

Geometrical isomers: Molecules having the same atoms linked together in the same sequence, but having different orientations in space. *11*

Gram atomic weight (gram-atom): The quantity of an element whose weight is equal to its atomic weight expressed in grams. *1*

Gram equivalent weight: That weight of a substance that will react with one gram-atom of hydrogen, or that is electrolyzed by one faraday of electrical charge. *4*

H

Half-reaction (half-cell reaction): The reaction of either the oxidant or the reductant in a redox reaction that shows the number of electrons gained or lost. *11*

Halocarbon: A hydrocarbon that, in addition to carbon and hydrogen, contains one or more halogen atoms. *14*

Halogens: The family name given the elements of Group VII (F, Cl, Br, I, and At). *3*

Heat content (enthalpy): The amount of heat "contained" in a substance. The heat content, *H,* of an element in its most stable form under standard conditions is defined to be zero. The standard heat content of any compound is the heat change of the reaction for forming the compound from its elements under standard conditions. For any reaction, $\Delta H = H_{products} - H_{reactants}$. *10*

Heat of fusion, ΔH_{fus}: The heat energy required to transform one mole of solid into one mole of liquid at one atmosphere of pressure; ΔH_{fus} is expressed commonly in units of cal mole^{-1}. *8*

Heat of solution ($\Delta H_{solution}$): The amount of heat released (exothermic) or absorbed (endothermic) when a solute is dissolved. $\Delta H_{solution}$ usually is expressed in cal mole^{-1}, or kcal mole^{-1}. *9*

Heat of sublimation, ΔH_{sub}: The heat energy required to transform one mole of substance from the solid phase to the vapor phase at one atmosphere pressure; ΔH_{sub} is expressed usually in units of cal mole^{-1}. *8*

Heat of vaporization, ΔH_{vap}: The heat energy required to transform one mole of substance from the liquid phase to the vapor phase at one atmosphere pressure; ΔH_{vap} usually is expressed in units of cal mole^{-1}. *8*

Heisenberg uncertainty principle: It is not possible to know with great accuracy both the momentum and the location of an atomic particle. The limits of accuracy are determined by the relationship $(\Delta mv)(\Delta x) > h/4\pi$. *5*

Heterogeneous equilibrium: An equilibrium involving substances in different phases. *10*

Heterogeneous mixture: A nonuniform

mixture with different phases; usually separable by mechanical means. *1*

Homogeneous mixture: A uniform mixture that usually can be separated by some physical change. *1*

Hybrid atomic orbital: An orbital that is strongly directed in space and therefore useful for making a localized σ bond to an orbital on an adjacent atom. Hybrid orbitals are constructed by appropriate mixing of atomic *s, p,* and *d* orbitals. Important hybrids are *sp, sp², sp³, dsp², dsp³* (or *sp³d*), and *d²sp³* (*sp³d²*). *7*

Hydride: A compound containing hydrogen and one other element. Most often used for metal-hydrogen compounds, such as NaH and CaH$_2$. *3*

Hydrocarbon: A compound containing only hydrogen and carbon. *14*

Hydrogen bond: The special name for the dipolar interaction between molecules that have a hydrogen atom attached to a highly electronegative atom such as N, O, or F. A bond of the type $-\overset{\delta-}{O}\cdots\overset{\delta+}{H}-\overset{\delta-}{O}-$ (illustrated by the dashed line). *8*

Hydrolysis: A term literally meaning "splitting by water," now used loosely to refer to reactions with water. *12*

I

Ideal gas: The imaginary gas whose properties are predicted exactly by the kinetic molecular theory. *2*

Inert: Resistant to chemical change. The term is applied especially to substances that can undergo spontaneous chemical reaction, but for which the reaction proceeds at an immeasureably slow rate. *10*

Ion: An atom or molecule with an electric charge. *4*

Ion product of water: The equilibrium constant, K_w, for the dissociation of water to H^+ and HO^- ions

$$K_w = [H^+][HO^-].\quad 10$$

Ionic bond: The type of bond formed when two oppositely charged particles (ions) are bound together through their mutual electrostatic attraction. *6*

Ionic solid: A solid consisting of discrete positive and negative ions held together by electrostatic attraction; NaCl and KBr are ionic solids. *8*

Ionization constant: Equilibrium constant for an ionization reaction. *10*

Ionization energy: The energy required to remove an electron from an atom; that is, the energy associated with the process $M \rightarrow M^+ + e^-$. *5*

Isomerization reaction: The conversion of a molecule into an isomer. *11*

Isomers: Molecules with the same chemical composition but different arrangements of the atoms. Example: urea $[(H_2N)_2C{=}O]$ and ammonium cyanate ($NH_4^+OCN^-$) both have the chemical composition CON_2H_4. *10*

Isotopes: Forms of the same element (same atomic number) that are essentially identical chemically, but which have different atomic weights. *4*

K

K_a (or K_{HA}): The equilibrium constant for the dissociation of an acid. The solvent base with which the acid reacts does not appear in the equilibrium expression. *12*

K_b (or K_B): The equilibrium constant for the dissociation of a base. The solvent acid with which the base reacts does not appear in the equilibrium expression. *12*

Ketone: A compound of the type $\overset{R}{\underset{R'}{>}}C{=}O$. *14*

Kinetic molecular theory: A model based on tiny particles in rapid motion, which can be used to predict the properties of gases. *2*

Kinetics: The way in which the rate of a reaction depends on the concentrations of the substances present. *13*

L

Lanthanides: A group of heavy, similar elements beginning with cerium and ending with lutetium. *3*

Law of combining volumes: The ratios of volumes of reacting gases are small whole numbers (Gay–Lussac). *2*

Law of conservation of energy: Energy can neither be created nor destroyed, but it can be transformed from one form into another. For example, chemical energy can be transformed into heat energy. *10*

Law of partial pressures: Each component of a gas mixture exerts its own partial pressure independent of the others, and the total pressure is the sum of the partial pressures of the gases present. *2*

Le Chatelier's principle: A system tends to respond to an imposed stress in such a way as to relieve the stress. *10*

Lewis acid: An electron-pair acceptor. *11*

Lewis base: An electron-pair donor. *11*

Lewis formula: A simple diagram showing the placement of shared and unshared valence electrons in a molecule. *6*

Localized molecular orbital: Molecular orbital extending over two atoms in a molecule. *7*

M

Magnetic quantum number (m): The quantum number that describes the different spatial orientations an orbital may have. It is limited to the values $+l \cdots 0 \cdots -l$. *5*

Malleability: The ability to be hammered into thin sheets. *3*

Melting point: The temperature at which the solid and liquid phases of a substance are in equilibrium with each other. The normal melting point is the melting point at a pressure of one atmosphere. *8*

Metals: Elements that are good conductors of heat and electricity and are ductile, malleable, and have a lustrous appearance. *3*

Methyl group: The group —CH_3. *7*

Molar volume: The volume occupied by one mole of a gas at STP; 22.4 liters. *2*

Molarity: The number of moles of solute dissolved in one liter of solution. *9*

Mole (gram molecular weight): The amount of a substance whose weight is equal to its molecular weight expressed in grams. Avogadro's number of particles. *1*

Mole fraction: The number of moles of a substance divided by the total number of moles of all substances present. *2*

Molecule: The smallest unit of a pure substance. Molecules are composed of atoms bonded together in definite ways. *1*

Molecular formula: The formula that gives the actual number of atoms of each element in one molecule of a substance. *1*

Molecular orbital: A wave function for an electron in a molecule; usually an additive or subtractive combination of atomic orbitals. *7*

Molecular orbital theory: Model of molecular electronic structure based on assigning electrons to molecular orbitals in a way analogous to the buildup of atomic structure by feeding electrons into atomic orbitals. *7*

Molecular solid: A solid consisting of discrete molecules held together through van der Waals or dipole–dipole attractive forces. Solid benzene is an example of a molecular solid bound primarily through van der Waals forces. Solid hydrogen fluoride is held together mainly by dipole–dipole forces. *8*

Molecular weight: The relative weight of a molecule on the atomic weight scale; the sum of the atomic weights of the atoms in the molecule. *1*

N

Neutron: A nuclear particle that has an atomic mass of 1.0 (roughly the same as the proton's mass) and is electrically neutral. *4*

Nitration: The replacement of hydrogen or other electrophilic group with —NO_2. *14*

Noble gases: The members of Group 0. So called because they are chemically unreactive. *3*

Nonmetals: Those elements that do not have metallic properties. Nonmetals are nonconducting, nonductile, nonmalleable, and usually dull in appearance. *3*

Nucleophilic: Nucleus-seeking; nucleophiles are Lewis bases. *14*

Nucleus: The small, dense part of the atom composed of protons and neutrons. *4*

O

Octet rule: For atoms other than hydrogen in molecular combination, the total number of shared-pair and unshared-pair electrons should be equal to eight, which represents the stable, closed valence shell of a noble gas element. *6*

Orbital overlap: The sharing of a common region of space by two or more atomic orbitals of different atoms. *7*

Orbital shape quantum number (l): The quantum number that is related to the shape of the orbital and is limited to the values $0 \cdots (n-1)$. *5*

Orbitals: Wave functions. *5*

Organic chemistry: The chemistry of carbon-containing compounds. *14*

Oxidation: The increase in the oxidation number of an atom; loss of electrons. *11*

Oxidation number: A somewhat arbitrary number assigned to an atom according to definite rules. The number is useful in determining whether a substance is oxidized or reduced in a reaction and in balancing redox equations. A change in oxidation number in a reaction corresponds to loss or gain of electrons. *11*

Oxidation–reduction (redox) reaction: A reaction in which the oxidation numbers of at least two different atoms change. *11*

Oxidizing agent (oxidant): A substance that can remove electrons from another substance (oxidize it) and is itself reduced (gains electrons). *11*

P

Paramagnetic: Drawn into a magnetic field. Paramagnetism is characteristic of compounds with unpaired electrons. *7*

Partition coefficient (distribution coefficient): The equilibrium constant that describes the distribution of a substance between two phases. *10*

Pauli exclusion principle: No two electrons in the same atom can have the same four quantum numbers. *5*

Period: A row in the periodic table; distinguished from a family, which occupies a column. *3*

Periodic table: A tabular arrangement of the elements in order of increasing atomic numbers; the columns present families with similar chemical properties. *3*

pH: A measure of acidity. $pH = -\log[H^+]$　*12*

Phase: The physical state of matter: solid, liquid, or gaseous. *8*

Phase diagram: A graph showing the temperature dependence of the vapor pressure above the solid and liquid phases of a substance, and the phase of that substance under particular temperature and pressure conditions. *8*

Phase transition: The change of a substance from one phase to another. *8*

Photochemical: Having to do with the transfer of light energy to and from compounds and the effects of the energy transfer on reactions. *14*

Photoelectric effect: The emission of electrons from a metal surface produced by beaming light of a critical wavelength on the surface of the metal. *4*

Photon: The packet or unit of electromagnetic radiation containing energy equal to the product of the frequency of the radiation times Planck's constant (or $h\nu$). *4*

Pi (π) orbital: A molecular orbital that has principal lobes on either side of the bond axis. *7*

pK: A measure of the free energy of a reaction. $pK = -\log K = \Delta G^0/2.303RT$. *12*

Planck's constant: A fundamental physical constant, h, relating the energy of a photon of electromagnetic radiation to its frequency. The value of h is 6.626×10^{-27} erg sec. *4*

Polyene: A molecule with many conjugated double bonds. A molecule of the type $\cdots$

$$\left(-\overset{|}{C}=\overset{|}{C}-\overset{|}{C}=\overset{|}{C}-\overset{|}{C}=\overset{|}{C}-\overset{|}{C}=\right) \cdots \quad 7$$

Polyfunctional: Having more than one functional group. *14*

Principal quantum number (n): The quantum number that describes the major energy levels in the atom and can have integer values from one to infinity. *5*

Probability density: The probability of finding the electron in a given volume. It is proportional to the square of the wave function, which describes the wave properties of the electron. *5*

Proton: A nuclear particle that has an atomic mass of 1.0 (roughly the same as the neutron's mass) and has unit positive charge. *4*

Protonic acid (base): A Brønsted acid (base). *12*

Q

Quantized energy: Energy that can assume only discrete values. *5*

Quantum theory: The idea that energy is packaged in discrete units, which we refer to as quanta. *4*

R

R—: Symbol used to denote a carbon-containing fragment of a molecule. *14*

Radioactivity: The spontaneous decomposition of atomic nuclei accompanied by the emission of α, β, or γ rays. *4*

Rare earths: Common name for the lanthanides. *3*

Rate constant: A temperature-dependent constant, k, which is characteristic of each reaction. The proportionality constant linking reaction rate to concentrations. *13*

Rate-determining step: The slowest elementary reaction in the mechanism of an overall reaction. *13*

Rate law: The mathematical relationship that describes how the rate depends on the concentrations of the substances present. *13*

Reaction coordinate: The measure of the progress of a reaction from reactants to products; usually a complicated function of bond angles and bond lengths. The reaction coordinate is the abcissa in an energy profile diagram. *13*

Reaction mechanisms: The series of elementary reactions through which a reaction proceeds from starting materials to products; the detailed description of the bond-making and bond-breaking processes that occur in the reaction. *13*

Reaction order: The sum of the exponents to which the reactant concentrations are raised in the rate law. *13*

Reducing agent (reductant): A substance that can reduce another and is itself oxidized. *11*

Reduction: The decrease in oxidation number of an atom; gain of electrons. *11*

Reflux: Boiling a liquid and returning the condensed vapor; reflux allows a liquid to boil continuously. *8*

Reversible reaction: A reaction that rapidly attains the equilibrium state. *10*

Rydberg constant: The proportionality constant, R_H, which relates the wavelengths of the several lines in the emission spectrum of the element hydrogen. The value of R_H is 109,678 cm^{-1}. *4*

S

Saturated compound: Having the maximum number of bonded atoms. Carbon compounds having no double or triple bonds are said to be saturated. *14*

Saturated solution: A solution that has undissolved solute in equilibrium with dissolved solute. In other words, a solution that cannot dissolve any more solute at a given temperature. *9*

Sigma (σ) orbital: A molecular orbital that is symmetrical around a bond axis. *7*

Simplest formula: The formula that gives the relative numbers of atoms of the various elements in a substance. *1*

Solubility: The amount of solute that can be dissolved to make a saturated solution. *9*

Solubility equilibrium: The condition that exists when the number of molecules dissolving equals the number of molecules coming out of solution. The equilibrium condition is represented with two arrows, solid $\rightleftarrows$ solution. *9*

Solubility product constant: The equilibrium constant, K_{sp}, describing a reaction of the type

$$AB(s) \rightleftarrows A^+ + B^-$$
$$\text{where} \quad K_{sp} = [A^+][B^-]. \quad 10$$

Solvation: The interaction of a solute molecule with solvent molecules. *9*

Spectroscope: An instrument that separates light into its various components or bands. *4*

Spin quantum number (s): The quantum number that derives from the fact that the electron behaves as if it were a spinning charged particle. It is limited to values of $\pm\frac{1}{2}$. *5*

Spontaneous reaction: A reaction capable of producing a substantial amount of product; that is, a reaction with $K_{eq} > 1$ or $\Delta G^0 < 0$. *10*

Standard state: The set of conditions 25°C, 1 atm (1 atm partial pressure for a gas), and for solutions, the concentration 1M. *10*

Standard temperature and pressure (STP): Standard conditions for reporting the properties of gases and other substances. The standard temperature is 0°C = 273°K = 32°F, and the standard pressure is 760 torr = 1 atm. *2*

Steric effects: Effects on rates or mechanisms due to the physical arrangements of atoms in space. *13*

Strong acid (base): An acid (base) in water with $K_a(K_b) > 1$. *12*

Structural isomers: Molecules having the same atoms linked together in different ways. *11*

Sublime: To pass directly from the solid phase to the vapor phase without passing through the liquid phase. *8*

Substitution (replacement) reaction: An exchange reaction in which there is no change in the number of bonds to each atom. *11*

Supercooling: Cooling a liquid below the normal freezing point. The supercooled state is unstable; a slight disturbance usually will bring about the phase transition from the supercooled liquid to the stable solid. *8*

T

Titration: The process of adding a measurable volume of one solution of known concentration to a measured volume of a second solution of unknown concentration for the purpose of determining the unknown concentration. *9*

Trace analysis: Finding the amount of a trace of one substance in a sample. *10*

Transition metals: The group of metals in the center (B groups) of the periodic table. *3*

Transition state: A short-lived state through which reactants pass on their way to products. The activation energy is the energy required to achieve this transition state. *13*

Triple point: The temperature at which the liquid, solid, and vapor phases are in equilibrium. *8*

U

Unimolecular reaction: An elementary process that involves only one molecule. *13*

Unsaturated compounds: Having multiple bonds. *14*

V

Valence: The number of bonds an element forms in a molecule. *1*

Valence electrons: Those electrons in an atom that have the highest values of n and are in the outermost orbitals. *5*

Van der Waals attraction: The intermolecular attraction due to the continuous fluctuation of electron clouds throughout molecules, which produce transient dipole moments. *8*

W

Wave function: The mathematical function that describes the wavelike behavior of a small particle such as an electron. *5*

Wave number: The reciprocal of the wavelength. The wave number, $\bar{\nu} = \lambda^{-1}$, is proportional to frequency, ν, and often is used to describe wave properties. *4*

Wavelength: The distance between equivalent points on a wave. *4*

Weak acid (base): An acid (base) in water with $K_a(K_b) < 10^{-2}$. *12*

Weight percent: The ratio of the weight of one component of a mixture or solution to the total weight of the mixture or solution, multiplied by 100. *9*

X

x Ray: High-energy electromagnetic radiation emitted by elements under bombardment by high-energy electrons. The x-ray frequency is characteristic of the atomic number of the element. *4*

APPENDIX 6
MOLALITY, WEIGHT PERCENT,
MOLE FRACTION, AND
NORMALITY

In Chapter 9, we discussed molarity and formality, two ways to express the concentrations of solutions. Following is a brief description of four other useful measures of concentration.

Molality

The molal concentration of a solution is the number of moles (or formula weights) of solute dissolved in 1000 g (1000 g = 1 kg) of solvent. The usual symbol for molal concentration is m.

Example A–1.   A 25.0-g sample of sugar (mol wt = 342) is dissolved in 150 g of water. What is the molality of the solution?
Solution.

g mol wt of sugar = 342 g

$$\text{moles sugar} = \frac{25.0 \text{ g}}{342 \text{ g mole}^{-1}} = 0.0731 \text{ mole}$$

We have determined that there are 0.0731 mole of sugar in 150 g of water. We need to know how many moles of sugar would be dissolved in 1000 g of water, maintaining this same concentration. This may be treated as a simple ratio and proportion problem; that is, if there are 0.0731 mole of sugar dissolved in 150 g of water, how many (x) moles of sugar will be dissolved in 1000 g of water, or

$$\frac{0.0731 \text{ moles}}{150 \text{ g H}_2\text{O}} = \frac{x \text{ moles}}{1000 \text{ g H}_2\text{O}}$$

$$x = \frac{0.487 \text{ mole}}{1000 \text{ g H}_2\text{O}} = 0.487 \text{ molal}$$

An alternative approach is to recognize that molality is simply the number of moles of solute in one kilogram of solvent. After converting grams of sugar to moles of sugar, divide the number of moles by the fraction of a kilogram of solvent used; this will give the molality of the solution

$$\frac{0.0731 \text{ mole}}{0.150 \text{ kg}} = 0.487 \text{ molal}$$

Molal concentrations are useful because some physical properties of dilute solutions are proportional to the molal concentrations of solutes in the solutions. These properties are called *colligative* properties. The amount by which the freezing point of a solvent is lowered by the addition of some solute is an example of a colligative property.

Weight percent

The weight percent is the percentage of the total weight of the solution represented by a given constituent.

Example A–2. Calculate the weight percent of a $0.487 m$ solution of sugar in water.

Solution. A $0.487 m$ solution has 0.487 mole of sugar (mol wt $= 342$ g mole^{-1}) dissolved in 1000 g of solvent. The weight of sugar present is

$$0.487 \text{ mole} \times 342 \text{ g mole}^{-1} = 167 \text{ g sugar}$$

The total weight of the solution is 1000 g of solvent plus 167 g of sugar, or 1167 g. The weight percent of sugar is

$$\frac{167 \text{ g}}{1167 \text{ g}} \times 100 = 14.3\%$$

Weight percent is used as a measure of concentration when the molecular weight of a solute either is unknown, or of no interest in the use intended for the solutions. The concentrations of solutions made by the addition of ethylene glycol (antifreeze) to water in the radiators of automobiles are usually measured in weight percent.

Mole fraction

In the discussion of Dalton's law of partial pressures (Section 2–5), we introduced the concept of mole fraction as a means of expressing the relative amounts of the components of a gaseous mixture. This is also a useful notation for liquid solutions, for which the mole fraction of solute, X, can be calculated from the relationship

$$X = \frac{\text{(moles solute)}}{\text{(moles solute} + \text{moles solvent)}}$$

This method of expressing concentration is not used widely, but it does have application in certain situations. For example, when a solute is dissolved in a solvent, the vapor pressure above the solution is less than it is above the pure solvent, at the same temperature. The extent of this vapor pressure lowering depends on the mole fraction of the solute.

Mole fraction can be calculated easily. Consider the solution described in the previous example, in which 25.0 g of sugar is dissolved in 150 g of water. The mole fraction of sugar present is

$$X_{\text{sugar}} = \frac{\text{moles sugar}}{\text{moles sugar} + \text{moles water}}$$

$$\text{moles of sugar} = \frac{25.0 \text{ g}}{342 \text{ g mole}^{-1}} = 0.0731 \text{ mole}$$

$$\text{moles of water} = \frac{150 \text{ g}}{18.0 \text{ g mole}^{-1}} = 8.33 \text{ moles}$$

$$X_{\text{sugar}} = \frac{0.0731 \text{ mole}}{(0.0731 + 8.33) \text{ moles}} = \frac{0.0731}{8.40} = 0.870 \times 10^{-2}$$

Normality

The normal concentration symbol, N, is the number of *equivalent weights* of a solute per liter of solution. This method of specifying concentrations is similar to molarity, but instead of moles, we use the gram equivalent weight (usually simply referred to as the *equivalent weight*) of solute. The equivalent weight of a compound is the molecular weight divided by the number of *reaction equivalents* per mole

$$\text{equivalent weight} = \frac{\text{molecular weight}}{n}$$

where n is the number of reaction equivalents. The term "reaction equivalents" will be defined more precisely in the examples in this section. To calculate the equivalent weight, we must ascertain n from the reaction in which the substance will be used. It is important that we always work with a *balanced equation for the reaction*.

Normality is used most often in giving the concentrations of solutions of acids and bases and of oxidizing and reducing agents. Consider the reaction of sulfuric acid with sodium hydroxide

(a base)

$$2NaOH \ + \ H_2SO_4 \ \rightarrow \ Na_2SO_4 \ + \ 2H_2O$$

| sodium hydroxide | sulfuric acid | sodium sulfate | water |

or the reaction of barium hydroxide (a base) with hydrochloric acid

$$Ba(OH_2) \ + \ 2HCl \ \rightarrow \ BaCl_2 \ + \ 2H_2O$$

| barium hydroxide | hydrochloric acid | barium chloride | water |

The common feature of these reactions is neutralization of hydroxide ions (HO^-) from the bases by hydrogen ions (H^+) from the acids to form water. *The number of reaction equivalents of an acid is the number of hydrogen ions that can be supplied by one molecule of the acid.* Sulfuric acid supplies two hydrogen ions, so its equivalent weight is its molecular weight divided by two

$$\frac{98.0 \text{ g } H_2SO_4 \text{ mole}^{-1}}{2 \text{ equiv mole}^{-1}} = 49.0 \text{ g } H_2SO_4 \text{ equiv}^{-1}$$

Hydrochloric acid supplies only one hydrogen ion, so its equivalent weight is the same as its molecular weight.

Similarly, *the number of reaction equivalents of a base is the number of hydroxide ions it can supply.* In the examples given, the equivalent weight of sodium hydroxide is equal to its molecular weight, whereas the equivalent weight of barium hydroxide is one half its molecular weight.

It is helpful to think of an equivalent weight as that quantity of an acid (or base) that can furnish Avogadro's number, or one mole, of H^+ (or HO^-) ions. Sulfuric acid, H_2SO_4, can furnish two hydrogen ions. We say that H_2SO_4 has two acidic hydrogen ions. Clearly, if we used only 49 g (0.5 mole) of H_2SO_4, there would be 2×0.5 mole, or 1 mole, of hydrogen ions. By defining equivalent weight in terms of the quantity of acid or base needed to furnish a mole of hydrogen or hydroxide ions, it follows that

one equivalent of base $=$ one equivalent of acid

since

$6.022 \times 10^{23} \ HO^-$ will neutralize $6.022 \times 10^{23} \ H^+$

Example A–3. What weight of barium hydroxide is required to prepare two liters of a $0.00500N$ solution?

Solution. The formula weight of $Ba(OH)_2$ is 171 g. There are two hydroxide ions (HO^-) per formula unit of $Ba(OH)_2$, and the equivalent weight is

$$\frac{171 \text{ g}}{\text{mole}} \times \frac{1 \text{ mole}}{2 \text{ equiv}} = 85.5 \text{ g equiv}^{-1}$$

Two liters of $0.00500N$ solution require 2.00 liters $\times$ 0.00500 equiv liter^{-1} = 0.0100 equiv weight $Ba(OH)_2$ = 0.0100 equiv $\times$ 85.5 g equiv^{-1} = 0.855 g of $Ba(OH)_2$.

APPENDIX 7
ANSWERS
TO SOME PROBLEMS

Chapter 1

3 75.8 cm
5 Greece
7 60.6% Cl; 39.4% Na
11 6.1 g mole^{-1}
13 (a) 16.043   (c) 259.60   (f) 231.3
14 (a) 2.50×10^{-2} mole O_2 (c) 8.75×10^{-5} mole N_2
    (f) 28.7 moles U
17 (b) 12×10^{20} molecules O_2
18 (b) 6.35×10^{-3} g Na
20 $AlCl_3$, Al_2Cl_6
24 3.67 g CO_2;   2.33 g CO
    Twice as much oxygen in one compound as the other.

Chapter 2

3 1.9 liters
6 1.25 liters
8 4.67
11 284 ml
13 24.6 atm
16 1.86×10^{-6} atm
20 8.72 g liter^{-1}
22 0.3 atm O_2; 0.1 atm H_2; 0.5 atm N_2
25 0.81 liter
28 0.907 g acetylene
    0.0349 mole CaC_2
30 The hexafluoride of ^{235}U diffuses 1.005 times faster than ^{238}U hexafluoride.

Chapter 3

2 RbH, BaH_2, GaH_3, GeH_4, SbH_3, TeH_2, HI

4 Li, Na, K, Rb, Cs, Fr

$2Li + 2H_2O \rightarrow 2LiOH + H_2$

6 Relative combining power of calcium to sodium is 2:1.

10 Calculated, 86.00; actual, 85.47.

Chapter 4

4 0.068 mole Al

6 176 coulombs

9 1.60×10^{-19} coulomb

12

Carbon-12	*Carbon-13*
6 protons	6 protons
6 neutrons	7 neutrons
6 electrons	6 electrons

15 purple

18 $\lambda = 6000$ Å (yellow)

20 9.9×10^{-9} erg photon^{-1}; 1.4×10^5 kcal mole^{-1}; yes

24 3.03×10^{-12} erg; 4.08×10^{-12} erg

Chapter 5

2 energy $= 1.94 \times 10^{-11}$ erg atom^{-1}

1025 Å, ultraviolet

3 $\Delta x = 3.6 \times 10^{-32}$ cm; insignificant

5 n = principal quantum number; gives the major energy levels within the atom. $n = 1, 2, 3, \cdots \infty$.

l = orbital shape quantum number; gives the general shape of the orbital in space. $l = 0, 1, 2, 3, \cdots (n-1)$.

m = magnetic quantum number; gives the orientation of the orbital in space. $m = +l \cdots 0 \cdots -l$.

s = spin quantum number; gives the spin orientation of the electron. $s = \pm\frac{1}{2}$.

8   $Ar = 1s^2 2s^2 2p^6 3s^2 3p^6$

     6 s electrons, 12 p electrons. The 18th electron will be a $3p$ electron: $n = 3$, $l = 1$, $m = +1$, 0, or -1 and $s = +\frac{1}{2}$ or $-\frac{1}{2}$.

11   The number of positive charges in the nuclei of the atoms increases from lithium ($Z = 3$) to neon ($Z = 10$). The electrons, which equal the number of protons, are added to the $n = 2$ shell from Li to Ne. These electrons are not effective in screening each other from the increasingly positive charge of the nucleus in this series of elements. Sodium ($Z = 11$) has its last electron in the $3s$ shell, well beyond the filled shell of neon. The Na $3s$ electron is well shielded by the 10 electrons of the underlying orbitals and is, on the average, farther from the nucleus than the Li $2s$ electron. Thus the IE of Na is less than the IE of Li.

13   $n = 4$; contains 32 electrons when filled

     $l = 3$; f electrons

15   Al $1s^2 2s^2 2p^6 3s^2 3p^1$

     Si $1s^2 2s^2 2p^6 3s^2 3p_x{}^1 3p_y{}^1$

     P $1s^2 2s^2 2p^6 3s^2 3p_x{}^1 3p_y{}^1 3p_z{}^1$

     S $1s^2 2s^2 2p^6 3s^2 3p_x{}^2 3p_y{}^1 3p_z{}^1$

17   np^5

19   $O(1s^2 2s^2 2p^4) \xrightarrow{+2e^-} O^2(1s^2 2s^2 2p^6)$

     $Cl(1s^2 2s^2 2p^6 3s^2 3p^5) \xrightarrow{+e^-} Cl^-(1s^2 2s^2 2p^6 3s^2 3p^6)$

     $Na(1s^2 2s^2 2p^6 3s^1) \xrightarrow{-e^-} Na^+(1s^2 2s^2 2p^6)$

     $Ca(1s^2 2s^2 2p^6 3s^2 3p^6 4s^2) \xrightarrow{-2e^-} Ca^{2+}(1s^2 2s^2 2p^6 3s^2 3p^6)$

Chapter 6

1   K·   ·C̈·   :N̈·   :Ö·   :F̈·   :N̈e:

5   (a)   H has a valence of 2 and is surrounded by 4 electrons.

     (b)   Too many electrons; the double bond should be a single bond.

     (d)   H has 8 electrons; too many electrons in the molecule.

7

Polar	Nonpolar	Ionic
IF	S_2	NaI
NO		MgO
		Cs_3N

9   BrCl < HCl < HF < NaI < CsI

11   33% ionic character; B has the partial negative charge.

13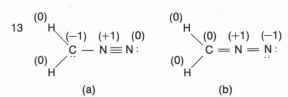

(a)   (b)

17 $EN_{Br} = 2.99$

19 (b) bent

(d) planar

(e) pyramidal

21 H—C≡C—H   A triple bond is shorter than a double bond, which is shorter than a single bond. Acetylene will have the largest C—C bond energy and ethane the least. The higher the bond order, the higher the bond energy.

Chapter 7

1 NF_3, pyramidal

$CH_3^{\oplus}$, planar

$CH_3^{\ominus}$, pyramidal

SO_2, bent

PF_6^-, octahedral

2 Silicon has low-lying $3d$ orbitals that can be used in conjunction with its $3s$ and $3p$ valence orbitals to construct six sp^3d^2 hybrid orbitals. In this model, the electronic structure of SiF_6^{2-} would consist of six $sp^3d^2\sigma$ electron-pair Si—F bonds. Carbon has no low-lying d orbitals for electron-pair sharing. Therefore, CF_6^{2-} would not be stable.

8 C_2H_4; the π electrons in the carbon–carbon double bond should have a lower IE than the σ electrons in the carbon–carbon single bond.

9 There are two more electrons in O_2 than in N_2. The two additional electrons in O_2 are in antibonding orbitals, resulting in a lower bond energy than in N_2.

13 Sulfur is surrounded by four pairs of bonding electrons and one unshared pair of electrons. The two possible structures for the five groups around the central sulfur atom in a trigonal bipyramidal model (sp^3d hybrid orbitals for sulfur) are (a) and (b).

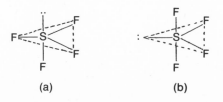

(a)   (b)

Structure (b) is preferred according to VSEPR theory, because the lone pair of electrons makes a 90° angle with only two S—F bonding pairs, as opposed to three such 90° interactions in Structure (a).

Chapter 8

2 (a) vapor   (c) liquid

3 1510 cal mole^{-1}; 755 cal melts 34.4 g of propionitrile.

7 Hydrogen bonds between molecules in the liquid phase give rise to anomalously high boiling points. HF has the lowest molecular weight of the hydrogen halide series, yet it has the highest boiling point.

   HI(−50.8°C) > HBr(−67°C) > HCl(−83°C) < HF(19°C).

13 The model for metal bonding describes the valence electrons as loosely bound and highly mobile. Under the influence of an electrical potential, the electrons move; that is, current flows.

14 The high ΔH_{vap} for water reflects the presence of intermolecular hydrogen bonds.

17 LiH(ionic); Si(covalent); Br_2(molecular); $SrCl_2$(ionic); N_2(molecular); $CHCl_3$(molecular); B_2O_3(covalent); CH_3OH(molecular); CaO(ionic); Kr(molecular); SF_6(molecular); Ba(metallic); $(CH_3)_2CO$(molecular); Cd(metallic).

Chapter 9

1 2.46 metric tons of seawater

3 37.3 g NaBr

5 4.82M

8 0.167F

13 1.88M

15 4.8 g

17 80 ml

18 35 ml

21 1.2 liters

22 0.105M

Chapter 10

1 (a) $K = \dfrac{[SO_3]^2}{[SO_2]^2[O_2]}$ (f) $K = [Ba^{2+}][SO_4^{2-}]$ (h) $K = [CO_2]$

3 $H_3PO_4 \rightleftarrows 3H^+ + PO_4^{3-}$ $K = \dfrac{[H^+]^3[PO_4^{3-}]}{[H_3PO_4]}$

   Or, in three steps:

   (1) $H_3PO_4 \rightleftarrows H^+ + H_2PO_4^-$ $K_1 = \dfrac{[H^+][H_2PO_4^-]}{[H_3PO_4]}$

   (2) $H_2PO_4^- \rightleftarrows H^+ + HPO_4^{2-}$ $K_2 = \dfrac{[H^+][HPO_4^{2-}]}{[H_2PO_4^-]}$

(3) $HPO_4^{2-} \rightleftarrows H^+ + PO_4^{3-}$ $\qquad K_3 = \dfrac{[H^+][PO_4^{3-}]}{[HPO_4^{2-}]}$

$K = K_1 \times K_2 \times K_3$

5 1.0×10^{-3} mole CuCl liter^{-1}

6 2.2×10^{-5} mole liter^{-1}

9 -6.9 cal deg^{-1}

12 $[A] = [B] = 0.167$ mole liter^{-1}; $[C] = [D] = 0.833$ mole liter^{-1}

15 (a) favor Cl_2 and H_2O formation

　　(b) favor $NH_4CO_2NH_2$ formation

　　(c) favor N_2O_4 formation

17 $[HO^-] = 4.0 \times 10^{-12}$ mole liter^{-1}

Chapter 11

1 (a) substitution

　　(b) addition

　　(d) redox

3 substitution or elimination

5 $KMnO_4$ 　　　$K(1+)\ Mn(7+)O(2-)$

　$BrCl$ 　　　　$Br(1+)Cl(1-)$

　$Na_2C_2O_4$ 　　$Na(1+)C(3+)O(2-)$

7 (a) $ClO_2 + 1e^- \xrightarrow{\text{reduced}} ClO_2$

　　$ClO_2 \xrightarrow{\text{oxidized}} ClO_3^- + 1e^-$

9 (a) $2e^- + Cl_2 \rightarrow 2Cl^-$

　　(c) $C_2O_4^{2-} \rightarrow 2CO_2 + 2e^-$

12 (a) redox

　　(b) substitution

　　(c) redox

　　(d) substitution

Chapter 12

1 H_2O (acid or base) 　　　HCl (acid)

　S^{2-} (base) 　　　　　　$(CH_3)_3NH^+$ (acid)

　$H_2PO_4^-$ (acid or base) 　OAc^- (base)

3 $[H_3O^+] = 1.5 \times 10^{-3}F$

5 5.6×10^{-2}% dissociated

7 2.0×10^{-11} mole liter$^{-1} = [H_3O^+]$

9 (a) pH $= 1.49$; pOH $= 12.51$

　　(b) pH $= 8.88$; pOH $= 5.12$

11 $K_a = 1.6 \times 10^{-4}$

13 pH $= 5.0$; pOH $= 9.0$

15 $[H_3O^+] = 4.9 \times 10^{-10}$ mole liter^{-1}

Chapter 13

1 (a) second order in NO; first order in O_2; third order overall

  (c) first order in $(CH_3)CCl$; first order overall

3 rate $= k[NO]^2[H_2]$; third order reaction

5 Some molecular collisions do not result in chemical reactions; only those which achieve the activation energy. The more complex the structure the more important the geometry of the molecular collisions and the fewer collisions that result in a chemical event.

9 A catalyst lowers the activation energy, E_a, for a reaction system, allowing more of the collisions to be effective in producing a chemical reaction. But it affects the reverse reaction as well as the forward reaction, leaving the equilibrium position unaffected.

Chapter 14

2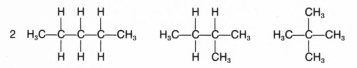

4 (a) ketone

  (b) aldehyde

  (c) amine

  (d) ester

  (e) amide

  (f) acid

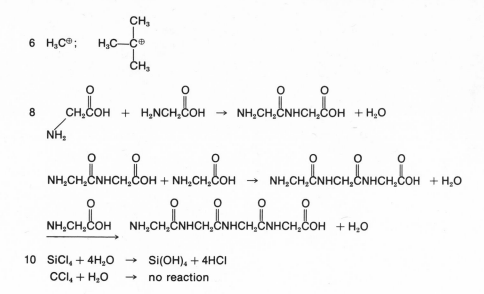

6   $H_3C^{\oplus}$; $H_3C—\overset{\overset{\displaystyle CH_3}{|}}{\underset{\underset{\displaystyle CH_3}{|}}{C^{\oplus}}}$

8   $\underset{\underset{\displaystyle NH_2}{|}}{CH_2}\overset{\overset{\displaystyle O}{\|}}{C}OH$ + $H_2NCH_2\overset{\overset{\displaystyle O}{\|}}{C}OH$ → $NH_2CH_2\overset{\overset{\displaystyle O}{\|}}{C}NHCH_2\overset{\overset{\displaystyle O}{\|}}{C}OH$ + H_2O

$NH_2CH_2\overset{\overset{\displaystyle O}{\|}}{C}NHCH_2\overset{\overset{\displaystyle O}{\|}}{C}OH$ + $NH_2CH_2\overset{\overset{\displaystyle O}{\|}}{C}OH$ → $NH_2CH_2\overset{\overset{\displaystyle O}{\|}}{C}NHCH_2\overset{\overset{\displaystyle O}{\|}}{C}NHCH_2\overset{\overset{\displaystyle O}{\|}}{C}OH$ + H_2O

$NH_2CH_2\overset{\overset{\displaystyle O}{\|}}{C}OH$ $NH_2CH_2\overset{\overset{\displaystyle O}{\|}}{C}NHCH_2\overset{\overset{\displaystyle O}{\|}}{C}NHCH_2\overset{\overset{\displaystyle O}{\|}}{C}NHCH_2\overset{\overset{\displaystyle O}{\|}}{C}OH$ + H_2O
$\xrightarrow{\hspace{3cm}}$

10   $SiCl_4 + 4H_2O$ → $Si(OH)_4 + 4HCl$
     $CCl_4 + H_2O$ → no reaction

INDEX

ATOMIC MASSES OF ELEMENTS
REFERRED TO ^{12}C = 12.0000

NAME	SYMBOL	ATOMIC NUMBER	ATOMIC WEIGHT	NAME	SYMBOL	ATOMIC NUMBER	ATOMIC WEIGHT
Actinium	Ac	89	(227)	Gallium	Ga	31	69.72
Aluminum	Al	13	26.9815	Germanium	Ge	32	72.59
Americium	Am	95	(243)	Gold	Au	79	196.967
Antimony	Sb	51	121.75	Hafnium	Hf	72	178.49
Argon	Ar	18	39.948	Helium	He	2	4.0026
Arsenic	As	33	74.9216	Holmium	Ho	67	164.930
Astatine	At	85	(210)	Hydrogen	H	1	1.00797^a
Barium	Ba	56	137.34	Indium	In	49	114.82
Berkelium	Bk	97	(249)	Iodine	I	53	126.9044
Beryllium	Be	4	9.0122	Iridium	Ir	77	192.2
Bismuth	Bi	83	208.980	Iron	Fe	26	55.847^b
Boron	B	5	10.811^a	Krypton	Kr	36	83.80
Bromine	Br	35	79.909^b	Lanthanum	La	57	138.91
Cadmium	Cd	48	112.40	Lawrencium	Lw	103	(257)
Calcium	Ca	20	40.08	Lead	Pb	82	207.19
Californium	Cf	98	(251)	Lithium	Li	3	6.939
Carbon	C	6	12.01115	Lutetium	Lu	71	174.97
Cerium	Ce	58	140.12	Magnesium	Mg	12	24.312
Cesium	Cs	55	132.905	Manganese	Mn	25	54.9380
Chlorine	Cl	17	35.453^b	Mendelevium	Md	101	(256)
Chromium	Cr	24	51.996^b	Mercury	Hg	80	200.59
Cobalt	Co	27	58.9332	Molybdenum	Mo	42	95.94
Copper	Cu	29	63.54	Neodymium	Nd	60	144.24
Curium	Cm	96	(247)	Neon	Ne	10	20.183
Dysprosium	Dy	66	162.50	Neptunium	Np	93	(237)
Einsteinium	Es	99	(254)	Nickel	Ni	28	58.71
Erbium	Er	68	167.26	Niobium	Nb	41	92.906
Europium	Eu	63	151.96	Nitrogen	N	7	14.0067
Fermium	Fm	100	(253)	Nobelium	No	102	(253)
Fluorine	F	9	18.9984	Osmium	Os	76	190.2
Francium	Fr	87	(223)	Oxygen	O	8	15.9994^a
Gadolinium	Gd	64	157.25	Palladium	Pd	46	106.4